ENGINEERING REFERENCE BOOK ON ENERGY AND HEAT

ENGINEERING REFERENCE BOOK ON ENERGY AND HEAT

Edited by

Verein Deutscher Ingenieure (VDI)
Gesellschaft Energietechnik (GET)

Association of German Engineers
Society for Energy Technologies

SPRINGER-VERLAG BERLIN HEIDELBERG GMBH

Die Deutsche Bibliothek – CIP-Einheitsaufnahme

Engineering reference book on energy and heat / ed.
by Verein Deutscher Ingenieure, Gesellschaft
Energietechnik. –
Düsseldorf : VDI-Verl., 1992
   Dt. Ausg. u.d.T.: Wärmetechnische Arbeitsmappe
   ISBN 978-3-642-51125-7   ISBN 978-3-642-51123-3 (eBook)
   DOI 10.1007/978-3-642-51123-3
NE: Gesellschaft Energietechnik

Preface

The suggestion for and the production of the "Wärmetechnische Arbeitsmappe" ("Engineering Reference Book on Energy and Heat") came from Dipl.-Ing. W. Goldstern. In 1932 he selected suitable diagrams and made them available in the form of worksheets. The first edition in 1934 was followed by further twelve ones, a fact which indicates that the "Wärmetechnische Arbeitsmappe" is a basic source for relevant engineering work, which cannot be given up.

The great success of the "Wärmetechnische Arbeitsmappe" in German speaking countries is the reason for translating it into the English language, and to make it available in many more countries.

The professional reasons for reviewing the twelfth edition were due to the following three points, which have been taken into consideration for the new edition:

- The units used in the diagrams and equations were changed to SI-units and units prescribed by Law.

- To simplify data processing, the fundamentals of calculation were added to the explanation for all diagrams and the included examples. The front page shows the graphs, while on the back of the worksheet the corresponding equations and flow sheets are complemented for faster computer programming.

- The 13th edition of the "Wärmetechnische Arbeitsmappe" is divided into 14 chapters, which include completely new subjects.

The 13th edition of the "Wärmetechnische Arbeitsmappe" is focussing more closely on the anticipated needs of engineers working in research and industry.

As in previous editions the requirements for the worksheets and the information contained was considerably increased. The concept was established by a committee, specially founded for this purpose, which also submitted many suggestions for the design of the worksheets. The following members of the committee have also contributed to the new chapters and worksheets:

Chapter 1: *Bohn, Th.,* Prof. Dr. techn., Essen,
Chapter 2: *Knoche, K. F.,* Prof. Dr.-Ing., Aachen,
                  Renz, U., Prof. Dr.-Ing., Aachen,
                  Schneider, F., Dr.-Ing., Aachen,
Chapter 3: *Renz, U.,* Prof. Dr.-Ing., Aachen,
                  Odenthal, H.-P., Dr.-Ing., Aachen,
Chapter 4: *Brandt, F.,* Prof. Dr.-Ing., Darmstadt,
Chapter 5: *Brandt, F.,* Prof. Dr.-Ing., Darmstadt,
Chapter 6: *Hömig, H. E.,* Dr.-Ing., Stuttgart,
Chapter 7: *Mellgren, A.,* Dipl.-Ing., Mannheim,
Chapter 8: *Sauer, E.,* Priv.-Doz. Dr.-Ing., Düsseldorf,
Chapter 9: *Beck †, K.,* Dr.-Ing., Trier,
                  Bohn, Th., Prof. techn., Essen,
                  Werner, K., Dipl.-Ing., Essen,
                  Stephan, K., Prof. Dr.-Ing., Stuttgart,
                  Auracher, H., Dr.-Ing., Stuttgart,
Chapter 10: *Goldstern, W.,* FIMechE FInst CE, Timperley/GB,
                  Bitterlich, W., Dr.-Ing., Essen,
Chapter 11: *Ritgen, G.,* Dipl.-Ing., Duisburg,
                  Gaffal, K., Dipl.-Ing., Frankenthal,
                  Hasenrahm, H., Dipl.-Ing., Moers,

Chapter 12: *Hentschel, W.,* Dipl.-Ing., Köln,
Oberländer, G., Dipl.-Ing., Köln,
Chapter 13: *Kunst, B.,* Prof. Dipl.-Ing., Köln,
Korek, J., Dipl.-Ing., Köln,
Chapter 14: *Ritgen, G.,* Dipl.-Ing., Duisburg,
Hoene, E., Dipl.-Ing., Karlsruhe,
Arens-Fischer, F., Dipl.-Ing., Essen,
Vesper, H., Dipl.-Ing., Erlangen.

They and many more, who have assisted this work deserve our thanks for their co-operation.

The management of the VDI-Gesellschaft Energietechnik supported the project effectively; whereby the activities of Ms. U. Stricker-Berghoff, Dipl.-Ing., must be specially mentioned.

Essen, April 1992
VDI-Gesellschaft Energietechnik
Committee "Wärmetechnische Arbeitsmappe"
The Chairman
Prof. Dr. *Th. Bohn*

Extract from the Preface to the First German Edition (1934)

Since April 1932 the "Archiv für Wärmewirtschaft und Dampfkesselwesen" ("Archives for Heat Economics and Steam Boiler Engineering") regularly publishes diagrams, in which information for heat engineering calculations is provided in ready usable form. They include data presented in graphs, which are important for solving individual problems and which are based on most recent and reliable sources. The use of each sheet is clearly illustrated by an example. Obviously, the accuracy of the presentation is limited by the available space and the range required; but it is generally adequate for practical calculations in operation and design. References of the sources used and the relevant literature facilitate further study of the subject. The worksheets have mainly been designed by Dipl.-Ing. *W. Goldstern,* who has initiated the publication of such sheets in the "Archiv". The compilation of these worksheets in the Reference Book cannot claim to cover any subject completely.

Contents Working sheet no.

8. Cooling Systems

9. Pipework

10. Energy Storage

11. Driven Machines

12. Gas Turbine Plants

13. Cogeneration

14. Thermal Measurement Technique

Comment

1° This book contains diagrams drawn in Germany. Therefore, you will find in those numbers a German ", =comma" instead of the English ". =decimal point".

2° The working sheet no 2.1.4.2 is added at the end of the book a second time on waxed tissue paper. You will need this sheet for the calculations with sheet no 2.1.4.1.

An increasing use of the International System Units (SI-Units) requires the use of conversion factors, that combine these units with that of the technical system. The largest part of the following tables (which in addition contain the most important British units) has been taken from INGENIEURWISSEN, vol. 4 "Die Umstellung auf das internationale Einheitensystem in der Mechanik und Wärmetechnik" by *H. W. Hahnemann* (VDI-Verlag, Düsseldorf).

Naturally, many of the conversion factors, given in these tables, have to be rounded to the required accuracy before use.

1. Force

Unit	Newton N	Kilopond kp	pound lb
1 N =	1	0.10197	0.2247
1 kp =	9.80665	1	2.2046
1 lb =	4.45	0.4536	1

In France also 1 sthene (sn) = 10^3 N

2. Pressure

Unit	bar	kp/cm²	Torr	atm	pound per square foot lb/ft²	pound per square inch lb/in²
1 bar = 10^5 N/m² =	1	1.0197	750.06	0.98692	2089	14.5038
1 kp/cm² = 1 at =	0.9807	1	735.56	0.96784	2048	14.223
1 Torr = 1 mm QS (bei 0 °C) =	$1.333 \cdot 10^{-3}$	$1.3595 \cdot 10^{-3}$	1	$1.3158 \cdot 10^{-3}$	2.7841	0.01934
1 atm =	1.01325	1.03323	760	1	$2.116 \cdot 10^{-3}$	14.696
1 lb/ft² =	478.8	$0.4883 \cdot 10^{-3}$	0.3591	472.58	1	1/144
1 lb/in² =	0.06895	0.07031	51.715	0.06806	144	1

In France also 1 pièze (pz) = 10^3 N/m² und 1 hectopièze (hpz) = 100 pz = 1 bar

3. Energy

Unit	Joule J	Kilopond-meter kpm	Kilocalorie kcal	Kilowatt-hour kWh	Horse power-hour PSh	Britisch thermal unit Btu
1 J = 1 Nm = 1 Ws = 1 kg m²/s² =	1	0.101972	$2.38844 \cdot 10^{-4}$	$2.77778 \cdot 10^{-7}$	$3.77673 \cdot 10^{-7}$	$9.47817 \cdot 10^{-4}$
1 kpm =	9.80665	1	$2.34228 \cdot 10^{-3}$	$2.72407 \cdot 10^{-6}$	$3.70370 \cdot 10^{-6}$	$9.29491 \cdot 10^{-3}$
1 kcal =	$4.1868 \cdot 10^3$	426.935	1	$1.16300 \cdot 10^{-3}$	$1.58124 \cdot 10^{-3}$	3.96832
1 kWh =	$3.6 \cdot 10^6$	$3.670978 \cdot 10^5$	859.845	1	1.35962	$3.41214 \cdot 10^3$
1 PSh =	$2.647796 \cdot 10^6$	$2.70000 \cdot 10^5$	632.416	0.735499	1	$2.50963 \cdot 10^3$
1 Btu =	$1.055056 \cdot 10^3$	107.5857	0.251996	$2.93071 \cdot 10^{-4}$	$3.98466 \cdot 10^{-4}$	1

4. Specific Heat Capacity

Unit	Joule per Kilogram and Kelvin J/(kg·K)	Kilocalorie per Kilogram and Kelvin kcal/(kg·K)	Kilowatt-hour per Kilogram and Kelvin kWh/(kg·K)	British thermal unit per pound and degree [1] Btu/(lb·deg)
1 J/(kg·K) =	1	$2.38844 \cdot 10^{-4}$	$2.77778 \cdot 10^{-7}$	$2.38844 \cdot 10^{-4}$
1 kcal/(kg·K) =	4186.8	1	$1.16300 \cdot 10^{-8}$	1
1 kWh/(kg·K) =	$3.6 \cdot 10^6$	859.845	1	859.845
1 Btu/(lb·deg) =	4186.8	1	$1.16300 \cdot 10^{-3}$	1

1) degrees are in Fahrenheit

5. Heat – Flux Density

Unit	Watt per square meter W/m^2	Kilowatt per square centimeter kW/cm^2	Kilocalorie per square-meter and hour $kcal/(m^2 \cdot h)$	British thermal unit per square inch and second $Btu/(in^2 \cdot sec)$	British thermal unit per square foot and second $Btu/(ft^2 \cdot sec)$	British thermal unit per square foot and hour $Btu/(ft^2 \cdot hr)$
1 W/m^2 =	1	$0.1 \cdot 10^{-6}$	0.860	$0.612 \cdot 10^{-6}$	$88.06 \cdot 10^{-6}$	0.317
1 kW/cm^2 =	$10 \cdot 10^6$	1	$8.6 \cdot 10^6$	6.12	880.6	$3.17 \cdot 10^6$
1 $kcal/(m^2 \cdot h)$ =	1.163	$11.63 \cdot 10^{-8}$	1	$71.17 \cdot 10^{-8}$	$1.024 \cdot 10^{-4}$	0.3687
1 $Btu/(in^2 \cdot sec)$ =	$1634 \cdot 10^3$	$16.34 \cdot 10^{-2}$	$1.405 \cdot 10^6$	1	144	$51.84 \cdot 10^4$
1 $Btu/(ft^2 \cdot sec)$ =	$11.35 \cdot 10^3$	$1.135 \cdot 10^{-3}$	$9.765 \cdot 10^3$	$6.944 \cdot 10^{-3}$	1	3600
1 $Btu/(ft^2 \cdot hr)$ =	3.154	$31.54 \cdot 10^{-8}$	2.713	$1.929 \cdot 10^{-6}$	$2.778 \cdot 10^{-4}$	1

6. Thermal Conductivity

Unit	Watt per meter and Kelvin $W/(m \cdot K)$	Kilocalorie per meter, hour and Kelvin $kcal/(m \cdot h \cdot K)$	British thermal unit per square foot, hour, and degree[1]) per inch $Btu\ in/(ft^2 \cdot hr \cdot deg)$	British thermal unit per foot, hour and degree[1]) $Btu/(ft \cdot hr \cdot deg)$	British thermal unit per inch, hour and degree[1]) $Btu/(in \cdot hr \cdot deg)$
1 $W/(m \cdot K)$ = $J/(m \cdot s \cdot K)$ =	1	0.86	6.935	0.5779	0.04815
1 $kcal/(m \cdot h \cdot K)$ =	1.163	1	8.064	0.6719	0.05599
1 $Btu\ in/(ft^2 \cdot hr \cdot deg)$ =	0.1442	0.1240	1	0.08333	$6.944 \cdot 10^{-3}$
1 $Btu/(ft \cdot hr \cdot deg)$ =	1.731	1.488	12	1	0.08333
1 $Btu/(in \cdot hr \cdot deg)$ =	20.77	17.858	144	12	1

1) degree in Fahrenheit

7. Heat Transfer Coefficient and Heat Transmission Figure

Unit	Watt per square meter and Kelvin $W/(m^2 \cdot K)$	Kilocalorie per square meter, hour and Kelvin $kcal/(m^2 \cdot h \cdot K)$	British thermal unit per square foot, hour and degree $Btu/(ft^2 \cdot hr \cdot deg)$
1 $W/(m^2 \cdot K)$ = 1 $J/(m^2 \cdot s \cdot K)$ =	1	0.859845	0.1761
1 $kcal/(m^2 \cdot h \cdot K)$ =	1.163	1	0.2048
1 $Btu/(ft^2 \cdot hr \cdot deg)$ =	5.681	4.886	1

8. Radiation Coefficient

Unit	Watt per square meter and Kelvin[4] $W/(m^2 \cdot K^4)$	Kilocalorie per square meter, hour and Kelvin[4] $kcal/(m^2 \cdot h \cdot K^4)$	British thermal unit per square foot, hour and degree[4] $Btu/(ft^2 \cdot hr \cdot deg^4)$
1 $W/(m^2 \cdot K^4)$ = 1 $J/(m^2 \cdot s \cdot K^4)$ =	1	0.859845	$3.020 \cdot 10^{-2}$
1 $kcal/(m^2 \cdot h \cdot K^4)$ =	1.163	1	$3.512 \cdot 10^{-2}$
1 $Btu/(ft^2 \cdot hr \cdot deg^4)$ =	33.11	28.49	1

9. Dynamic Viscosity

Unit	Pascal-second $Pa\ s$	Poise P	Kilogram per meter and hour $kg/(m \cdot h)$	Kilopond-second per square meter $kp\ s/m^2$	Kilopond-hour per square meter $kp\ h/m^2$	pound-mass per foot and second $\dfrac{\text{lb-mass}}{\text{ft} \cdot \text{s}}$	pound-force second per square foot $\dfrac{\text{lb-force} \cdot \text{s}}{\text{ft}^2}$
1 $Pa\ s$ = 1 $N\ s/m^2$ = 1 $kg/(m \cdot s)$ =	1	10	3600	0.10197	$2.833 \cdot 10^{-3}$	0.6721	$2.0885 \cdot 10^{-2}$
1 P =	0.1	1	360	0.010197	$2.833 \cdot 10^{-6}$	0.06721	$2.0885 \cdot 10^{-3}$
1 $kg/(m \cdot h)$ =	$2.778 \cdot 10^{-4}$	$2.778 \cdot 10^{-3}$	1	$2.833 \cdot 10^{-5}$	$78.68 \cdot 10^{-10}$	$1.867 \cdot 10^{-4}$	$5.801 \cdot 10^{-6}$
1 $kp\ s/m^2$ =	9.807	98.07	$3.5304 \cdot 10^4$	1	$2.778 \cdot 10^{-4}$	6.5919	0.20482
1 $kp\ h/m^2$ =	$0.35304 \cdot 10^5$	$0.35304 \cdot 10^6$	$1.2709 \cdot 10^8$	3600	1	$2.3730 \cdot 10^{-4}$	$0.73728 \cdot 10^3$
1 lb-m./(ft $\cdot$ s) =	1.488	14.882	5357	0.1518	$4.214 \cdot 10^{-4}$	1	0.03108
1 lb-force s/ft^2 =	47.88	478.8	$1.724 \cdot 10^5$	4.882	$1.3558 \cdot 10^{-3}$	32.174	1

Depreciation of Investment

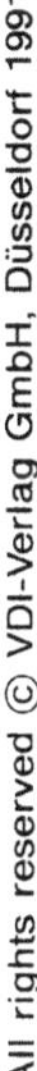

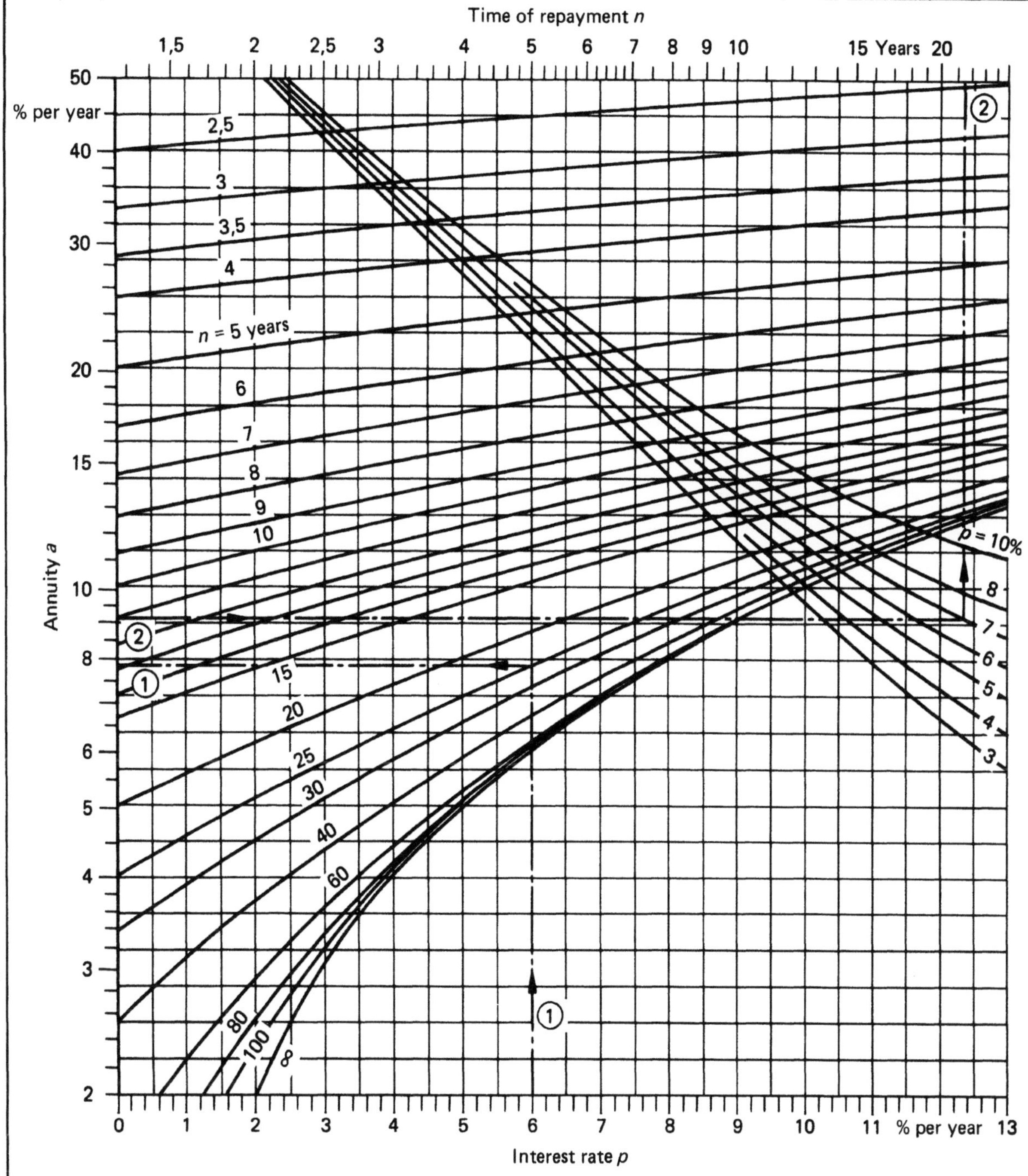

Example ①: Interest rate p % per year $p = 6.0$
Runtime for amortization,
Number of years $n = 25$
Results in
Annuity a % per year $a = 7.8$.

Example ②: Annuity a % per year $a = 9$
Interest rate p % per year $p = 7$
Results in
Runtime for amortization,
Number of years $n = 21.5$.

Basic equation: $a = \frac{q^n p}{q^n - 1}$, with $q = 1 + \frac{p}{100}$.

Explanation

The annuity a in % per year is the yearly amortization rate related to the investment costs, which have to be paid at the end of each year for depreciation and interest.

Example ① calculates the annuity, if interest rate (p %/year) and runtime for amortization (number of years n) are known.

Example ② calculates the runtime for amortization, if (for example for economic calculations) annuity and interest rate are known.

Hütte, 28th edition 1955, vol. 1 p. 65/66

W. Lenz, W. Goldstern

Name	Symbol	Name	Symbol
Steam		Combustable waste	
Steam containing oil		Other substances	
Cycle water		Control impulse	
Water containing oil		Pipe with heating or cooling	
Raw water		Pipe heated by steam	
Sludge water, waste water		Pipe heated by electricity	
Solutions, chemicals		Thermal insulation	
Oil		Intersection of two pipelines without junction	
Liquid metals		Intersection of two pipelines with junction	
Air		Branch point	
Inflammable gases		Hopper	
Non-inflammable gases		Reducer	
Solid fuels		Exhaust	
Shut-off fittings, general		Shut-off fitting magnetically operated	
Shut-off fittings with safety device		Shut-off fitting hydraulically operated	
Non-return fittings, general		Shut-off fitting piston operated	
Shut-off valve		Shut-off fitting membrane operated	
Non-return valve		Dead weight safety valve	
Non-return ball valve		Spring loaded safety valve	
Closure ball operated valve		Shut-off fitting, closed	
Pressure reducing valve		Shut-off fitting, opened	
Steam reducing valve		Sound absorber	
Stop valve		Blind disk	
Straight through valve		Changeable disk	
Isolating flap		Pressure test valve	
Non-return valve		Throttle disk	
Shut-off fitting operated by hand		Rupture disk	
Shut-off fitting operated by motor		Steam trap	

Name	Symbol
Cross flow heat exchanger (the heat absorbing fluid is flowing through serrated line)	
Flue gas heated heat exchanger	
Condensate heated economizer	
Tubular feedwater heater	
Condensate cooler	
Non cross flow heat exchanger (the heat absorbing fluid is flowing through serrated line)	
Flue gas heated air heater	
Water cooled condenser	
Air cooled condenser	
Mixing feedwater heater	
Spray attemperator	
Deaerator, mixing preheater	
Steam generator without superheater	
Conveyor, general	
Feeder for solid fuel	
Clutch, general	
Hydraulic clutch	
Gear, general	
Separator, general	
Electrostatic precipitator	
Cyclone	
Flash tank	
Sieve	
Coarse rake	
Fine rake	
Filter device, general	
Air filter, gas filter	
Filter for liquids	

Name	Symbol
Steam generator, heated by electricity	
Steam regenerator	
Steam generator with superheater	
Steam generator with flue gas heated reheater	
Steam generator, gas fired	
Steam generator, oil fired	
Steam generator, coal dust fired	
Steam generator, coal dust fired with liquid ash removal	
Steam generator, stoker fired	
Steam generator, fired by waste	
Gasification plant	
Furnace, combustion chamber	
Heat consumer	
Heat consumer with heating surface	
Open tank, general	
Agitator, general	
Chemical water treatment plant, general	
Water treatment tank	
Dosing plant	
Cooling tower, general	
Wet cooling tower with natural draught	
Wet cooling tower with suction fan	
Wet cooling tower with pressurized fan	
Dry cooling tower with natural draught	
Dry cooling tower with suction fan	
Dry cooling tower with pressurized fan	
Wet-dry cooling tower with natural draught	

Name	Symbol	Name	Symbol
Storage tank, general		Generator, rotating	
Storage tank with trickle deaeration		Pump for liquids, general	
Storage tank with trickle deaeration and steam supply at the bottom		Rotary pump	
Drop accumulator		Reciprocating pump	
Drive machine with expansion of fluid		Ejector	
Steam turbine		Turbo compressor, fan	
Steam turbine, non-automatic extraction		Inlet vane control	
Steam turbine, automatic extraction		Outlet vane control	
Gas turbine		Size reduction machine, general	
Drive machine with reciprocating piston		Crusher	
Reciprocating steam engine		Mill, general	
Diesel engine, Otto engine		Hammer mill	
Electric motor		Ball mill	
Alternating current (a.c.) motor		Tube mill	
Direct current (d.c.) motor		Roller mill	
Spray attemperator with water injection and temperature regulation to achieve a constant steam temperature		Squirrel-cage motor, stator winding in wye	
Pressure reducing valve opens with sinking pressure in line b.		Squirrel-cage motor with start up capacitor	
Pressure reducing valve with attemperator opens with sinking pressure in line b, but also, when the pressure in line a reaches a certain level. The water injection valve automatically controls the steam temperature behind the attemperator.		Choke coil	
Generator, general		Transformer with two separated coils	
D.c. generator		Transformer with three separated coils	
A.c. three phase generator, general		Energy saving transformer	
Electric motor, general		Choke coil with continuous control	
D.c. motor, general		Transformer with step by step control	
A.c. motor, general		Closing switch	
Syncrone motor, general		Opening switch	
Motor with two winding rotor, stator winding in wye		Alternating switch	
		Circuit breaker	
		Current transformer with 1 kernel	
		Current transformer with 2 kernel	
		Voltage transformer	

Name	Symbol			
	Measurement		Identification	Measurement device
	Form A	Form B		
Flow metering, general	⌣⌢	⊖	F	▱
Level metering	∇	⊽	L	▽
Humidity metering	%	⊛	M	%
Pressure metering	→●←	⊙	P	→●←
pH-value metering	⌀	–	Q	–
Conductivity metering	⌀	–	Q	–
Number of revolutions metering	○	◎	S	○
Temperature metering	⊤	⊕	T	⊤
Expansion metering	⇄	⊕		⇄
Vibration metering	∿	⊘		∿

Name	Symbol	Name	Symbol
Main impulse, opens on rise of control variable	⊕	Limiting impulse, closes at upper limit of control variable	⇡
Main impulse, opens on drop of control variable	⊖	Limiting impulse, closes at lower limit of control variable	▽
Limiting impulse, opens at upper limit of control variable	⊕	Level control of water height	
Limiting impulse, opens at lower limit of control variable	⊖		

Bibliography

Handbuch Energie edited by Prof. Dr. *Thomas Bohn*. vol. 6: Fossil beheizte Dampfkraftwerke.
Technischer Verlag, Resch Verlag, TÜV Rheinland

General Remarks

In all steam diagrams, zero of enthalpy and entropy is defined at the triple point of water ($\vartheta = 0.01\,°\mathrm{C}$, $p = 0.00612$ bar) with water in the liquid state. The values of the diagrams of properties (oblique-angled h,s-, T,s-, h,p-, and h,ϱ-diagrams) have been calculated according to the equations of *Juza* [1], because these equations are valid even for pressures above 1000 bar.

In order to show the example of the liquid state, for the h,s-diagram the oblique-angled drawing according to *Bošnjaković* [2] has been chosen.

The diagrams for the compressibility factors, the specific heat capacity, the isentropic and isenthalpic exponent, thermal conductivity and dynamic viscosity have been calculated according to the equations of the Steam Tables [3].

Mollier (h, s)-Diagram for Water and Steam

The properties of water and steam have been calculated according to the equations of *Juza* [1], because they are valid for pressures up to 100 000 bar.

For the diagram the oblique-angled drawing according to *Bošnjaković* [2] has been chosen. Here, the lines of constant enthalpy are inclined in such a way, that the line of constant temperature of liquid water at $\vartheta = 0\,°\mathrm{C}$ will be parallel to the abscissa. Thus it is possible to show the whole range of properties of superheated steam up to liquid water and ice in a single diagram with an appropriate resolution.

Even the region of very high pressures is shown, which, for example, becomes more and more important for the manufacturing industry. In the region of high pressures, the diagram has been complemented by the curves of the different modifications of ice (I to VIII) [2]. Lines of constant pressure, constant temperature and constant specific volume are shown.

Zero for enthalpy and entropy have been chosen according to the VDI-Steam Tables [3] for liquid water at 1 bar and 0 °C.

Bibliography

[1] *Juza J.:* An Equation of State for Water and Steam. Academia, Nakladatelstiv Ceskoslovenske Akademie VED, Prag 1966
[2] *Bošnjaković F., Renz, U., Burow, P.:* Mollier Enthalpy, Entropy Diagram of Water. Tehnicka Knjiga, Zagreb
[3] *Schmidt E.:* Zustandsgrößen von Wasser und Wasserdampf, Springer-Verlag, Berlin 1969

Additional information of this book

(Engineering Reference Book on Energy and Heat; 978-3-642-51125-7;

978-3-642-51125-7_OSFO1) is provided:

http://Extras.Springer.com

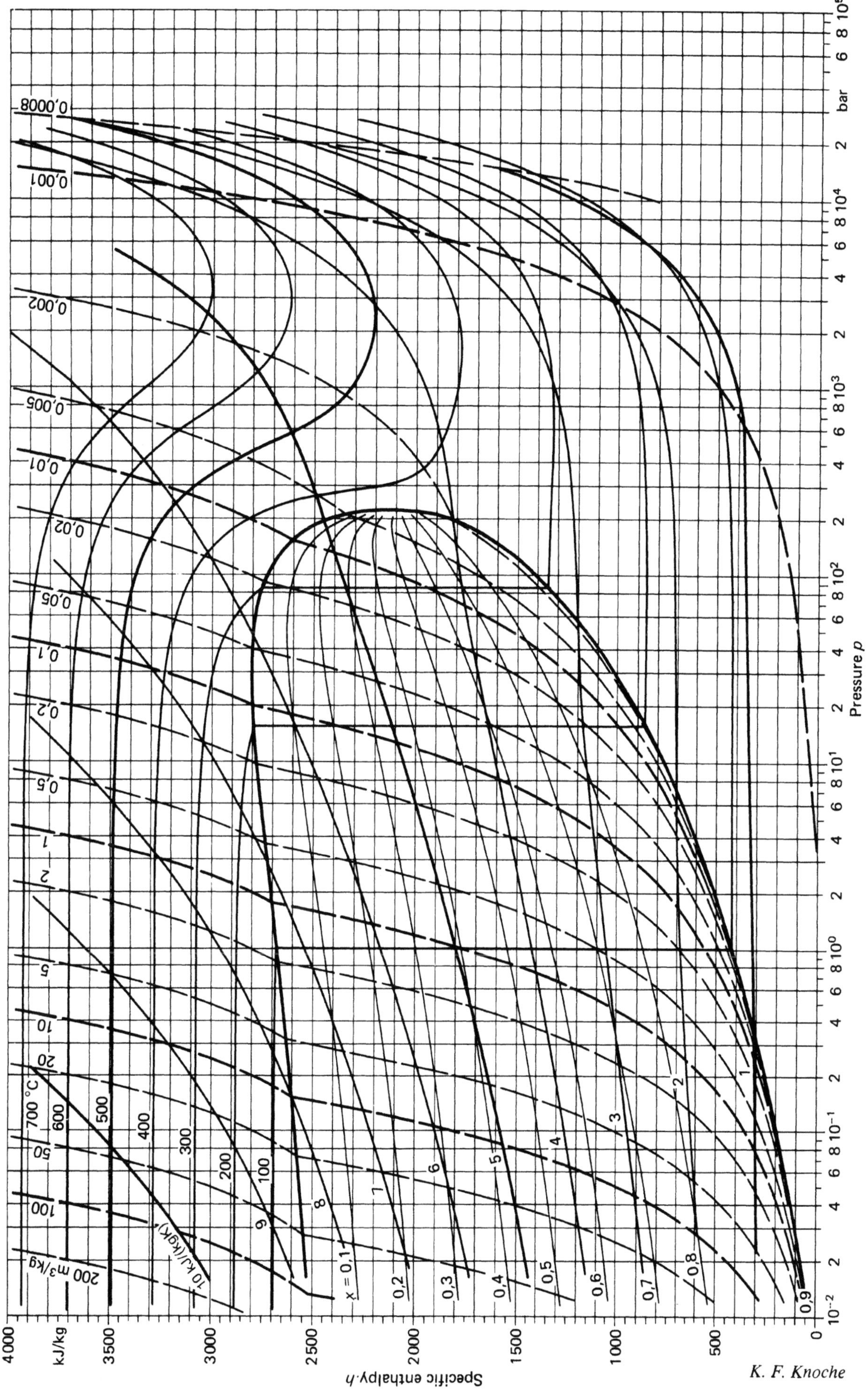

Pressure p
bar
Specific enthalpy h
kJ/kg
200 m³/kg
10 kJ/(kgK)
700 °C
x = 0,1

K. F. Knoche

Enthalpy-Density-Diagram for Water and Steam

The enthalpy-density-diagram is especially useful to examine one dimensional flows. The enthalpy has a linear and the density a logarithmic axis. The properties of state are the same as in the Mollier h, s-diagram. With the local velocity c, the volume flow $\dot{V}$ in a channel of the cross sectional area A is

$$\frac{\dot{V}}{\dot{V}_0} = \frac{A}{A_0} \frac{c}{c_0} \tag{1}.$$

In this equation $\dot{V}_0$, A_0 and c_0 are arbitrary reference values, for example $\dot{V}_0 = 1\,\text{m}^3/\text{s}$, $A_0 = 1\,\text{m}^2$, $c_0 = 1\,\text{m/s}$.

If the velocity c in equation (1) will be substituted by the specific total enthalpy of the fluid h_t and the specific static enthalpy h (equation 2)

$$h_t = h + 1/2\,c^2 \tag{2}$$

one obtains after drawing the logarithm:

$$\log \frac{\dot{V}}{\dot{V}_0} = \log \frac{A}{A_0} + \log \frac{c}{c_0} =$$
$$= \frac{A}{A_0} + \frac{1}{2} \log\left(\frac{h_t - h}{c_0^2/2}\right) \tag{3}.$$

These curves are shown in the transparent worksheet No. 2.1.4.2. The lines of constant cross section area A are called Fanno curves. At any point their gradient equals the square of the velocity:

$$\left[\frac{\partial(h_t - h)}{\partial \log \dfrac{\dot{V}}{\dot{V}_0}}\right]_A = \left[\frac{\partial \dfrac{c^2}{2}}{\partial \log \dfrac{\dot{V}}{\dot{V}_0}}\right]_A = c^2 \tag{4}.$$

By superimposing the specific total enthalpy of the fluid h_t in the transparent worksheet No. 2.1.4.2 with the equivalent value in the enthalpy-density-diagram, and besides, the line A = const. with the properties of flow h, ϱ (as shown in figure 1), according to the continuity equation of the specific mass flow

$$\log \frac{\dot{m}}{\dot{m}_0} = \log \frac{\varrho}{\varrho_0} + \log \frac{\dot{V}}{\dot{V}_0} =$$
$$= \log \frac{\varrho}{\varrho_0} + \log \frac{A}{A_0} + \log \frac{c}{c_0} \tag{5}$$

can be determined in the transparent worksheet at the axis for volume flow. If, on the contrary, the related mass flow $\dot{m}/\dot{m}_0$ is known, then this value has to be superimposed on the transparent worksheet with the line $\varrho_0 = $ const. and the specific total enthalpy of the fluid h_t in the h, p-diagram. Now it is possible to determine the square area of the nozzle A, which is correlated to the specific flow conditions of h and ϱ.

The throat area of the nozzle (laval condition L) is indicated by the point, where the Fanno curve A = const. and the expansion line touch each other. The equivalent velocity c_L may be determined from the velocity curves in the transparent worksheet.

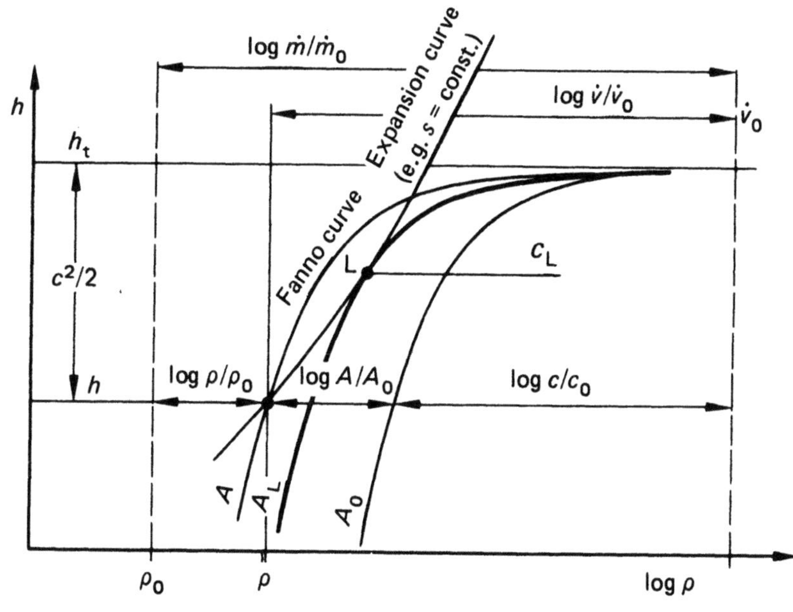

Figure 1:
Nozzle correction on one dimensional flow by superimposition of diagram 2.1.4.1 and 2.1.4.2.

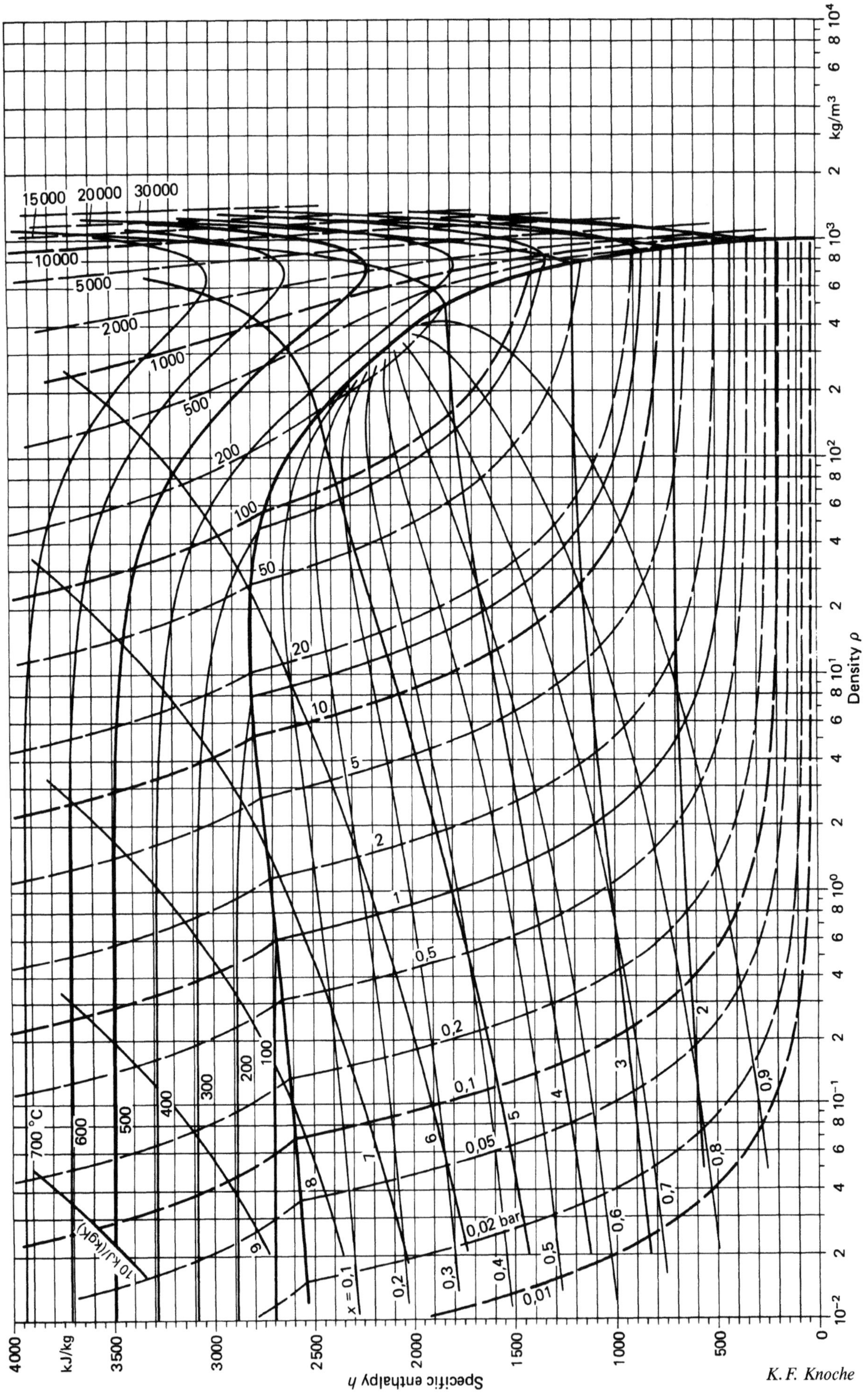
Specific enthalpy *h*
kJ/kg
Density *ϱ*
kg/m³
700 °C
600
500
400
300
200
100
10 kJ/(kgK)
x = 0,1
0,02 bar
0,01
15000
20000
30000
10000
5000
2000
1000

K. F. Knoche

Fanno Curves

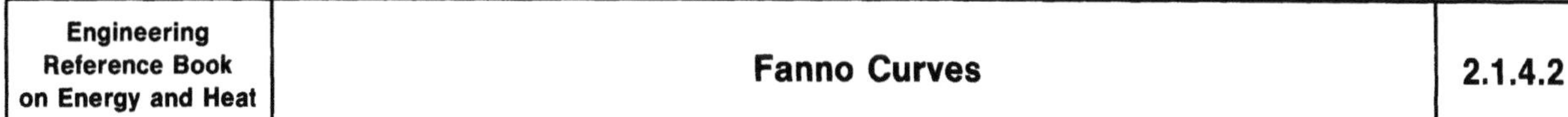

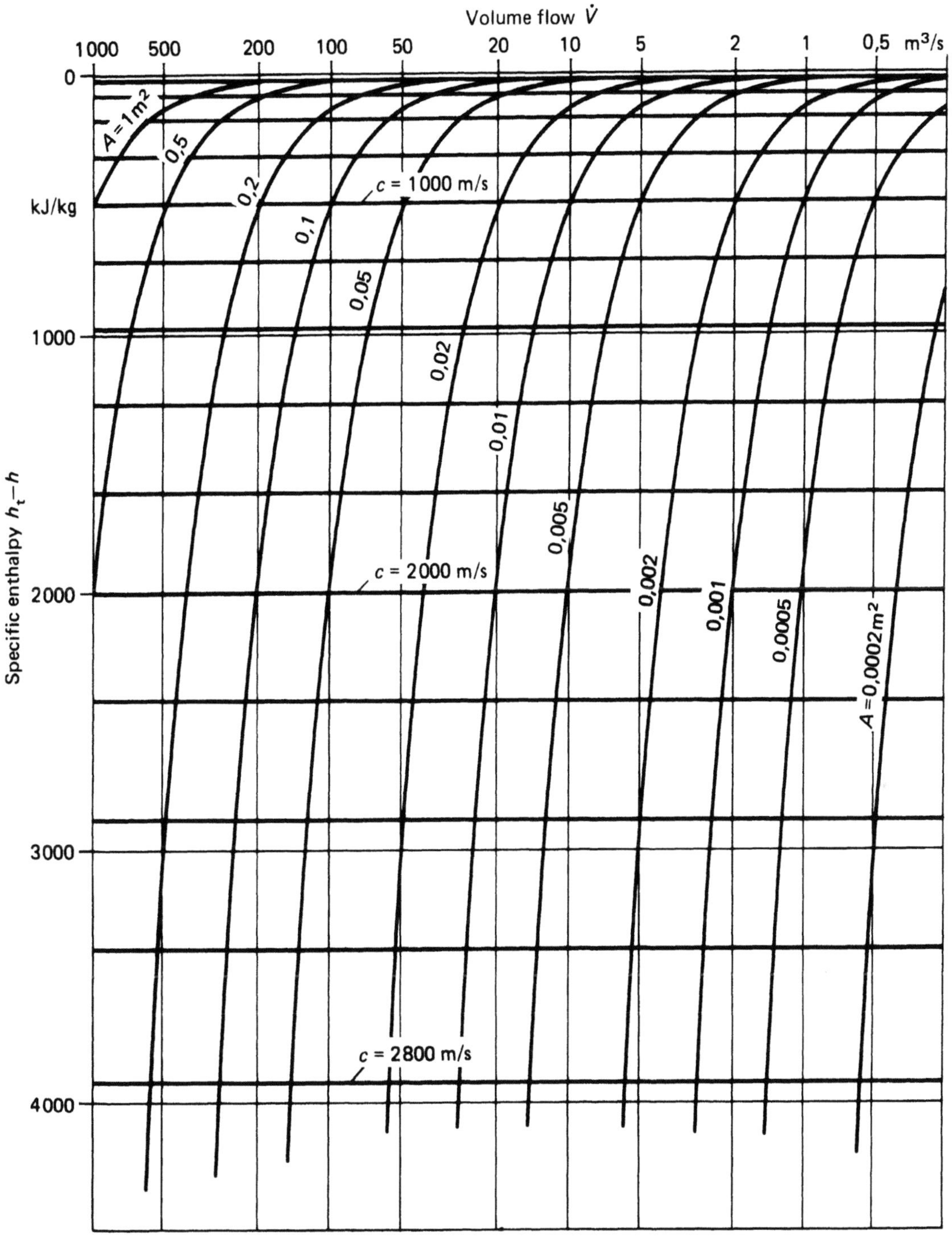
Volume flow $\dot{V}$
1000 500 200 100 50 20 10 5 2 1 0,5 m³/s
0
kJ/kg
1000
2000
3000
4000
Specific enthalpy $h_t - h$
$A = 1\,m^2$
0,5
0,2
0,1
0,05
0,02
0,01
0,005
0,002
0,001
0,0005
$A = 0{,}0002\,m^2$
c = 1000 m/s
c = 2000 m/s
c = 2800 m/s

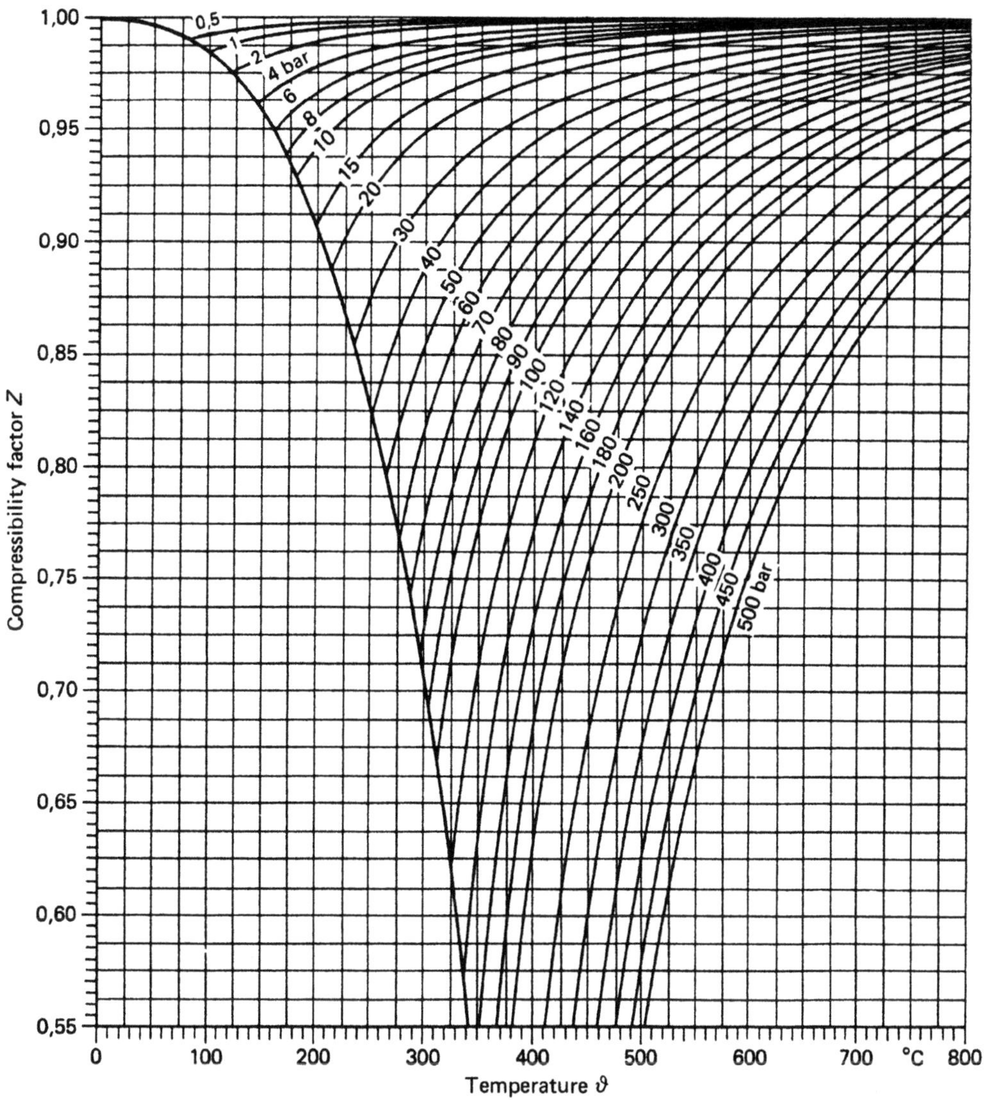

Calculated according to: Properties of Water and Steam in SI-Units; Professor Dr.-Ing. *Ernst Schmidt;* Springer-Verlag Berlin, Heidelberg, New York; 1969

K. F. Knoche

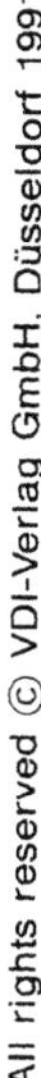

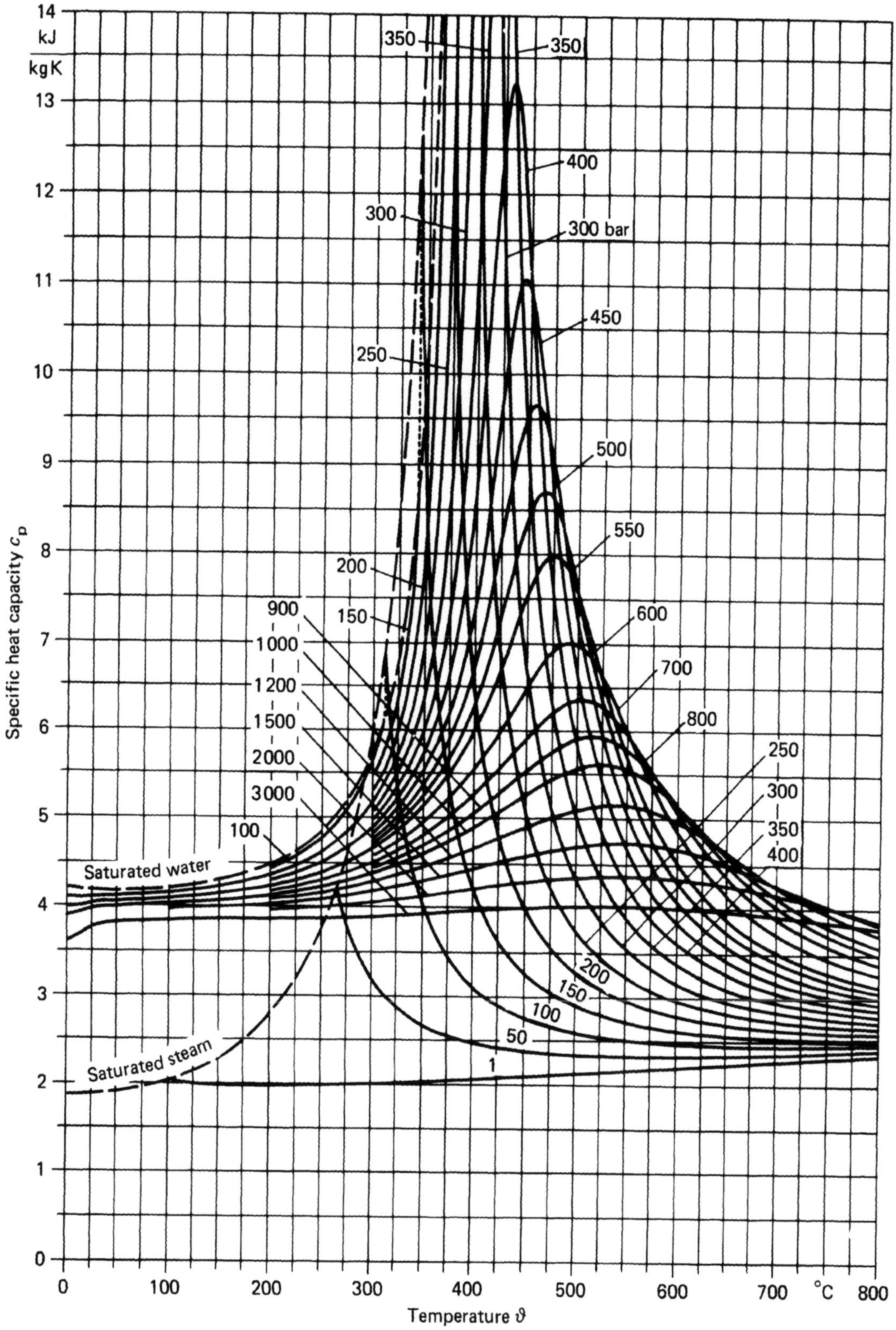

Bibliography

Haar, Lester, J.S. Gallagher, G.S. Kell: NBS/NRC Wasserdampftafeln; Springer-Verlag
Berlin, Heidelberg, New York, London, Paris, Tokyo; 1988

K.F. Knoche

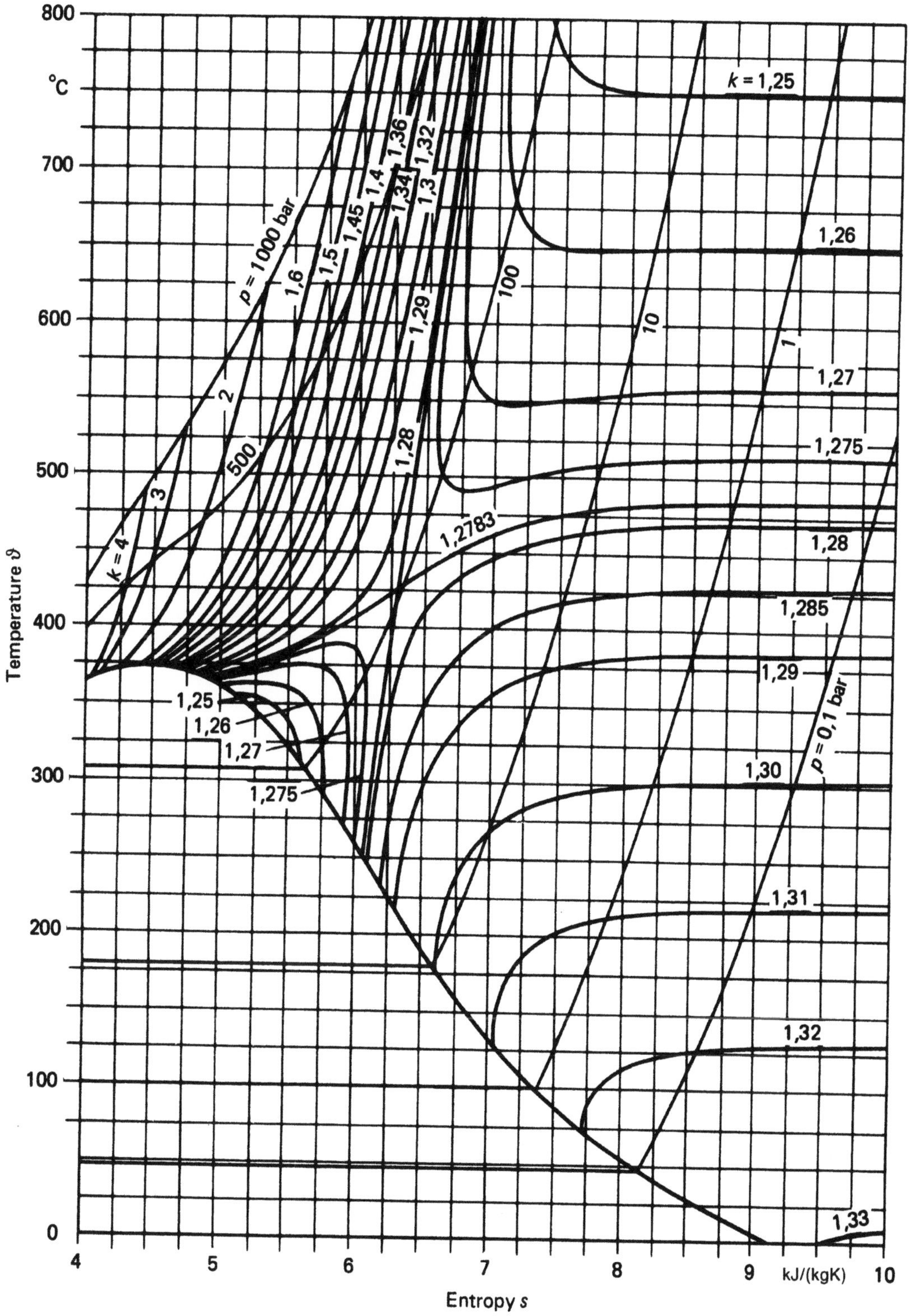

Bibliography

Properties of Water and Steam in SI-Units; Professor Dr.-Ing. *Ernst Schmidt*; Springer-Verlag
Berlin, Heidelberg, New York; 1969

K. F. Knoche

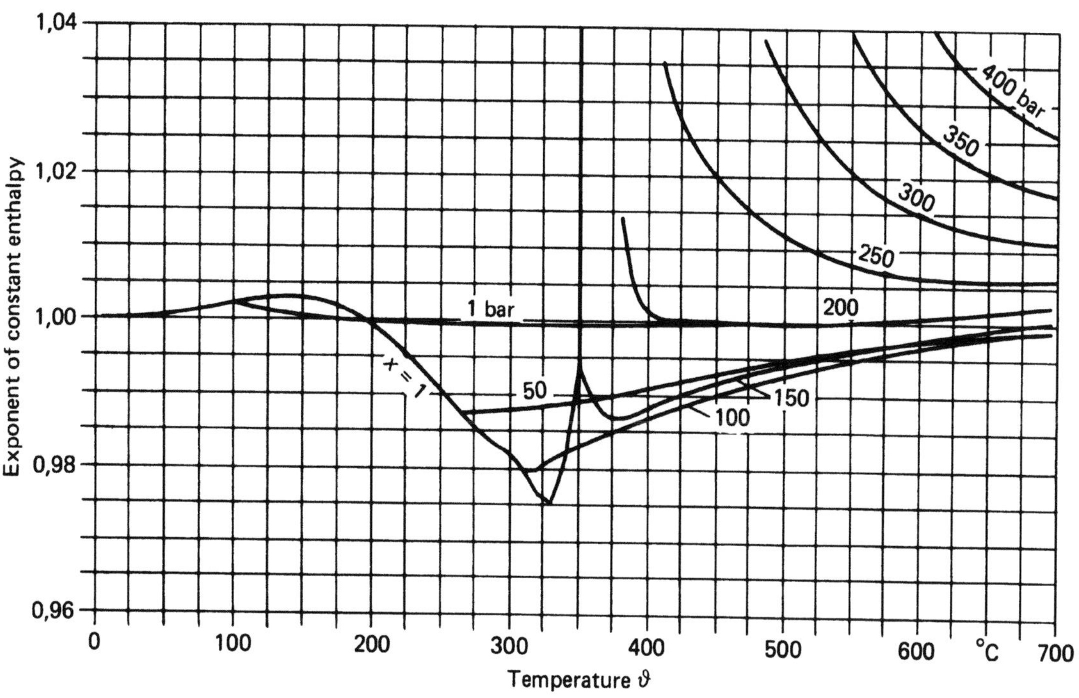

Bibliography

Properties of Water and Steam in SI-Units; Professor Dr.-Ing. *Ernst Schmidt;* Springer-Verlag
Berlin, Heidelberg, New York; 1969

K. F. Knoche

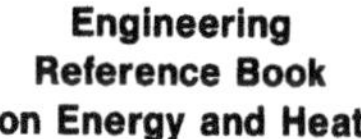

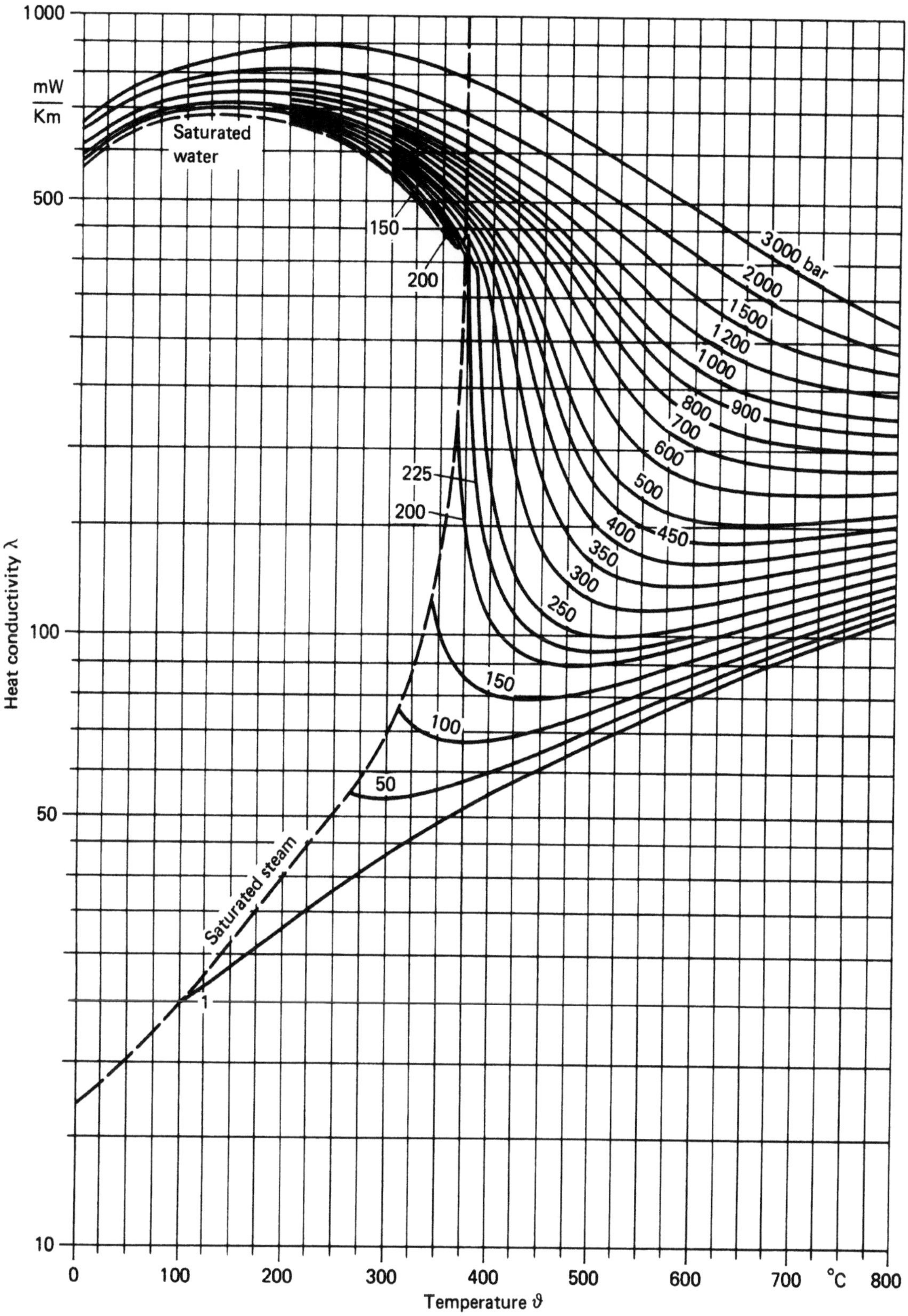

Bibliography

Haar, Lester, J.S. Gallagher, G.S. Kell: NBS/NRC Wasserdampftafeln; Springer-Verlag
Berlin, Heidelberg, New York, London, Paris, Tokyo; 1988

K. F. Knoche

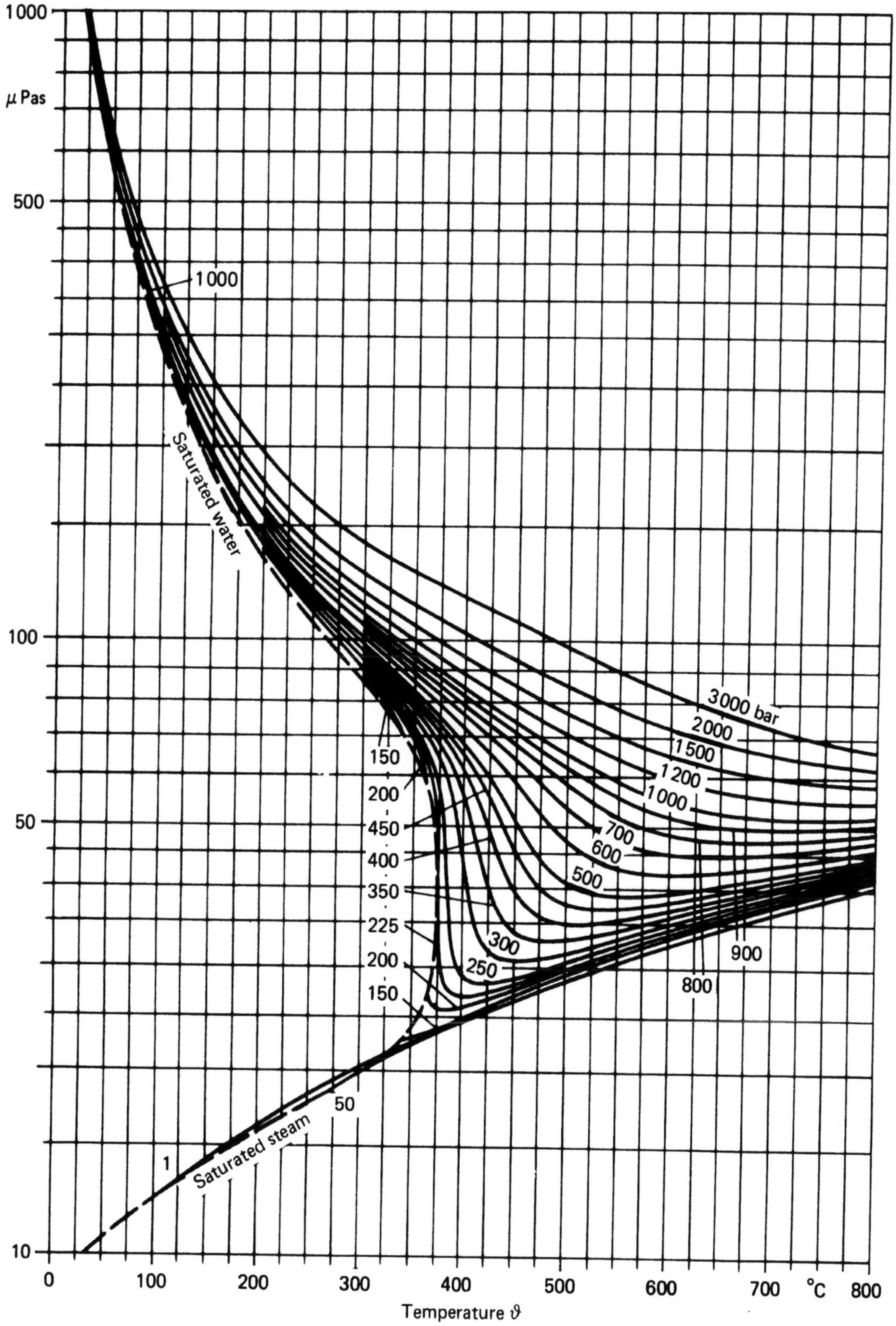

Bibliography

Haar, Lester, J.S. Gallagher, G.S. Kell: NBS/NRC Wasserdampftafeln; Springer-Verlag
Berlin, Heidelberg, New York, London, Paris, Tokyo; 1988

K.F. Knoche

Additional information of this book

(Engineering Reference Book on Energy and Heat; 978-3-642-51125-7; 978-3-642-51125-7_OSFO2) is provided:

http://Extras.Springer.com

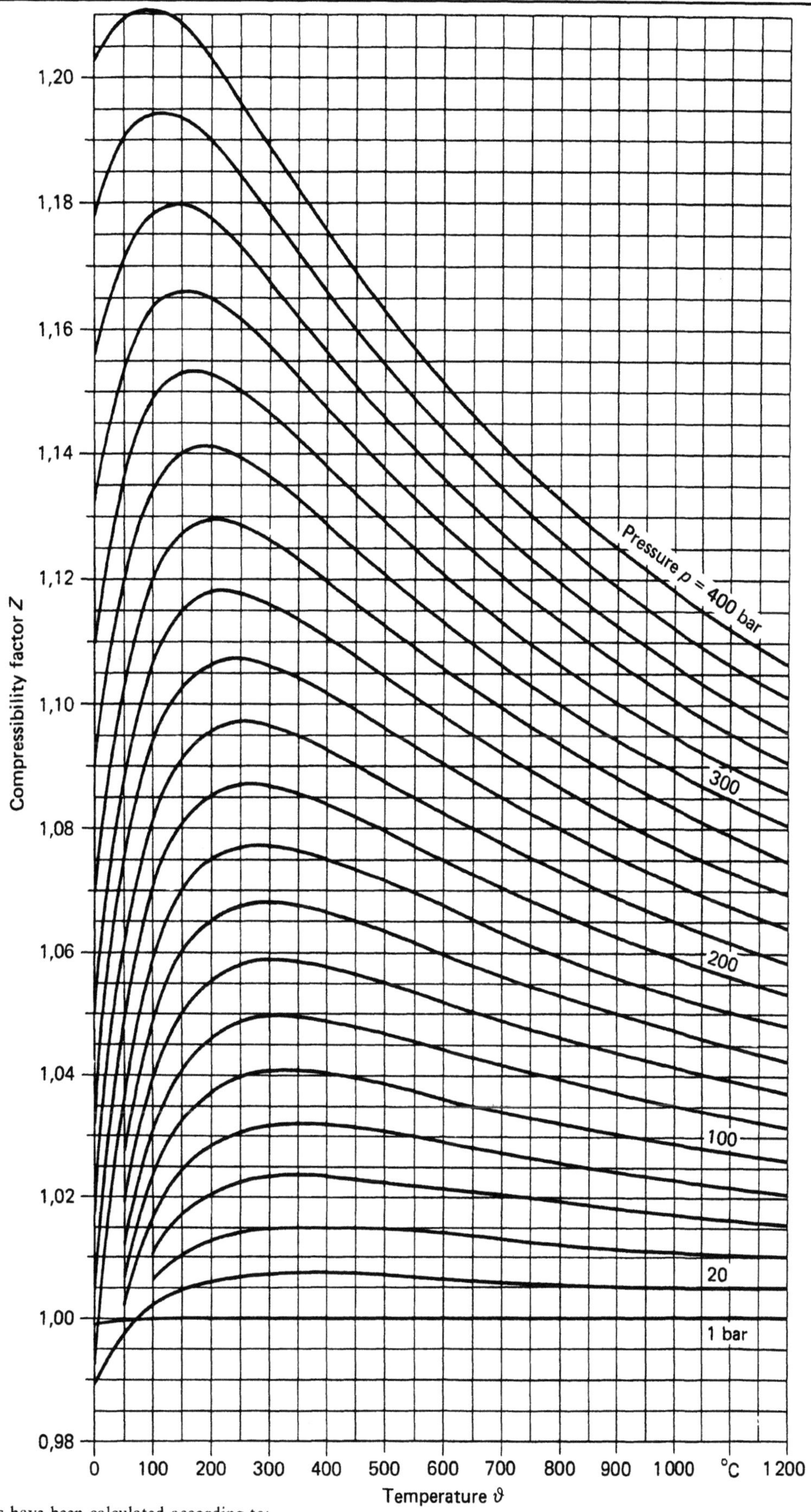

The curves have been calculated according to:

Baehr H. D.: Thermodynamische Eigenschaften der Gase und Flüssigkeiten. Vol. 1: „Die thermodynamischen Eigenschaften der Luft im Temperaturbereich zwischen −210 °C und +1250 °C bis zu Drücken von 4500 bar" by *H. D. Baehr* and *K. Schwier,* Springer-Verlag, Berlin 1961

H. Rögener, J. Tölle

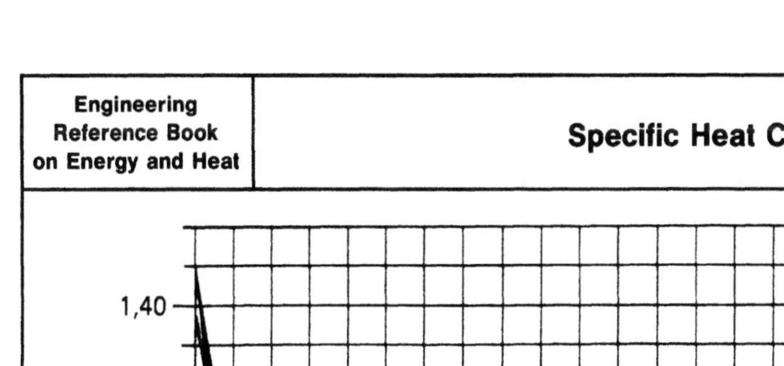

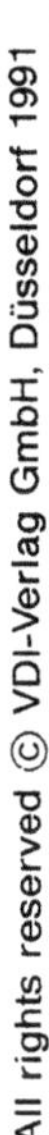

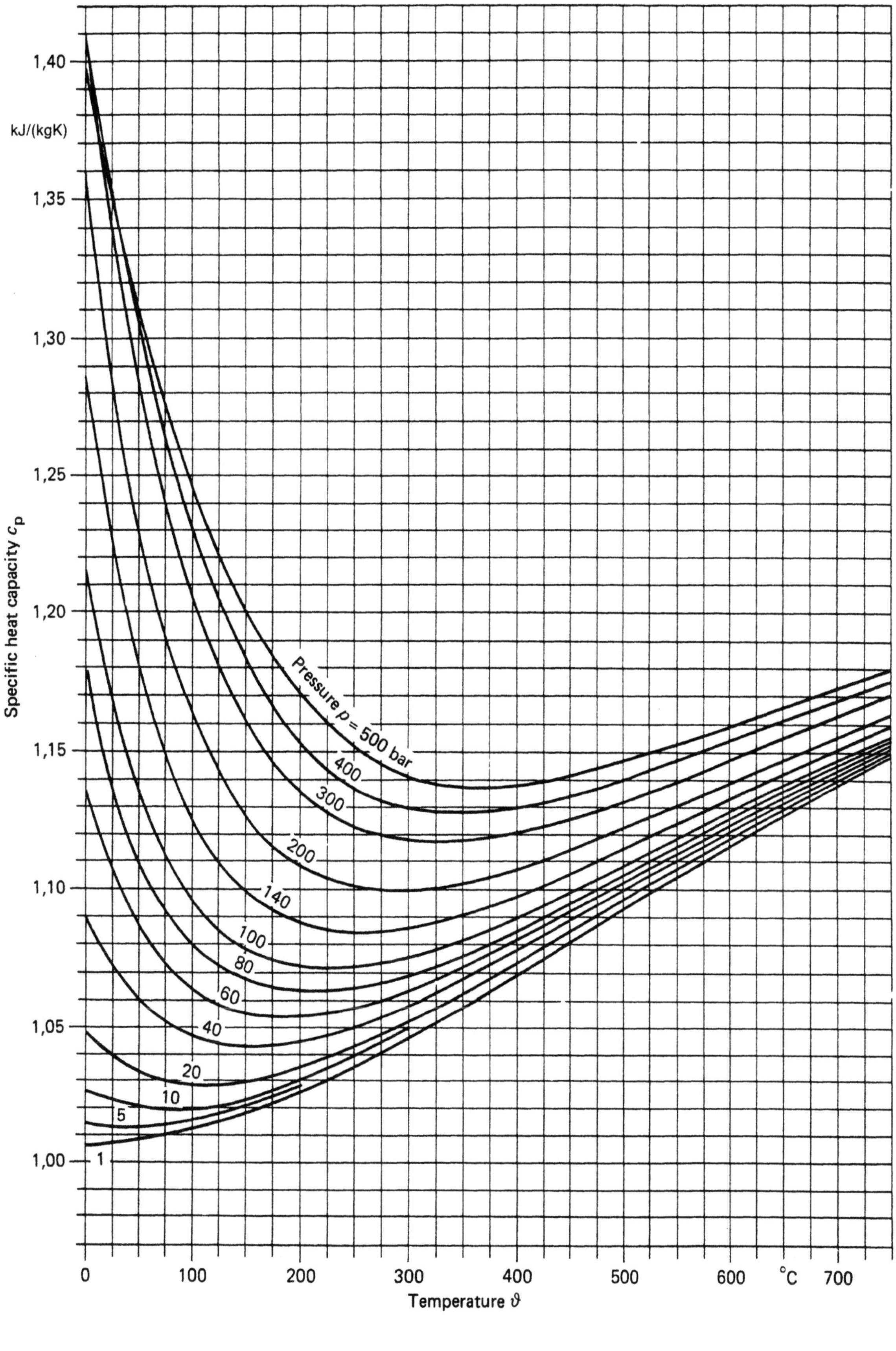

Bibliography

Baehr H.D.: Thermodynamische Eigenschaften der Gase und Flüssigkeiten. Vol. 1: „Die thermodynamischen Eigenschaften der Luft im Temperaturbereich zwischen −210 °C und +1250 °C bis zu Drücken von 4500 bar" by *H.D. Baehr* and *K. Schwier.* Springer-Verlag, Berlin 1961

H. Rögener, J. Tölle

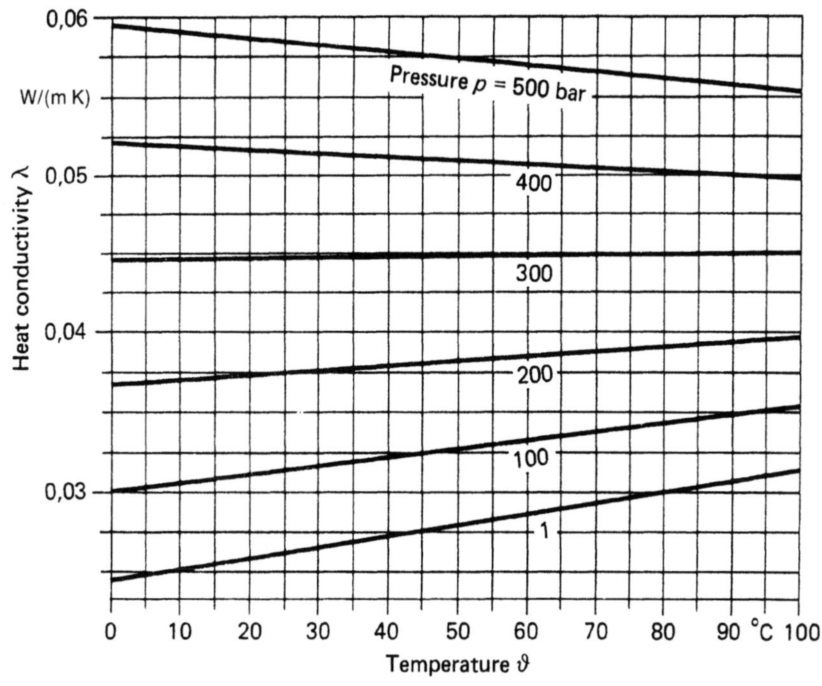

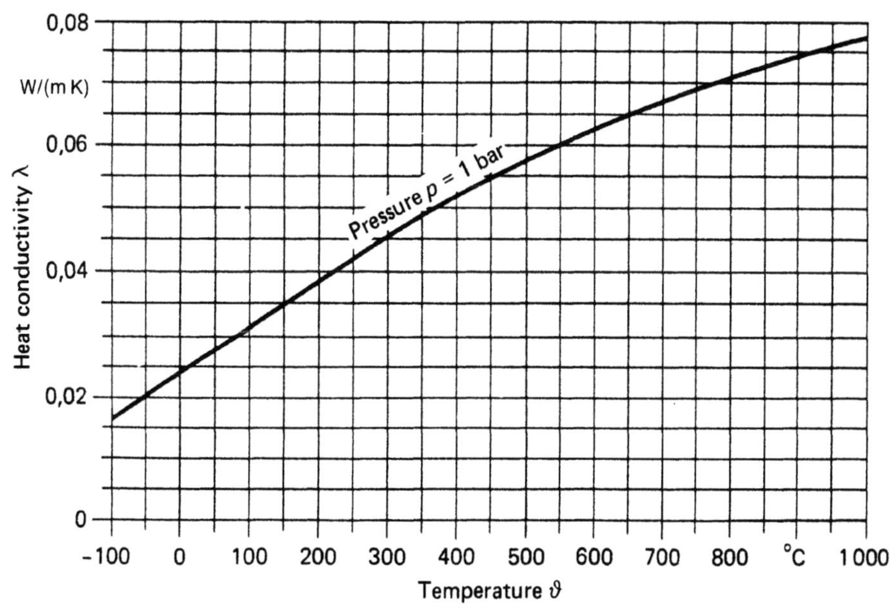

According to a data collection of the Physikalisch-Technische Bundesanstalt Braunschweig

H. Rögener, J. Tölle

Dynamic Viscosity of Air

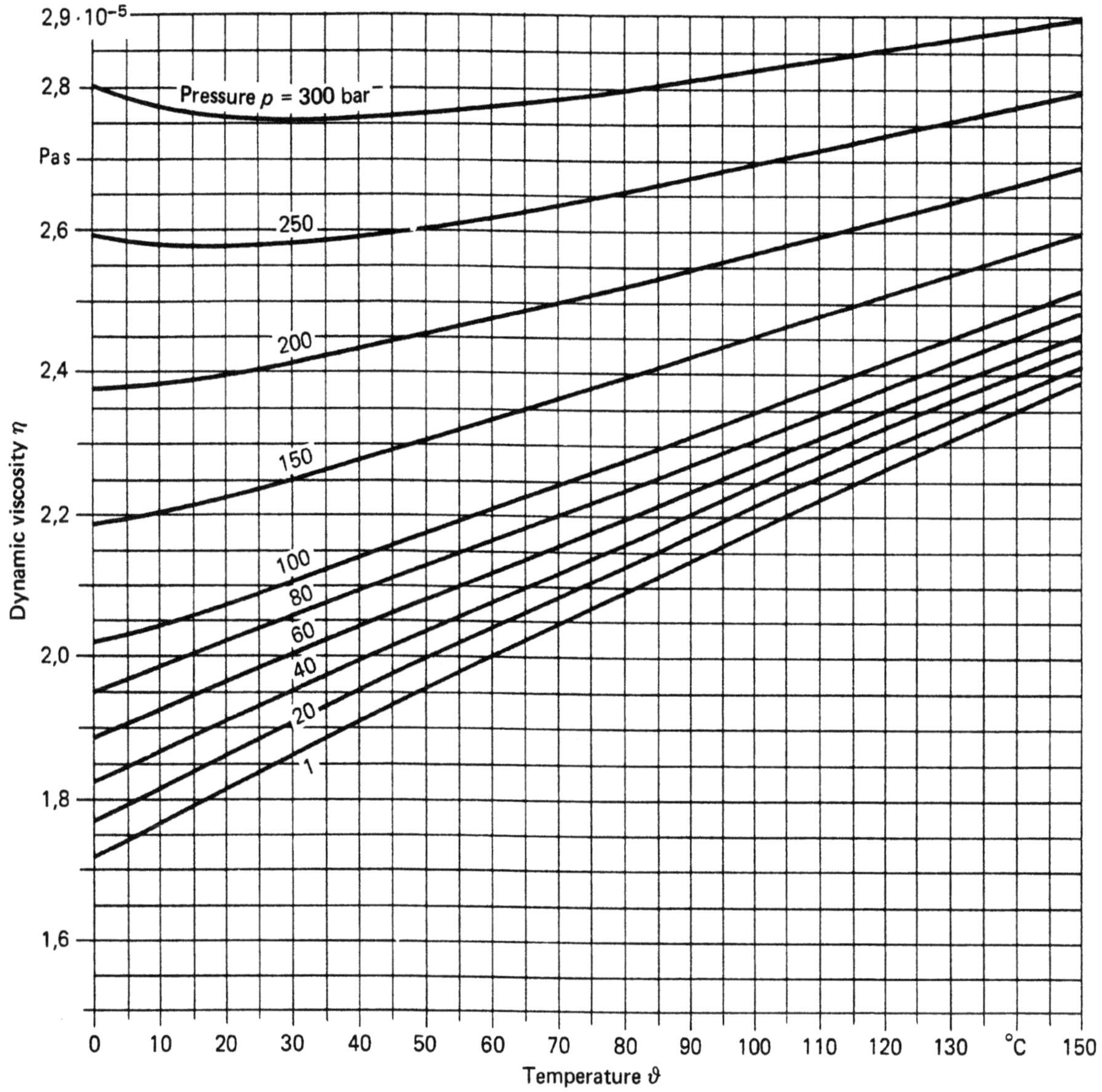

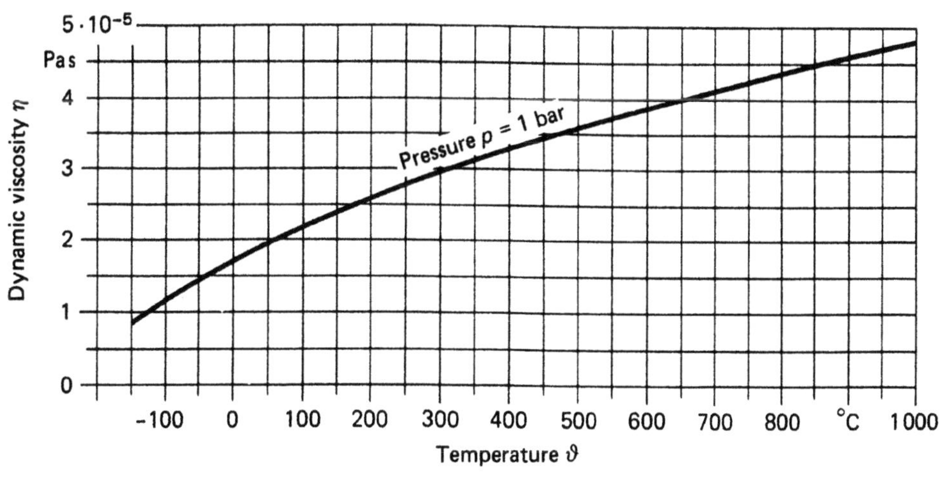

According to a data collection of the Physikalisch-Technische Bundesanstalt Braunschweig

H. Rögener, J. Tölle

Δh/Δx   MJ/kg
°C
Temperature ϑ
Absolute humidity x
Partial pressure of steam pD/p
h = 400 kJ/kg
v · p = 1,00 m³bar/kg
x · p = 0,01 bar⁻¹
ϑ
Δh/Δx
g/kg
MJ/kg

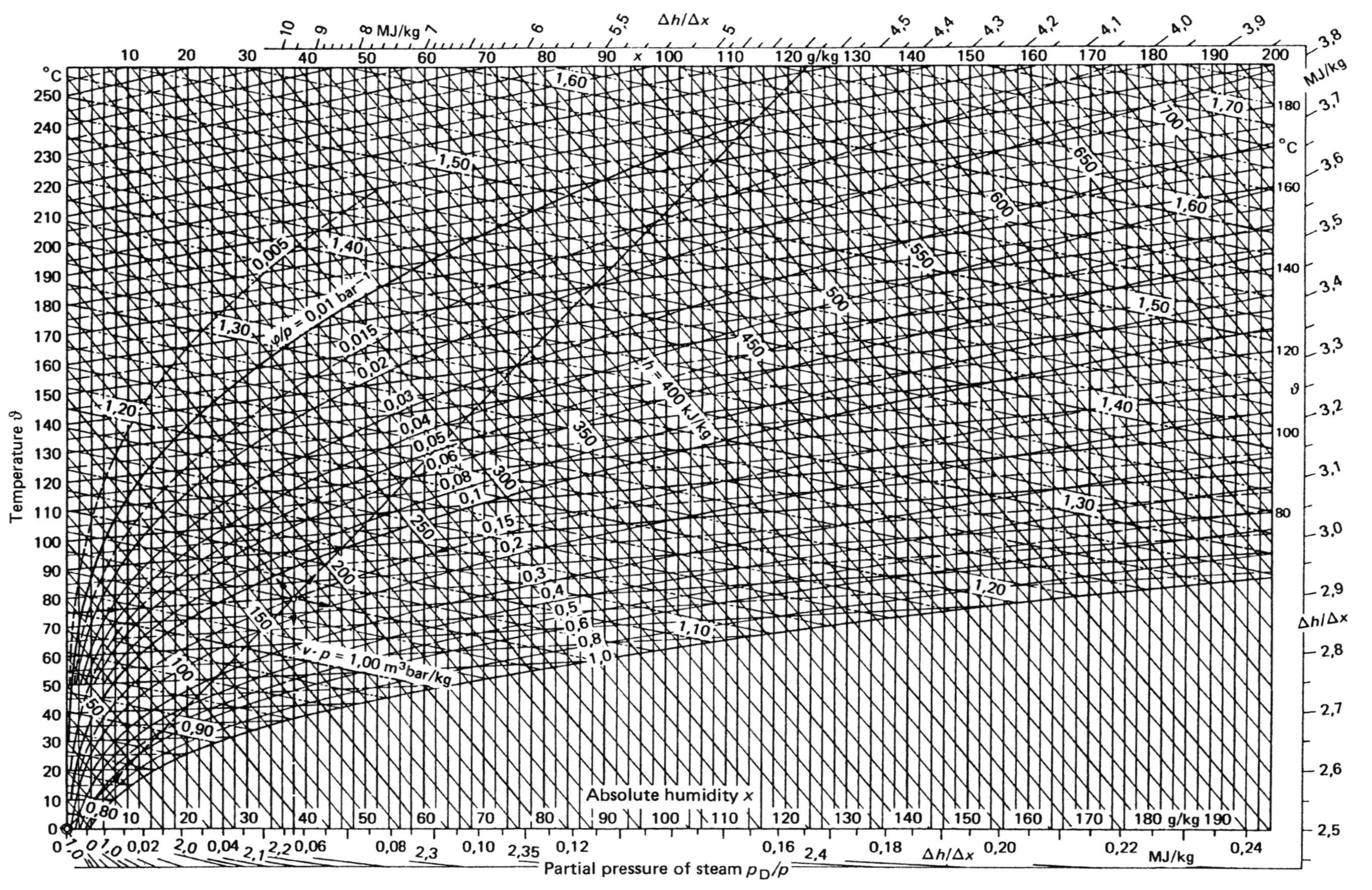

K. F. Knoche

Example:

Temperature of air	ϑ $= 75\,°C$	$(T = 348\ \mathrm{K})$
Pressure of air	p $= 2.0$ bar	
Relative humidity of air	φ $= 0.3\ (30\%)$	
Equivalent to	$\varphi/p = 0.15\ \mathrm{bar}^{-1}$	

results in

Specific water content	x $= 0.038$ kg/kg (38 g/kg)	
Specific volume	v $= 0.53\ \mathrm{m}^3/\mathrm{kg}$	$vp = 1.06\ \mathrm{m}^3\ \mathrm{bar/kg}$
Specific enthalpy	h $= 175$ kJ/kg	
Partial pressure of steam	p_D $= 0.116$ bar	$p_\mathrm{D}/p = 0.058$

Basic equations:
(in above used units)

$$x = \frac{p_\mathrm{D}\, 0.622}{p - p_\mathrm{D}},$$

$$v = (x + 0.622)\frac{R_\mathrm{D}\, T}{10^5\, p},$$

$$h = [c_{p\mathrm{mL}}]_0\, t + [[c_{p\mathrm{mD}}]_0 + r_0]\, x .$$

R_D	gas constant of steam $(461.51\ \mathrm{J/(kgK)})$
r_0	specific latent heat at $0\,°C$ (2502 kJ/kg)
$c_{p\mathrm{mL}}$	specific heat capacity of air at constant pressure
$c_{p\mathrm{mD}}$	specific heat capacity of steam at constant pressure

Explanation

The zero point is valid for the parameter $\Delta h/\Delta x$ at the outer axis.

Bibliography

Häußler W.: Lufttechnische Berechnungen im Mollier i, x-Diagram, Dresden 1969

(Barin I., Knacke O., Kubaschewski O.; Thermochemical Properties of Inorganic Substances. Berlin, Heidelberg, New York 1973)

$$Cp = A + B*T/1000 + C \cdot 10^5/T**2 + D*(T/1000)**2 \quad (kJ/(kmol*K))$$
$$H = HO* + A*T/1000 + B/2*(T/1000)**2 - C*100/T + D/3*(T/1000)**3 \quad (kJ/mol)$$
$$S = SO* + A*\ln(T) + B/1000*T - 5*C*(100/T)**2 + D/2*(T/1000)**2 \quad (kJ/(kmol*K))$$

Chemical symbol	Name	State of aggregate	A	B	C	D	Range	HO*	SO*
Al	Aluminium	sol	31.397	−16.404	−3.609	20.767	298− 933	−10.015	−148.598
Al	Aluminium	liq	31.769	0.000	0.000	0.000	933−2767	−0.791	−145.733
Al₂O₃	Aluminium oxide	S−A	103.921	26.285	−29.111	0.000	298− 800	−1718.300	−565.298
Au	Gold	sol	24.024	4.379	0.000	0.000	298− 900	−7.356	−90.644
Au	Gold	sol	4.572	18.501	54.809	0.000	900−1200	10.521	32.339
Bi	Bismut	sol	11.857	30.488	4.107	0.000	298− 544	−3.509	−17.551
Bi	Bismut	liq	19.029	10.379	20.754	−3.982	544−1200	10.138	−27.640
C	Graphite	sol	0.109	38.967	−1.482	−17.396	298−1100	−2.107	−6.550
CH₄	Methane	Gas	12.456	76.740	1.449	−18.016	298−2000	−81.333	94.098
C₄H₁₀	Butane	Gas	40.277	265.255	−12.669	−76.413	298−1500	−152.123	−2.048
C₆H₁₄	Hexane	Gas	65.691	377.846	−20.109	−109.615	342−1500	−210.529	−107.399
C₇H₁₆	Heptane	Gas	77.833	435.289	−23.459	−126.847	371−1500	−239.355	−158.773
C₈H₁₈	Octane	Gas	90.644	491.103	−27.013	−143.046	398−1500	−268.860	−213.363
C₉H₂₀	Nonane	Gas	102.953	548.019	−30.388	−164.056	424−1500	−298.050	−265.642
C₁₀H₂₂	Decane	Gas	115.472	604.620	−33.909	−176.675	447−1500	−316.706	−295.869
C₂H₂	Acetylene	Gas	43.656	31.673	−7.511	−6.314	298−2000	210.002	−61.115
C₂H₄	Ethylene	Gas	32.657	59.871	0.000	0.000	298−1200	40.112	15.480
C₃H₄	Propadiene	Gas	39.096	111.243	−9.437	−31.749	298−1500	−211.693	−24.743
C₅H₈	Pentadiene (1,2)	Gas	64.309	223.860	−17.928	−64.573	298−1500	−180.250	−106.583
C₄H₆	1-Butyne	Gas	49.739	164.403	12.024	45.515	298−1500	147.710	−36.629
C₈H₁₄	1-Octyne	Gas	99.102	392.588	−24.966	−113.747	298−1500	28.144	−244.275
C₉H₁₆	1-Nonyne	Gas	109.904	452.421	27.691	131.993	298−1500	17.128	−266.390
C₆H₆	Benzene	Gas	44.171	245.476	−26.339	−75.576	353−1500	50.095	−68.092
CH₃OH	Methanol	Gas	4.312	128.811	4.538	−44.133	338−1000	−207.129	177.671
C₂H₆O	Ethanol	Gas	39.147	149.385	−5.778	−42.337	351−1500	−257.070	7.854
CO	Carbon monoxide	Gas	28.428	4.103	−0.461	0.000	298−2500	−119.424	34.244
CO₂	Carbon dioxide	Gas	44.171	9.043	−8.541	0.000	298−2500	−410.198	−45.347
CaCl₂	Calcium chloride	sol	71.929	12.728	−2.721	0.000	298−1045	−819.244	−310.443
Ca(OH)₂	Calcium hydroxide	sol	105.365	11.953	−18.979	0.000	298−1000	−1025.167	−531.079
CaSO₄	Calcium sulfate	sol	70.255	98.808	0.000	0.000	298−1200	−1460.389	−324.393
CaSO₄* *½H₂O	Alpha calcium sulfate ½-hydrate	sol	69.375	163.285	0.000	0.000	298− 800	−1605.721	−313.269
CaSO₄* *2H₂O	Calcium sulfate dihydrate	sol	91.440	318.197	0.000	0.000	298− 800	−2065.386	−421.498
CaC₂	Calcium dicarbide	S−A	68.664	11.891	−8.667	0.000	298− 720	−83.351	−329.267
CaC₂	Calcium dicarbide	S−B	64.477	8.374	0.000	0.000	720−1275	−72.662	−290.652
CaCO₃	Calcium carbonate	sol	104.586	21.939	−25.958	0.000	298−1200	1248.324	−528.944
Cl₂	Chlorine	Gas	36.928	0.251	−2.847	0.000	298−3000	−11.970	11.099
Cr	Chromium	sol	17.727	22.981	−0.377	−9.039	298−1000	−6.351	−83.996
Cu	Copper	sol	24.870	3.789	−1.390	0.000	298−1357	−8.047	−110.465
CuSO₄	Copper (2) sulfate	sol	73.457	152.952	−12.318	−71.636	298−1078	−802.681	−358.511
Fe	Iron	S−A	28.194	−7.323	−2.897	25.058	298− 800	−9.270	−133.887
Fe	Iron	S−A	−263.630	255.981	619.646	0.000	800− 1000	244.421	1621.468
Fe₃O₄	Magnetite	S−A	86.323	209.055	0.000	0.000	298− 866	−1154.137	−407.551
Fe₂O₃	Hematite	S−A	98.348	77.874	−14.863	0.000	298− 953	−863.808	−504.367
FeSO₄	Iron (2) sulfate	sol	156.440	9.031	−118.733	19.891	298− 944	−1016.509	−840.638
H₂	Hydrogen	Gas	27.298	3.266	0.502	0.000	298−3000	−8.110	−25.539
HCl	Hydrogen chloride	Gas	26.544	4.605	1.089	0.000	298−2000	−100.123	34.926
H₂O	Water	Gas	30.019	10.718	0.335	0.000	298−2500	−251.292	14.805
H₂S	Hydrogen sulfide	Gas	29.391	15.407	0.000	0.000	298−1800	−29.957	33.787
H₂SO₄	Sulfuric acid	liq	157.005	28.320	−23.480	0.000	298− 553	−870.457	−759.121
H₂SO₄	Sulfuric acid	Gas	94.831	52.595	−26.080	−12.971	298−2000	−780.294	−280.735
HNO₃	Nitric acid	Gas	91.888	6.289	−94.869	17.622	298−2000	−194.050	−313.020
H₃PO₄	Orthophosphoric acid	sol	106.763	0.000	0.000	0.000	298− 316	−1311.595	−497.626
H₃PO₄	Orthophosphoric acid	liq	200.966	0.000	0.000	0.000	316−1200	−1328.451	−998.983
Hg	Mercury	liq	30.396	−11.472	0.000	10.161	298− 630	−8.637	−94.128
Hg	Mercury	Gas	20.800	0.000	0.000	0.000	630−3000	55.123	56.434
LiCl	Lithium chloride	sol	41.445	23.413	0.000	0.000	298− 883	−421.929	−183.754
LiCl	Lithium chloride	liq	73.432	−9.479	0.000	0.000	883−1701	−417.512	−349.242
LiBr	Lithium bromide	sol	30.220	41.391	5.949	0.000	298− 823	−359.282	−107.048
LiBr	Lithium bromide	liq	65.314	0.000	0.000	0.000	823−1562	−357.209	−287.562
Mn	Manganese	S−A	20.758	18.740	0.000	0.000	298− 600	−7.017	−91.829
Mn	Manganese	S−A	24.024	13.469	0.000	0.000	600− 980	−8.034	−109.560
Mo	Molybdenum	sol	25.586	2.847	−2.186	0.000	298− 700	−8.482	−119.219
Mo	Molybdenum	sol	33.934	−11.920	−9.211	6.963	700−1500	−12.510	−166.005
N₂	Nitrogen	Gas	27.884	4.271	0.000	0.000	298−2500	−8.499	31.497
NH₃	Ammonia	Gas	25.812	31.644	0.352	0.000	298− 800	−54.952	36.513
NH₄Cl	Ammonium chloride	S−1	38.895	160.354	0.000	0.000	298− 458	−333.474	−174.334
NH₄Cl	Ammonium chloride	S−2	34.667	111.788	0.000	0.000	458− 793	−322.539	−117.666
Na	Sodium	sol	14.800	44.259	0.000	0.000	298− 371	−6.376	−46.302
Na	Sodium	liq	37.493	−19.167	0.000	10.643	371−1156	−8.022	−150.775
NaCl	Sodium chloride	sol	45.971	16.329	0.000	0.000	298−1074	−425.819	−194.586

Chemical symbol	Name	State of aggregate	A	B	C	D	Range	HO*	SO*
NaOH	Sodium hydroxide	S−A	71.804	−110.950	0.000	235.926	298− 568	−446.861	−322.008
NaOH	Sodium hydroxide	S−B	86.039	0.000	0.000	0.000	568− 593	−452.074	−426.032
NaOH	Sodium hydroxide	liq	89.514	−5.862	0.000	0.000	593− 900	−446.740	−434.012
NaOH	Sodium hydroxide	liq	83.736	0.000	0.000	0.000	900−1663	−443.918	−399.986
Na₂SO₄	Sodium sulfate	S−5	82.379	154.464	0.000	0.000	298− 522	−1419.543	−365.633
Na₂SO₄	Sodium sulfate	S−1	145.148	54.634	0.000	0.000	522− 980	−1427.958	−685.727
Na₂CO₃	Sodium carbonate	S−1	11.024	244.199	24.510	0.000	298− 723	−1137.428	17.099
Na₂CO₃	Sodium carbonate	S−2	50.116	129.163	0.000	0.000	723−1123	−1138.341	−158.533
Ni	Nickel	sol	19.096	23.513	0.000	0.000	298− 500	−6.737	−85.905
Ni	Nickel	sol	−251.334	356.678	259.628	0.000	500− 631	138.755	1480.042
Ni	Nickel	sol	467.506	−679.191	0.000	0.000	631− 640	−149.758	−2533.512
Ni	Nickel	sol	−385.956	404.495	654.970	0.000	640− 700	276.861	2367.501
Ni	Nickel	sol	−10.881	54.705	56.513	−16.500	700−1400	16.400	98.189
Ni	Nickel	sol	36.216	0.000	0.000	0.000	1400−1726	−15.056	−184.022
O₂	Oxygen	Gas	29.977	4.187	−1.675	0.000	298−3000	−9.680	32.201
O₃	Ozone	Gas	44.376	15.604	−8.616	−4.350	298−2000	126.002	−23.139
Pb	Lead	sol	24.237	8.717	0.000	0.000	298− 600	−7.612	−75.852
Pb	Lead	liq	32.511	−3.090	0.000	0.000	600−1200	−5.656	−113.718
PbO₂	Lead dioxide	sol	63.258	31.037	−8.985	−14.001	298−1200	−297.774	−302.230
Pt	Platinum	sol	24.267	5.380	0.000	0.000	298−2043	−7.469	−98.193
S (rhombic)	Sulfur (rhombic)	sol	14.821	24.074	0.729	0.000	298− 368	−5.241	−59.298
S (monoclinic)	Sulfur (monoclinic)	sol	68.400	−118.625	0.000	0.000	368− 374	−15.085	−322.488
S (monoclinic)	Sulfur (monoclinic)	sol	13.691	29.986	0.000	0.000	374− 388	−5.028	−53.993
S	Sulfur	liq	−2065.738	3469.723	1132.056	0.000	388− 440	836.368	11387.301
S	Sulfur	liq	−25.577	57.816	88.685	0.000	440− 718	31.828	201.054
SO₂	Sulfur dioxide	Gas	43.459	10.634	−5.945	0.000	298−1800	−312.432	−5.828
SO₃	Sulfur trioxide	Gas	57.183	27.365	−12.920	−7.733	298−2000	−418.553	−84.047
Si	Silicon	sol	22.839	3.860	−3.542	0.000	298−1685	−8.164	−114.429
SiO₂	Quartz	S−A	43.945	38.841	−9.684	0.000	298− 847	−929.537	−225.895
SiO₂	Quartz	S−B	58.950	10.048	0.000	0.000	847−1696	−930.056	−301.157
SiC	Silicon carbide	sol	50.824	1.951	−49.237	8.210	298−3259	−105.097	−301.593
Sn	Tin	S−B	21.608	18.108	0.000	0.000	298− 505	−7.243	−77.272
Sn	Tin	liq	21.554	6.150	12.891	0.000	505− 800	3.894	−54.449
Sn	Tin	liq	28.470	0.000	0.000	0.000	800−2876	−1.285	−96.774
SnO₂	Tin (4) oxide	S−1	73.939	10.048	−21.604	0.000	298−1500	−610.858	−384.059
Ti	Titanium	S−A	22.173	10.291	0.000	0.000	298−1155	−7.063	−98.721
U	Uranium	S−A	27.411	−3.643	−0.959	27.290	298− 941	−8.570	−106.504
V	Vanadium	sol	26.507	2.633	−2.114	0.000	298− 600	−8.725	−124.055
V	Vanadium	sol	16.722	12.678	11.438	0.000	600−1400	−2.407	−65.624
W	Tungsten	sol	22.927	4.689	0.000	0.000	298−2500	−7.042	−99.332
Zn	Zinc	sol	20.750	12.519	0.833	0.000	298− 693	−6.460	−79.817
Zn	Zinc	liq	31.401	0.000	0.000	0.000	693−1180	−3.643	−130.348
ZnSO₄	Zinc sulfate	sol	76.409	76.200	0.000	0.000	298−1027	−1008.169	−347.404
Zr	Zirconium	S−A	21.989	11.639	0.000	0.000	298−1135	−7.072	−89.807
Zr	Zirconium	S−B	23.253	4.647	0.000	0.000	1135−2125	0.013	−87.236
Zr	Zirconium	liq	33.494	0.000	0.000	0.000	2125−4777	9.680	−145.969

Example for the calculation of an ideal mixture

The molar enthalpy of a natural gas has to be calculated at a temperature of $127\,°C$ $(400\ K)$. The natural gas (Wels, Austria) has the following composition:

$$96.3\%\ CH_4$$
$$0.8\%\ H_2$$
$$0.7\%\ CO$$
$$0.6\%\ CO_2$$
$$0.6\%\ O_2$$
$$1.0\%\ N_2$$

$$Hm = \sum_i \psi_i * HO*_i + \left(\sum_i \psi_i * A_i\right) * T/1000$$
$$+ \left(\sum_i \psi_i * B_i\right)\bigg/ 2 * (T/1000) ** 2$$
$$- \left(\sum_i \psi_i * C_i\right) * 100/T$$
$$+ \left(\sum_i \psi_i * D_i\right)\bigg/ 3 * (T/1000) ** 3$$

with $\psi_i = n_i\big/ \sum_i n_i$

$$HO* = \sum_i \psi_i * HO*_i = 0.963 * (-81.333)$$
$$+ 0.008 * (-8.110)$$
$$+ 0.007 * (-119.424)$$
$$+ 0.006 * (-410.198)$$
$$+ 0.006 * (-9.680)$$
$$+ 0.010 * (-8.499)$$

$$A = \sum_i \psi_i * A_i = 0.963 * 12.456 + 0.008 * 27.298$$
$$+ 0.007 * 28.428 + 0.006 * 44.171$$
$$+ 0.006 * 29.977 + 0.010 * 27.884$$
$$= 13.1362$$

$$B = \sum_i \psi_i * B_i = 0.963 * 76.740 + 0.008 * 3.266$$
$$+ 0.007 * 4.103 + 0.006 * 9.043$$
$$+ 0.006 * 4.187 + 0.010 * 4.271$$
$$= 74.0776$$

$$C = \sum_i \psi_i * C_i = 0.963 * 1.449 + 0.008 * 0.502$$
$$+ 0.007 * (-0.461) + 0.006 * (-8.541)$$
$$+ 0.006 * (-1.675) + 0.010 * 0.000$$
$$= 1.33488$$

$$D = \sum_i \psi_i * D_i = 0.963 * (-18.016) + 0.008 * 0.000$$
$$+ 0.007 * 0.000 + 0.006 * 0.000$$
$$+ 0.006 * 0.000 + 0.010 * 0.000$$
$$= 17.3494$$

$$Hm = -81.8288 + 13.1362 * 400/1000$$
$$+ 74.0776/2 * (400/1000) ** 2$$
$$- 1.33488 * 100/400$$
$$- 17.3494/3 * (400/1000) ** 3$$

General Explanation

Heat transfer in single-phase flow

The correlations of heat transfer for forced and natural convection, offered in this chapter, have been calculated into the form of dimensionless nomographs, which always lead to the Nusselt number (section 3.2 and 3.3). Utilizing these, it is possible to use them for all single-phase fluids within their range of validity.

Essential for the use of these nomographs and the conversion of the Nusselt number into the heat transfer coefficient is the knowledge of the properties of media. For this reason, section 3.1 contains additional nomographs of the media air, water and steam. The calculation of the heat transfer coefficient of the mentioned media may be divided into the following steps.

- *initial values*

  i.e. temperature, pressure, velocity, characteristic dimensions

- *calculation of the Reynolds resp. Grashof number*

	Reynolds number	*Grashof number*
Air	nomograph 3.1.1.1	nomograph 3.1.1.4
Water/steam	nomograph 3.1.2.1	nomograph 3.1.2.4

- *calculation of the Prandtl number*

Air	nomograph 3.1.1.2
Water/steam	nomograph 3.1.2.2

- *calculation of the Nusselt number*

  laws for different geometric – and flow conditions
  nomographs 3.2.1.1 to 3.3

- *calculation of the heat transfer coefficient*

Air	nomograph 3.1.1.3
Water/steam	nomograph 3.1.2.3

To determine the heat transfer coefficient of other media, the necessary properties of these have to be collected from other papers and the corresponding values must be determined by calculation.

Heat transfer coefficient for boiling water in a vessel

The nomograph in section 3.4 is valid only for boiling water in a vessel. Using this diagram, it is possible to determine the heat transfer coefficient directly.

1200
°C
1000
800
600
400
200
0
Temperature ϑ
300
200
150
100
80
60
40
30
20
15
10
8
6
4
3
2
1,5
p = 1 bar
6 8 10⁻⁷   2   4   6 8 10⁻⁶   2   4   6 8 10⁻⁵   2   4   6 8 10⁻⁴ m²/s
Kinematic viscosity ν
d, dₕ resp. l = 1 (characteristic length)
6
2 4
0,6
0,4
0,2
0,1
0,06
0,04
0,02
0,01
0,006
0,004
w = 0,01 m/s
0,02
0,05
0,1
0,2
0,5
1
2
5
10
20
50
100
10¹   2   4   6 10²   2   4   6 10³   2   4   6 10⁴   2   4   6 10⁵   2   4   6 10⁶   2   4   6 10⁷
Reynolds number Re

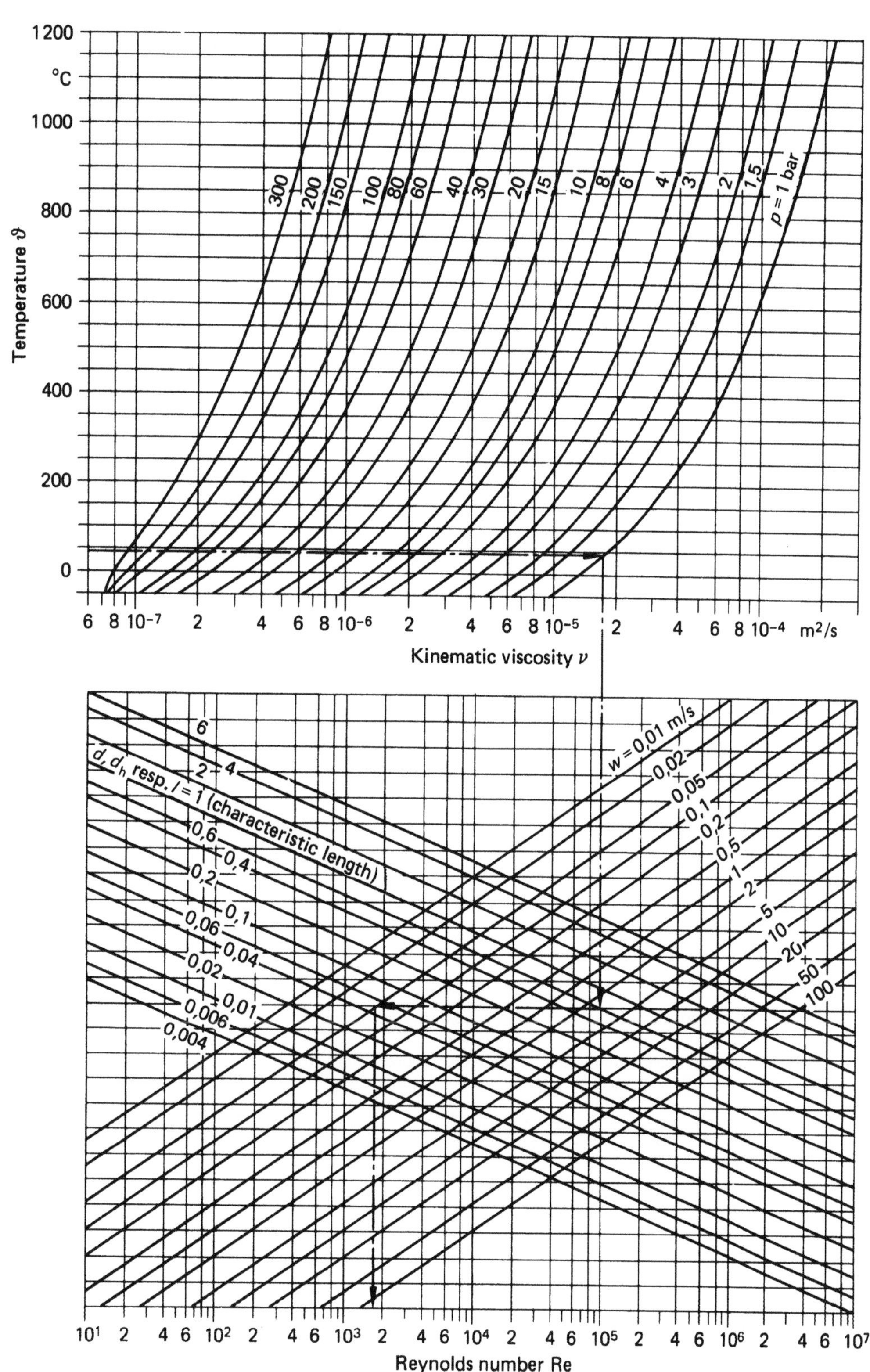

U. Renz

600
°C
500
400
300
200
100
0
Temperature ϑ
p = 1 bar
1,5
2
3
4
6
8
10
15
20
30
40
60
80
100
150
200
300
4 6 8 10⁰
2
4 6 8 10¹
2
4 6 8 10²
2 kg/m³ 4
Density ρ

500
°C
400
300
200
100
0
Temperature ϑ
p = 1 bar
300
200
100
50
0,65
0,70
0,75
0,80
0,85
0,90
0,95
Prandtl number Pr

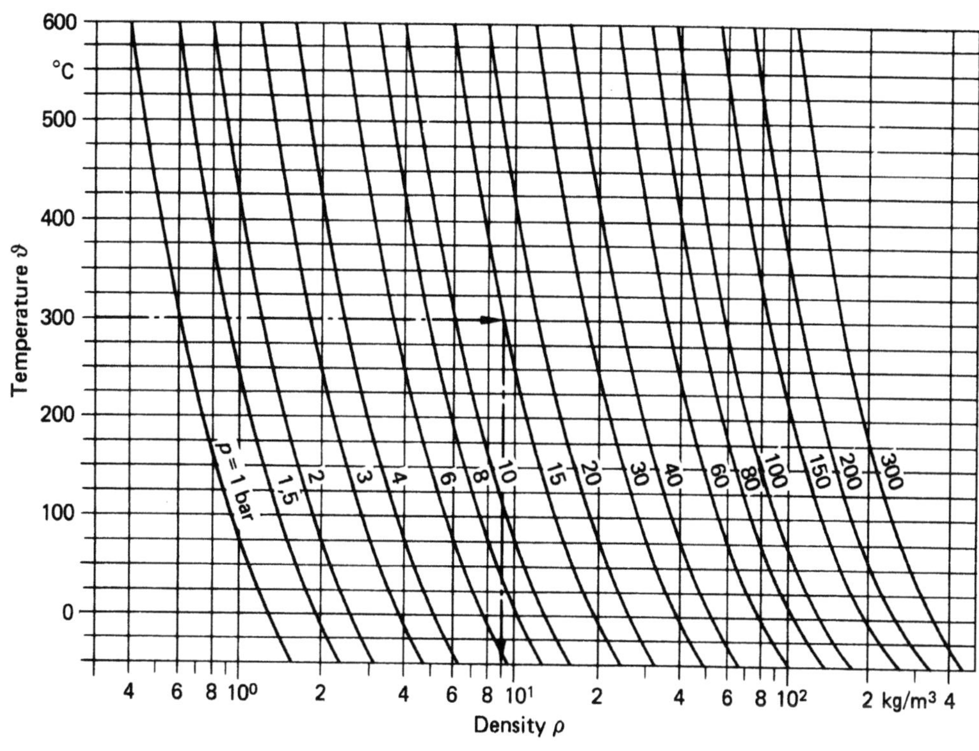

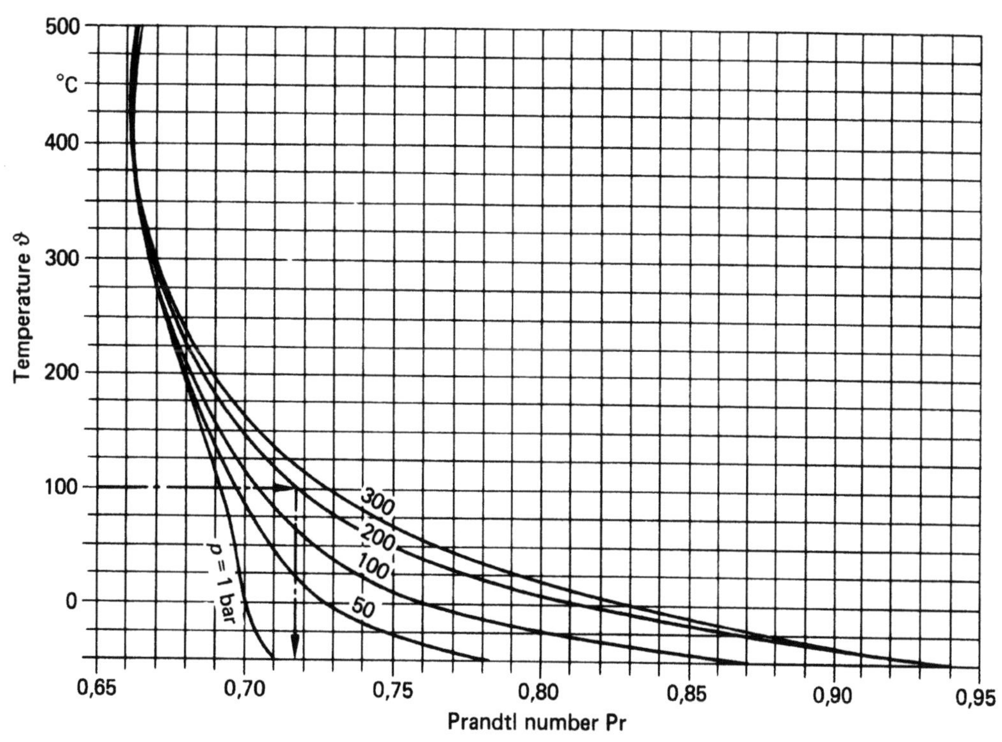

U. Renz

800
°C
600
400
Temperature ϑ
200
0
p = 1 bar
50
100
150
200
250
300
0,02
0,03
0,04
0,05
0,06
W/(m K)
0,07
Heat conductivity λ
d, d_h resp. l = 0,004 m (characteristic length)
0,006
0,01
0,02
0,04
0,06
0,1
0,2
0,4
0,6
1
2
4
6
Nu = 3,66
5
10
20
30
50
100
200
300
500
1000
2000
2
4
6
8 10⁰
2
4
6
8 10¹
2
4
6
8 10²
2
W/(m²K)
10³
Heat transfer coefficient α

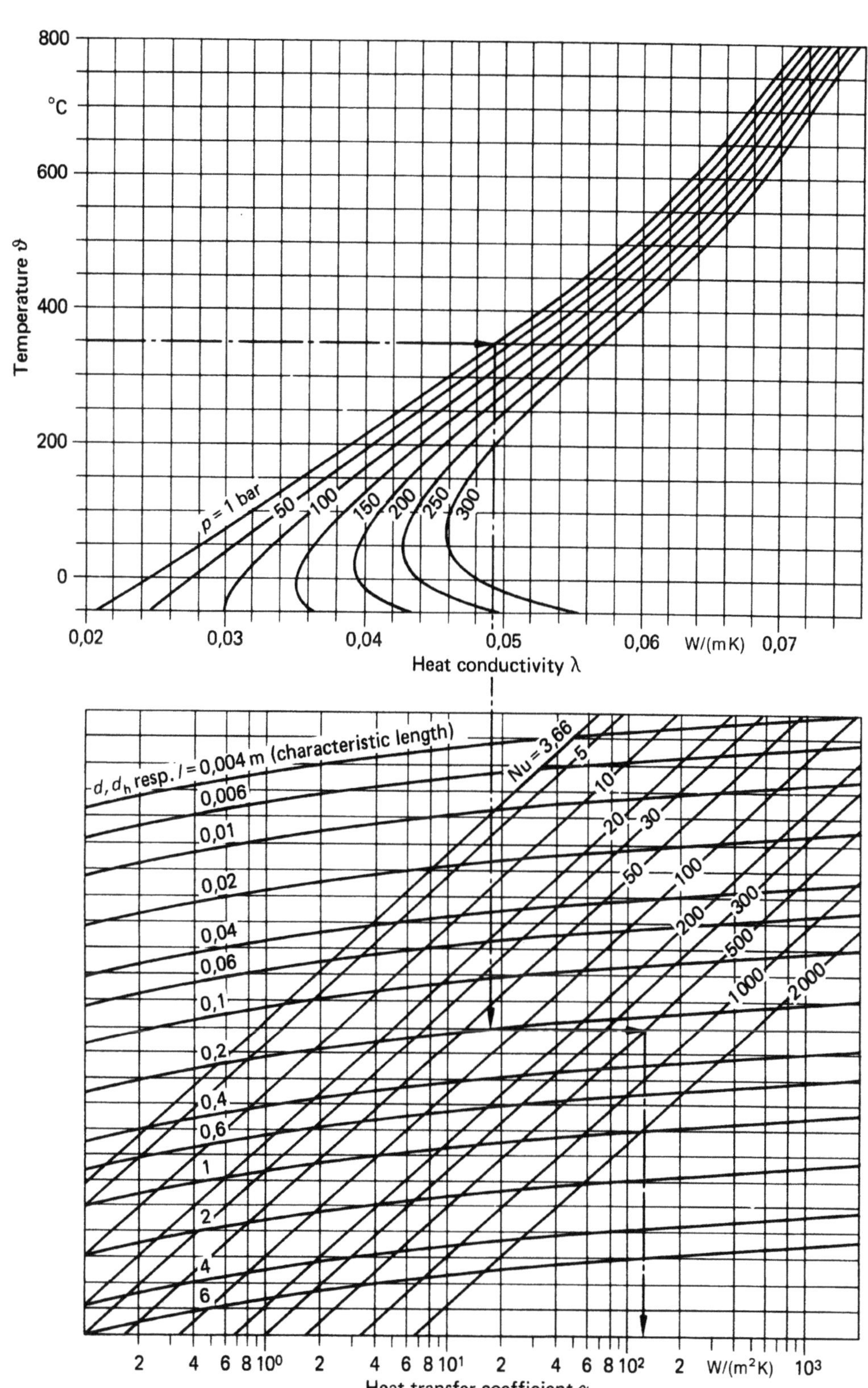

U. Renz

1200
°C
1000
800
600
400
200
0
Temperature ϑ
300
200
150
100
80
60
40
30
20
15
10
8
6
4
3
2
1,5
$p = 1$ bar
8 10⁻⁷
2
4
6 8 10⁻⁶
2
4
6 8 10⁻⁵
2
4
6 8 10⁻⁴ m²/s
Kinematic viscosity ν
20
10
5
3
2
$l = 1$ m
0,5
0,3
0,2
0,1
0,05
0,02
$(\rho_\infty - \rho_w)/\rho_w = 0,1$
0,2
0,5
1
2
5
10
10⁶
10⁷
10⁸
10⁹
10¹⁰
10¹¹
10¹²
10¹³
Grashof number Gr

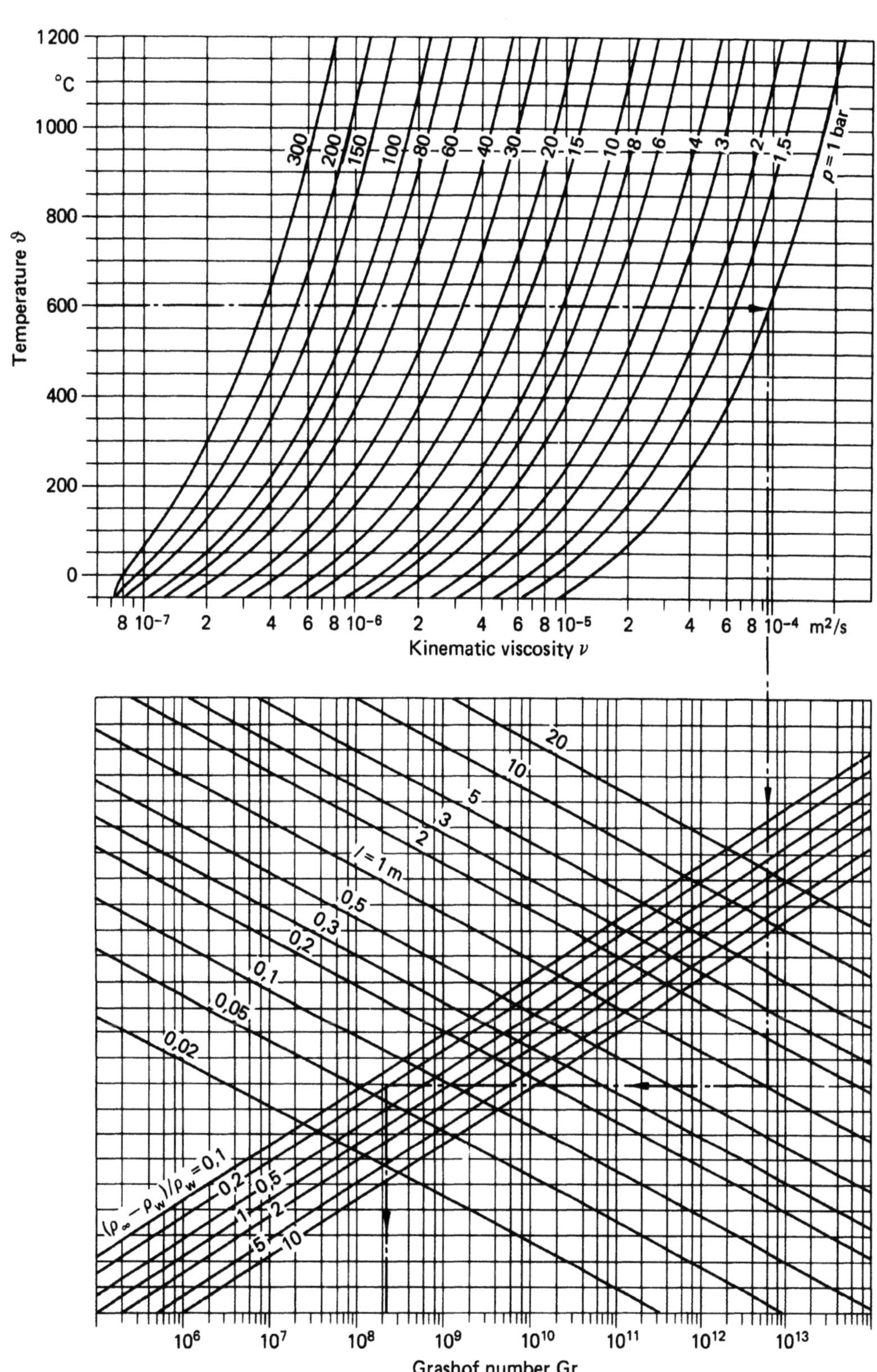

U. Renz

600
°C
500
400
Temperature ϑ
300
200
100
0
400
300
221.2
200
100
50
20
10
5
2
p = 1 bar
0.5
0.1
0.05
0.01
0.005
0.001
Liquid
Steam
10⁻⁷ 2 4 6 10⁻⁶ 2 4 6 10⁻⁵ 2 4 6 10⁻⁴ 2 4 6 10⁻³ 2 4 6 10⁻² 2 m²/s 10⁻¹
Kinematic viscosity ν
d, dₕ resp. l = 6m (characteristic length)
4
2
1
0.6
0.4
0.2
0.1
0.06
0.04
0.02
0.01
0.006
0.004
0.01
0.02
0.05
0.1
w = 0.2 m/s
0.5
1
2
5
10
20
50
100
10¹ 2 4 6 10² 2 4 6 10³ 2 4 6 10⁴ 2 4 6 10⁵ 2 4 6 10⁶ 2 4 6 10⁷
Reynolds number Re

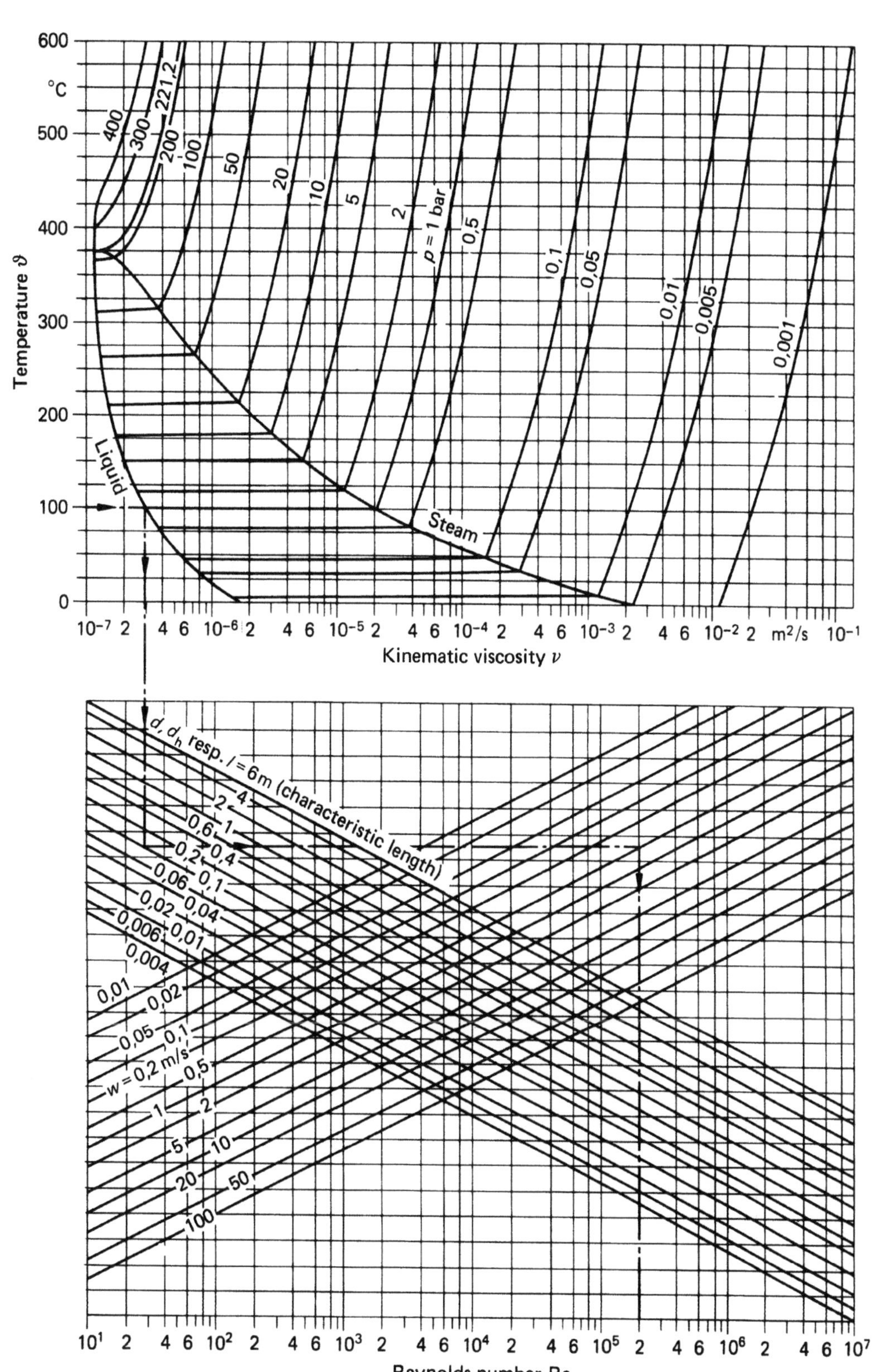

U. Renz

600
°C
500
400
300
200
100
0
Temperature ϑ
400
300
221,2
200
100
50
20
10
5
2
p = 1 bar
0,5
0,1
0,05
0,01
0,005
0,001
Liquid
Steam
10^{-4}
10^{-3}
10^{-2}
10^{-1}
10^0
10^1
10^2
kg/m³
10^3
Density ρ

600
°C
500
400
300
200
100
0
Temperature ϑ
400
300
221,2
200
100
50
20
10
5
2
1
0,5
p = 0,1 bar
0,05
0,01
Liquid
Steam
8
10^0
2
4
6
8
10^1
Prandtl number Pr

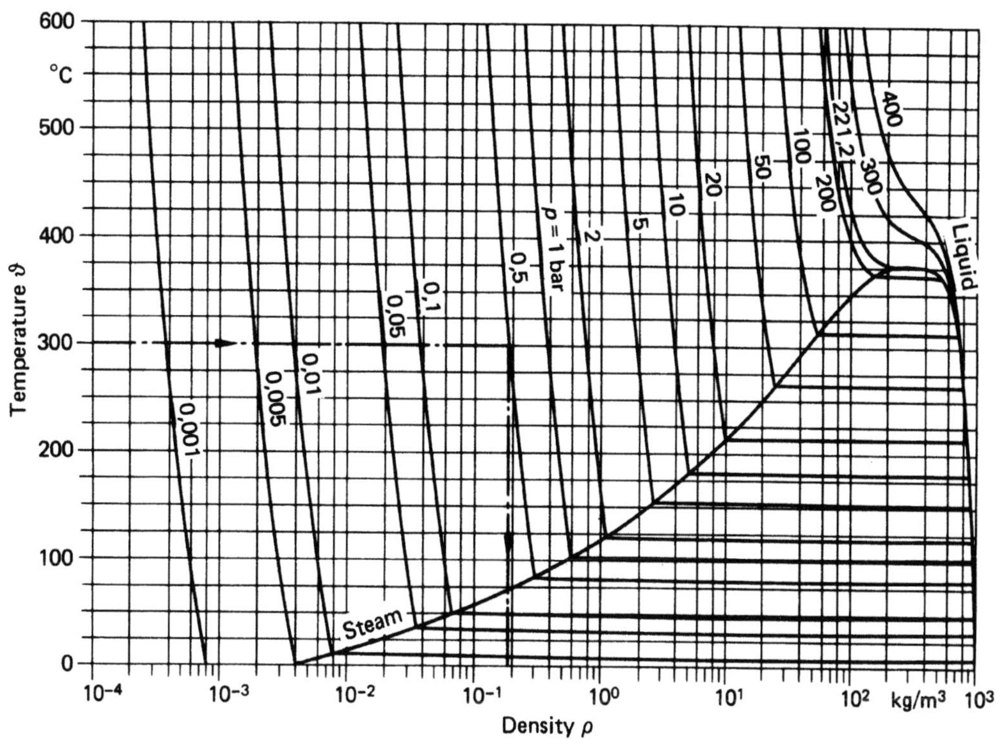

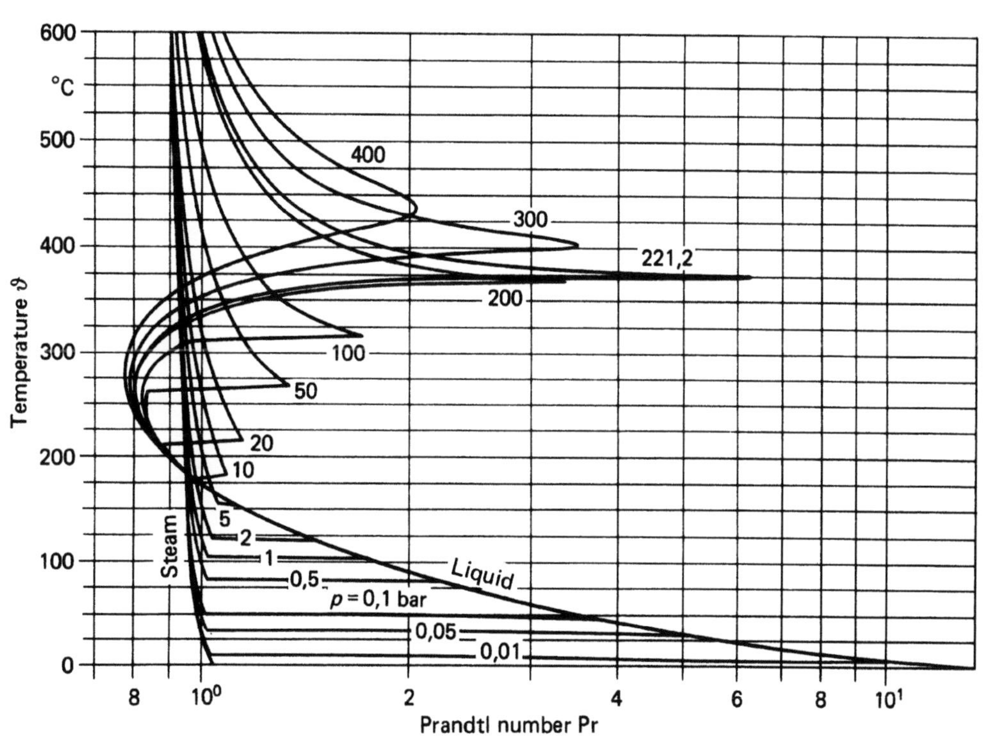

U. Renz

600
°C
500
400
300
200
100
0
Temperature ϑ
Steam
$p = 400\,bar$
300
221,2
200
100
50
20
10
5
2
1
0,5
0,1
0,05
0,01
Liquid
0
0,1
0,2
0,3
0,4
0,5
0,6 W/(m K) 0,7
Heat conductivity λ
d, d_h resp. $l = 0,004$ m (characteristic length)
0,006
0,01
0,02
0,04
0,06
0,1
0,2
0,4
0,6
1
2
4
6
10
20
30
50
100
$Nu = 200$
300
500
1000
2000
3000
5000
10000
20000
10^0
2
4
6
8
10^1
2
4
6
8
10^2
2
4
6
8
10^3
2 W/(m²K) 6 8 10^4
Heat transfer coefficient α

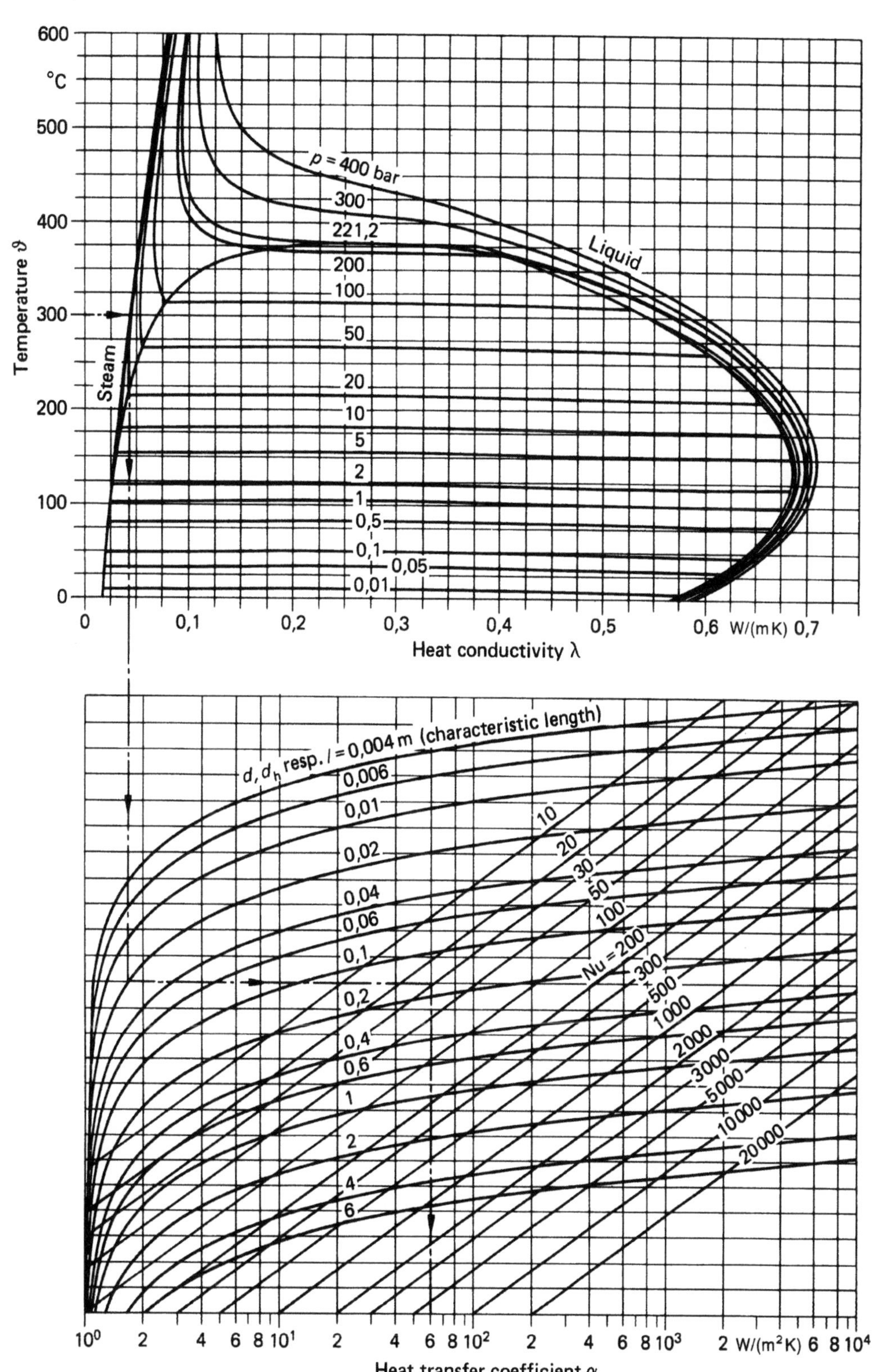

U. Renz

600
°C
500
400
300
200
100
0
Temperature ϑ
400
221,2
300
200
100
50
20
10
5
2
$p = 1$ bar
0,5
0,1
0,05
0,01
0,005
0,001
Liquid
Steam
10^{-7} 2 4 6 10^{-6} 2 4 6 10^{-5} 2 4 6 10^{-4} 2 4 6 10^{-3} 2 4 6 10^{-2} 2 m²/s 10^{-1}
Kinematic viscosity ν
$(\rho_\infty - \rho_w)/\rho_w = 0,1$
0,2
0,5
1
2
5
10
20
10
5
3
2
$l = 1$ m
0,5
0,3
0,2
0,1
0,05
0,02
10^3 10^4 10^5 10^6 10^7 10^8 10^9 10^{10} 10^{11} 10^{12} 10^{13} 10^{14} 10^{15}
Grashof number Gr

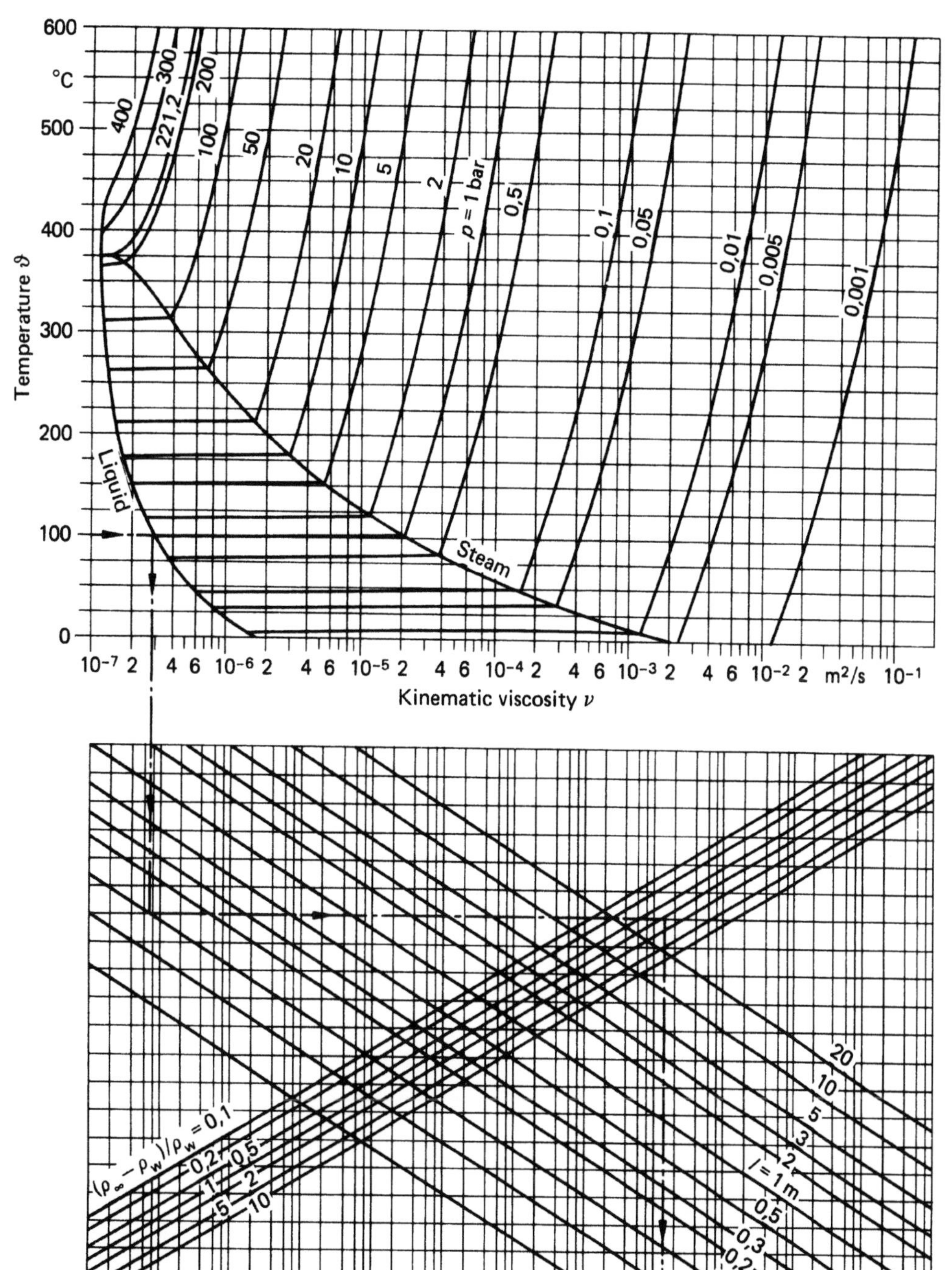

U. Renz

Reynolds number Re
2
10^3
8
6
4
2
10^2
8
6
4
2
10^1
Pr = 0,1
0,2
0,5
0,71
1
2
5
7
10
20
50
70
100
200
500
1000
d/L = 1,0
0,5
0,2
0,1
0,05
0,02
0,01
0,005
0,002
0,001
Pr/Pr$_w$ = 0,1
1
10
2
4
6
8
10^1
2
4
6
8
10^2
2
4
Nusselt number Nu

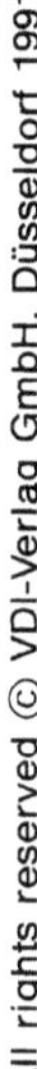

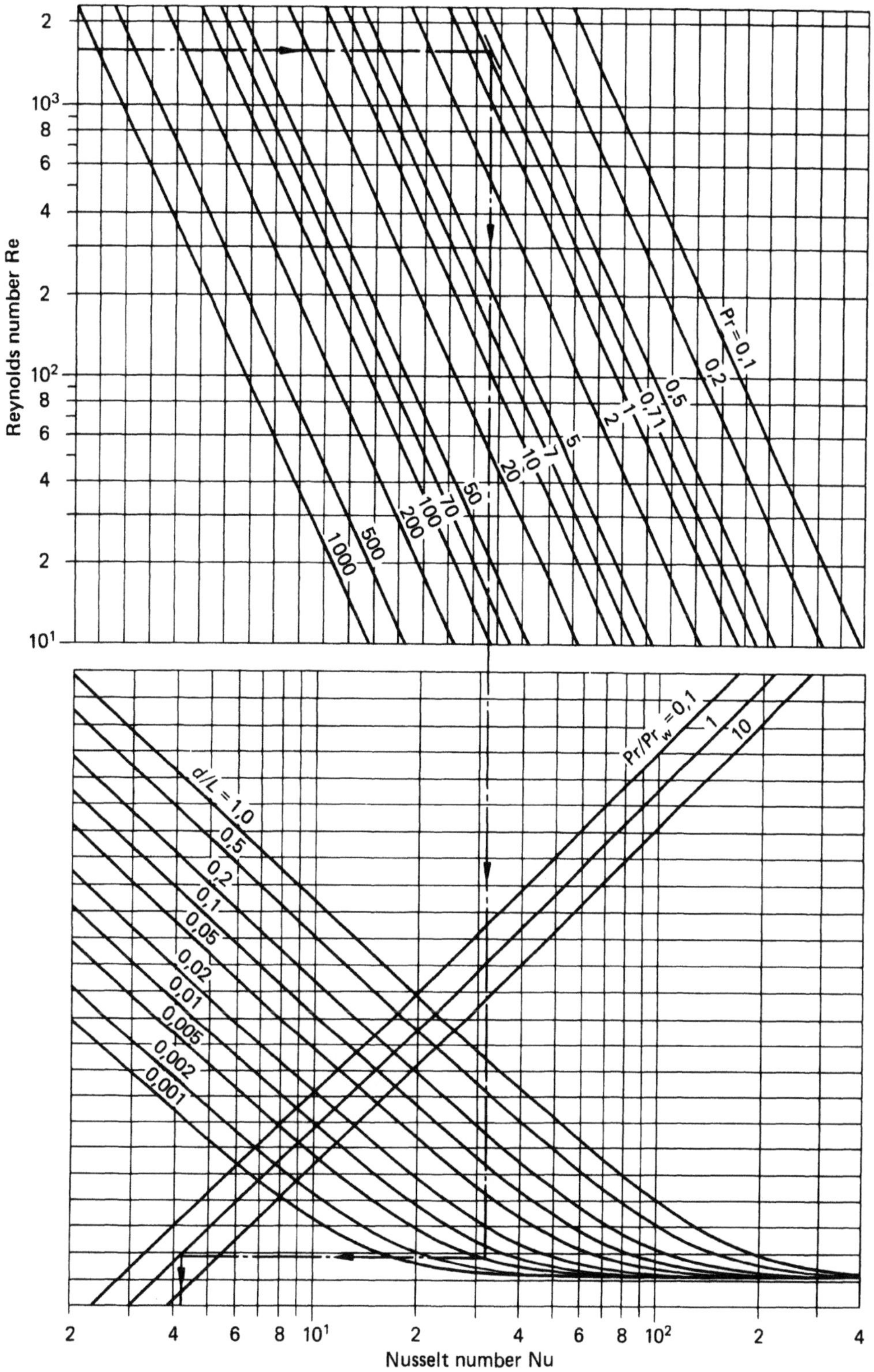

U. Renz

$$\mathrm{Nu} = \mathrm{Nu_o} \cdot K$$

$$\mathrm{Nu_o} = \left(3.66^3 + 1.61^3 \, \mathrm{Re} \cdot \mathrm{Pr} \cdot \frac{d}{L} \right)^{1/3} \quad [1]$$

Range of validity:

$$\mathrm{Re} < 2300$$

$$0.1 < \mathrm{Re} \, \mathrm{Pr} \, \frac{d}{L} < 10^4$$

Correction factor K:

The correction factor K considers the influence of the direction of heat flow (heating or cooling).

For liquid media this factor is:

$$K = (\mathrm{Pr}/\mathrm{Pr_w})^{0.11} .$$

For gases in the range of $0.5 < T/T_w < 2$ is:

$$\mathrm{Pr}/\mathrm{Pr_w} = 1$$

The *properties* of media have to be determined at the average pressure p and the average temperature ϑ_m of the fluid:

$$\vartheta_m = \frac{\vartheta_e + \vartheta_a}{2} .$$

Definitions:

$$\mathrm{Nu} = \frac{\alpha \cdot d}{\lambda} \quad \text{Nusselt number}$$

$$\mathrm{Re} = \frac{w \cdot d}{v} = \frac{\dot{m} \cdot d}{A \cdot \varrho \cdot v} = \frac{\dot{V} \cdot d}{A \cdot v} \quad \text{Reynolds number}$$

$$\mathrm{Pr} = \frac{\eta}{\lambda/c_p} \quad \text{Prandtl number}$$

$\mathrm{Pr_w}$	Prandtl number at the temperature of the tube wall ϑ_w
A in m^2	cross-sectional area of the tube
c_p in J/(kgK)	specific heat capacity
d in m	inner diameter of the tube
L in m	length of the tube
$\dot{m}$ in kg/s	mass flow
p in bar	average pressure
ϑ_m in °C	average temperature of the fluid
ϑ_e in °C	temperature at the entrance of the tube
ϑ_a in °C	temperature at the exit of the tube
T in K	average temperature of the fluid
T_w in K	temperature of the tube wall
$\dot{V}$ in m^3/s	volume flow
w in m/s	average flow velocity
α in W/(m^2 K)	heat transfer coefficient
λ in W/(mK)	thermal conductivity
v in m^2/s	kinematic viscosity
ϱ in kg/m^3	density
η in kg/(ms)	dynamic viscosity

Remarks:

The equation for $\mathrm{Nu_o}$ at laminar flow past a flat plate, recalculated for laminar tube flow

$$\mathrm{Nu_o} = 0.664 \, (\mathrm{Pr})^{1/3} \left(\mathrm{Re} \cdot \frac{d}{L} \right)^{1/2} \quad [1]$$

may result in a greater value for $\mathrm{Nu_o}$ for very short tubes ($d/L > 0.1$). In this case, the greater of the 2 values for $\mathrm{Nu_o}$ has to be used. This case has not been considered in the nomograph.

Example: Air

inner diameter of the tube	$d = 0.01\,\mathrm{m}$	$d/L =$
length of the tube	$L = 2\,\mathrm{m}$	0.005
average flow velocity	$w = 3\,\mathrm{m/s}$	
pressure	$p = 1\,\mathrm{bar}$	
average temperature of air	$\vartheta_m = 50\,°\mathrm{C}$	
temperature of the tube wall	$\vartheta_w = 80\,°\mathrm{C}$	

Solution:

1. Determination of the Reynolds number from the values of ϑ_m, p, w, d: $\mathrm{Re} = 1680$ (sheet No 3.1.1.1).

   remark: If the mass flow is known, the average flow velocity has to be calculated from

   $$w = \frac{\dot{m}}{A \cdot \varrho} = \frac{\dot{m}}{\varrho} \, \frac{4}{\pi \, d^2} .$$

   The density ϱ has to be determined with the average temperature of air ϑ_m and the pressure p from nomograph No 3.1.1.2.

2. Determination of the Prandtl number with the values of ϑ_m and p: $\mathrm{Pr} = 0.69$ (sheet No 3.1.1.2).

3. Examination of the range of validity of the law of heat transfer by the values of Re, Pr, d/L, T/T_w (sheet No 3.2.1.1).

4. Evaluation of the law of heat transfer: $\mathrm{Nu} = 4.2$ (sheet No 3.2.1.1).

   remark: $\mathrm{Pr}/\mathrm{Pr_w} = 1$.

5. Determination of the heat transfer coefficients by the values of d, Nu, ϑ_m, p: $\alpha = 12\,\mathrm{W}/(\mathrm{m^2\,K})$ (sheet No 3.1.1.3).

Bibliography

[1] VDI-Wärmeatlas, 4th edition 1984, VDI-Verlag Düsseldorf

Reynolds number Re
2
4
6
8
10^4
2
4
6
8
10^5
2
4
6
8
10^6
d/L = 1
0,2
0,001
Pr = 0,5
1,0
0,71
2,0
5,0
10
20
50
100
200
500
1000
Gases T/T$_w$ ≥ 1,0
0,75
0,5
Liquids Pr/Pr$_w$ = 0,1
1,0
10
Nusselt number Nu for liquids
10^0
10^1
10^2
10^3
10^4
10^5
Nusselt number Nu for gases

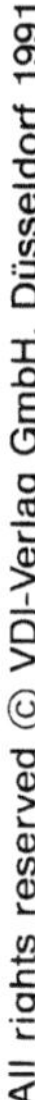

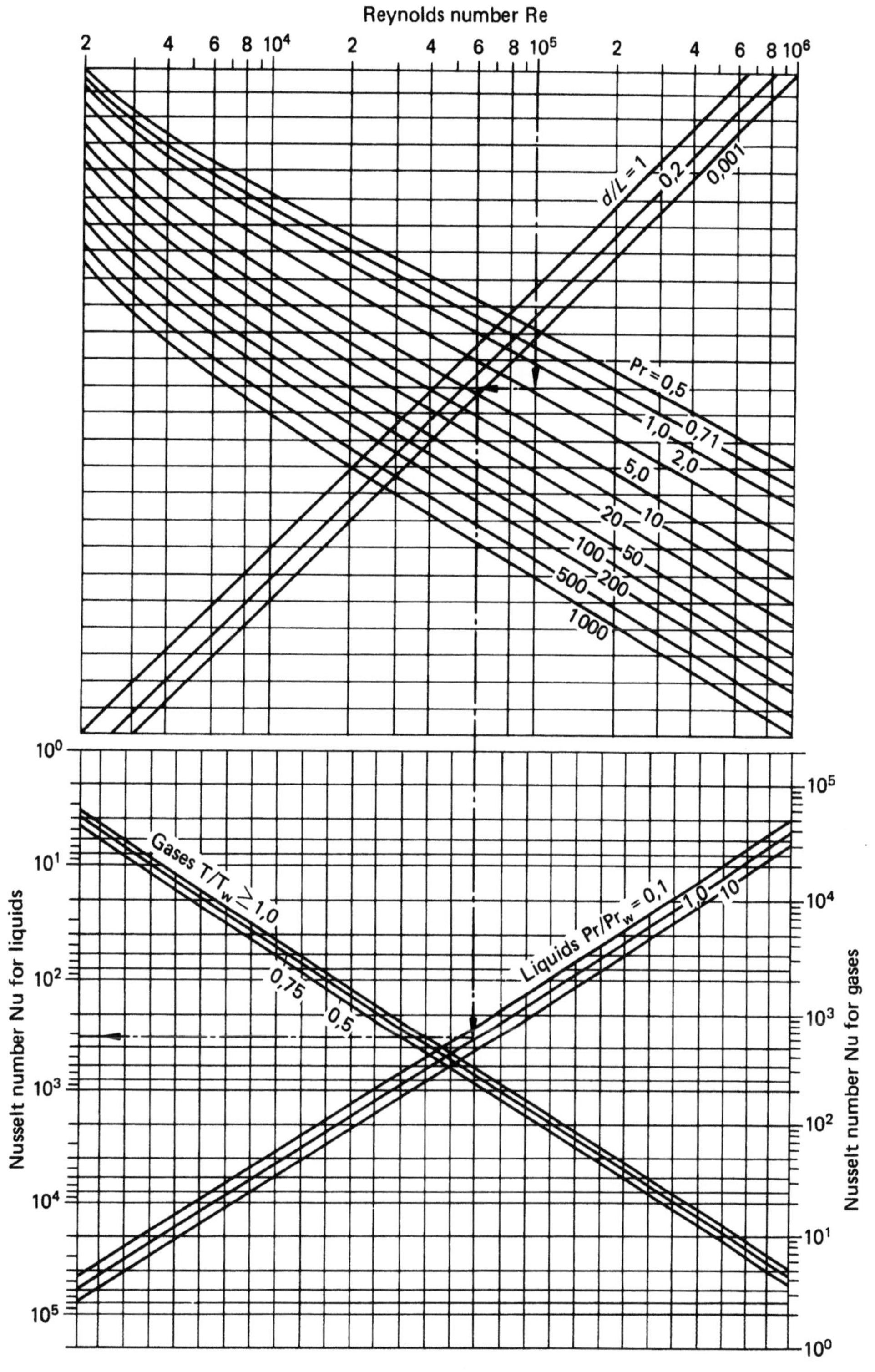

U. Renz

$$Nu = Nu_o \cdot K$$

$$Nu_o = \frac{\xi/8 \cdot (Re - 1000) \cdot Pr}{1 + 12.7\sqrt{\xi/8} \cdot (Pr^{2/3} - 1)} \quad 1 + \left(\frac{d}{L}\right)^{2/3} \quad [1]$$

with the resistance coefficient

$$\xi = (1.82 \cdot \log_{10} Re - 1.64)^{-2}$$

Range of validity:

$$2300 < Re < 10^6$$

$$\frac{d}{L} < 1$$

Correction factor K:

The correction factor K considers the influence of the direction of heat flow (heating or cooling).

For liquids this factor is:

$$K = (Pr/Pr_w)^{0.11} \quad \text{at} \quad 0.1 < Pr/Pr_w < 10$$

For cooling gases:

$$K = 1 \quad \text{at} \quad T/T_w > 1$$

For heating gases:

$$K = (T/T_w)^{0.45} \quad \text{at} \quad 0.5 < T/T_w < 1$$

The *properties* of the media have to be determined at the average pressure p and the average temperature ϑ_m of the flowing fluid:

$$\vartheta_m = \frac{\vartheta_e + \vartheta_a}{2}.$$

Definitions:

$$Nu = \frac{\alpha \cdot d}{\lambda} \quad \text{Nusselt number}$$

$$Re = \frac{w \cdot d}{v} = \frac{\dot{m} \cdot d}{A \cdot \varrho \cdot v} = \frac{\dot{V} \cdot d}{A \cdot v} \quad \text{Reynolds number}$$

$$Pr = \frac{\eta}{\lambda/c_p} \quad \text{Prandtl number}$$

Pr_w	Prandtl number at the temperature of the tube wall ϑ_w
A in m^2	cross-sectional area of the tube
c_p in J/(kgK)	specific heat capacity
d in m	inner diameter of the tube
L in m	length of the tube
$\dot{m}$ in kg/s	mass flow
p in bar	average pressure
ϑ_m in °C	average temperature of the fluid
ϑ_e in °C	temperature at the entrance of the tube
ϑ_a in °C	temperature at the exit of the tube
T in K	average temperature of the fluid
T_w in K	temperature of the tube wall
$\dot{V}$ in m^3/s	volume flow
w in m/s	average flow velocity
α in W/(m^{2}K)	heat transfer coefficient
λ in W/(mK)	thermal conductivity
v in m^2/s	kinematic viscosity
ϱ in kg/m^3	density
η in kg/(ms)	dynamic viscosity

Remarks:

In the transit region $(2300 < Re < 10^4)$ the following equations may result in a greater value for Nu_o for very short tubes $(d/L > 0.1)$.

1. $Nu_o = \left(3.66^3 + 1.61^3 \, Re \cdot Pr \cdot \dfrac{d}{L}\right)^{1/3}$

   (Nu_o at laminar tube flow: sheet No 3.2.1.1).

2. $Nu_o = 0.664 \, (Pr)^{1/3} \left(Re \, \dfrac{d}{L}\right)^{1/2}$

   (Nu_o at laminar flow past a flat plate, recalculated for laminar tube flow).

In these cases the greatest value for Nu_o has to be used [1].

These cases have not been considered in the nomogram.

Example: Water

inner diameter of the tube	$d = 0.04$ m	$d/L =$
length of the tube	$L = 4$ m	0.01
average flow velocity	$w = 0.83$ m/s	
average temperature of water	$\vartheta_m = 90$ °C	
pressure	$p = 2$ bar	
temperature of the tube wall	$\vartheta_w = 60$ °C	

Solution:

1. Determination of the Reynolds number from the values of ϑ_m, p, w, d: $Re = 10^5$ (sheet No 3.1.2.1).

   remark: If the mass flow is known, the average flow velocity has to be calculated from

   $$w = \frac{\dot{m}}{A \cdot \varrho} = \frac{\dot{m} \cdot 4}{\varrho \cdot \pi d^2}.$$

   The density ϱ has to be determined by the average temperature of water ϑ_m and the pressure p from nomograph No 3.1.2.2.

2. Determination of the Prandtl number at the average temperature of water (Pr), as well as at the temperature of the tube wall (Pr_w), with the values of p and ϑ_m resp. ϑ_w: $Pr = 2$, $Pr_w = 3$ (sheet No 3.2.1.2).

3. Examination of the range of validity of the law of heat transfer by the values of Re, Pr/Pr_w, d/L (sheet No 3.2.1.2).

4. Evaluation of the law of heat transfer: $Nu = 320$ (sheet No 3.2.1.2).

5. Determination of the heat transfer coefficient by the values of Nu, ϑ_m, p, d: $\alpha = 5400$ W/(m^2 K) (sheet No 3.1.1.3).

Bibliography

[1] VDI-Wärmeatlas, 4th edition 1984. VDI-Verlag Düsseldorf

Reynolds number Re
1000
500
200
100
50
20
10
5
2
1
0,71 Air
0,5
0,2
Pr = 0,1
1
0,5
0,2
0,1
0,05
0,02
0,01
0,005
0,002
$d_h/L = 0,001$
For gases in the range $0,5 < T/T_w < 2$
is $Pr/Pr_w = 1$
Liquids $Pr/Pr_w = 0,1$
1
10
$d_i/d_a = 0,1$
0,2
0,3
0,5
1,0
Nusselt number Nu

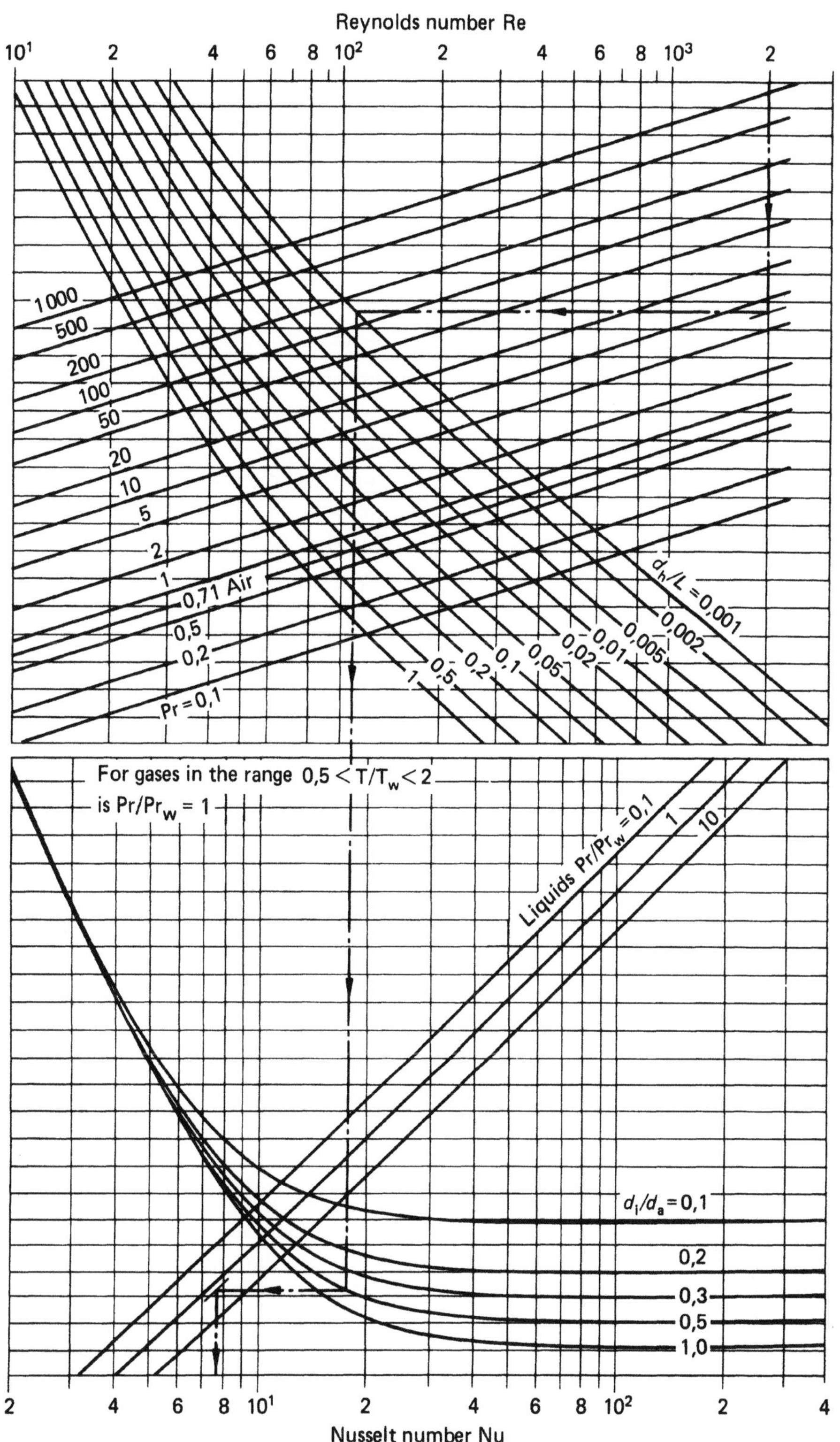

U. Renz

$$\mathrm{Nu} = \mathrm{Nu_o} \cdot K$$

$$\mathrm{Nu_o} = \mathrm{Nu_\infty} + f\!\left(\frac{d_\mathrm{i}}{d_\mathrm{a}}\right) \frac{0.19\left(\mathrm{RePr}\,\dfrac{d_\mathrm{h}}{L}\right)^{0.8}}{1 + 0.117\left(\mathrm{RePr}\,\dfrac{d_\mathrm{h}}{L}\right)^{0.467}}$$

$$\mathrm{Nu_\infty} = 3.66 + 1.2\left(\frac{d_\mathrm{i}}{d_\mathrm{a}}\right)^{-0.8}$$

$$f\!\left(\frac{d_\mathrm{i}}{d_\mathrm{a}}\right) = 1 + 0.14\left(\frac{d_\mathrm{i}}{d_\mathrm{a}}\right)^{-0.5} \quad [1]$$

Range of validity: concentric tubes

$$\mathrm{Re} < 2300$$

$$0.1 < \mathrm{Pr} < 1000$$

$$0 < \frac{d_\mathrm{i}}{d_\mathrm{a}} < 1$$

Correction factor K:

The correction factor K considers the influence of the direction of heat flow (heating or cooling).

For liquids this factor is:

$$K = (\mathrm{Pr}/\mathrm{Pr_w})^{0.11}.$$

For gases in the range $0.5 < T/T_\mathrm{w} < 2$ is:

$$\mathrm{Pr}/\mathrm{Pr_w} = 1$$

The *properties* of the media have to be determined at the average pressure p and the average temperature ϑ_m of the fluid:

$$\vartheta_\mathrm{m} = \frac{\vartheta_\mathrm{e} + \vartheta_\mathrm{a}}{2}.$$

The *characteristic length* on the calculation of Re and Nu is the hydraulic diameter of the circular gap:

$$d_\mathrm{h} = d_\mathrm{a} - d_\mathrm{i}.$$

Definitions:

$$\mathrm{Nu} = \frac{\alpha \cdot d_\mathrm{h}}{\lambda} \quad \text{Nusselt number}$$

$$\mathrm{Re} = \frac{w \cdot d_\mathrm{h}}{v} = \frac{\dot{m} \cdot d_\mathrm{h}}{\varrho \cdot A \cdot v} = \frac{\dot{V} \cdot d_\mathrm{h}}{A \cdot v} \quad \text{Reynolds number}$$

$$\mathrm{Pr} = \frac{\eta}{\lambda/c_\mathrm{p}} \quad \text{Prandtl number}$$

$\mathrm{Pr_w}$	Prandtl number at the temperature of the tube wall
A in m^2	cross-sectional area of the circular gap $= \pi(d_\mathrm{a}^2 - d_\mathrm{i}^2)/4$
c_p in J/(kgK)	specific heat capacity
d_a in m	inner diameter of the outer tube
d_h in m	hydraulic diameter of the circular gap
d_i in m	outer diameter of the inner tube
L in m	length of the tube
$\dot{m}$ in kg/s	mass flow
p in bar	average pressure
ϑ_a in °C	temperature at the exit of the circular gap
ϑ_e in °C	temperature at the entrance of the circular gap
ϑ_w in °C	temperature of the tube wall
T in K	average temperature of the fluid
T_w in K	temperature of the tube wall
$\dot{V}$ in m^3/s	volume flow
w in m/s	average flow velocity
α in W/(m^2 K)	heat transfer coefficient
λ in W/(mK)	thermal conductivity
v in m^2/s	kinematic viscosity
ϱ in kg/m^3	density
η in kg/(ms)	dynamic viscosity

Example: Water

average temperature of water	$\vartheta_\mathrm{m} = 20\,°\mathrm{C}$	
pressure	$p = 1$ bar	
inner diameter of the outer tube	$d_\mathrm{a} = 0.04\,\mathrm{m}$	$d_\mathrm{h} = d_\mathrm{a} - d_\mathrm{i}$ $= 0.02\,\mathrm{m}$
outer diameter of the inner tube	$d_\mathrm{i} = 0.02\,\mathrm{m}$	$d_\mathrm{h}/L = 0.001$
length of the tube	$L = 20\,\mathrm{m}$	$d_\mathrm{i}/d_\mathrm{a} = 0.5$
average flow velocity	$w = 0.1\,\mathrm{m/s}$	
temperature of the wall of the inner tube	$\vartheta_\mathrm{w} = 60\,°\mathrm{C}$	

Solution:

1. Determination of the Reynolds number from the values of ϑ_m, d_h, p, w: $\mathrm{Re} = 2000$ (sheet No 3.1.2.1).

   remark: If the mass flow is known, the average flow velocity has to be calculated from

   $$w = \frac{\dot{m}}{A \cdot \varrho}.$$

   The density ϱ has to be determined by the average temperature of water ϑ_m and the pressure p from nomograph No 3.1.2.2.

2. Determination of the Prandtl number at the average temperature of water (Pr), as well as at the temperature of the wall of the inner tube ($\mathrm{Pr_w}$): $\mathrm{Pr} = 7$, $\mathrm{Pr_w} = 3$ (sheet No 3.1.2.2).

3. Examination of the range of validity of the law of heat transfer by the values of Re, Pr, $d_\mathrm{i}/d_\mathrm{a}$ (sheet No 3.2.2.1).

4. Evaluation of the law of heat transfer: $\mathrm{Nu} = 7.8$.

   remark: $\mathrm{Pr}/\mathrm{Pr_w} = 2.3$ (sheet No 3.2.2.1).

5. Determination of the heat transfer coefficient by the values of Nu, d_h, ϑ_m, p: $\alpha = 230\,\mathrm{W/(m^2\,K)}$ (sheet No 3.1.2.3).

Bibliography

[1] VDI-Wärmeatlas, 4th edition 1984. VDI-Verlag Düsseldorf

Reynolds number Re
10^1 2 4 6 8 10^2 2 4 6 8 10^3 2
1000
500
200
100
50
20
10
5
2
1
0,71 Air
0,5
0,2
Pr = 0,1
$d_h/L = 0,001$
0,002
0,005
0,01
0,02
0,05
0,1
0,2
0,5
1
For gases in the range $0,5 < T/T_w < 2$
is $Pr/Pr_w = 1$
Liquids $Pr/Pr_w = 0,1$
1
10
$d_i/d_a = 1,0$
0,5
0,1
2 4 6 8 10^1 2 4 6 8 10^2 2 4
Nusselt number Nu

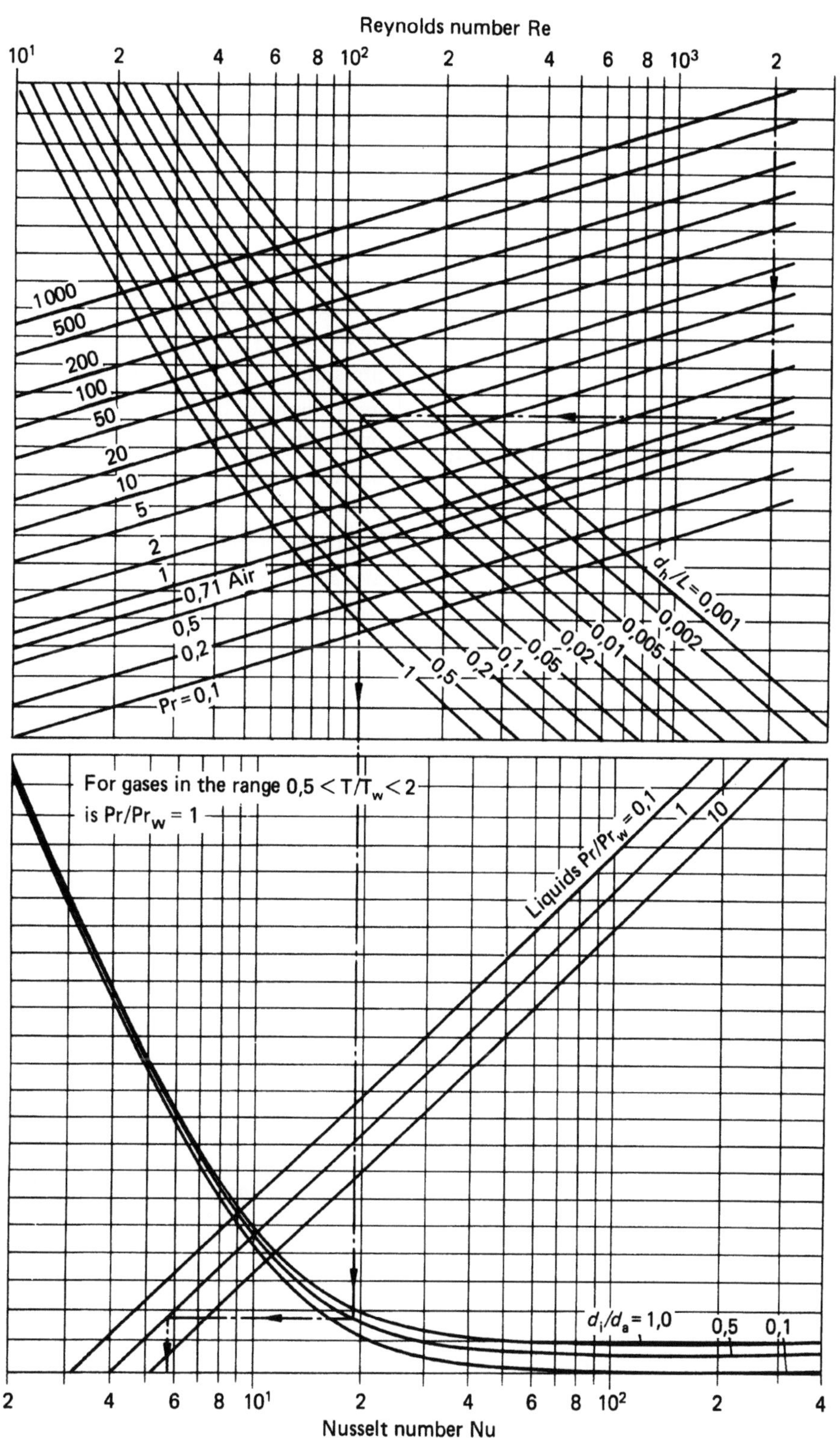

U. Renz

$$\mathrm{Nu} = \mathrm{Nu_o} \cdot K$$

$$\mathrm{Nu_o} = \mathrm{Nu_\infty} + f\!\left(\frac{d_i}{d_a}\right)\frac{0.19\left(\mathrm{RePr}\,\dfrac{d_h}{L}\right)^{0.8}}{1 + 0.117\left(\mathrm{RePr}\,\dfrac{d_h}{L}\right)^{0.467}}$$

$$\mathrm{Nu_\infty} = 3.66 + 1.2\left(\frac{d_i}{d_a}\right)^{0.5}$$

$$f\!\left(\frac{d_i}{d_a}\right) = 1 + 0.14\left(\frac{d_i}{d_a}\right)^{1/3} \quad [1]$$

Range of validity: concentric tubes

$$\mathrm{Re} < 2300$$

$$0.1 < \mathrm{Pr} < 1000$$

$$0 < \frac{d_i}{d_a} < 1$$

Correction factor K:

The correction factor K considers the influence of the direction of heat flow (heating or cooling).

For liquids this factor is:

$$K = (\mathrm{Pr}/\mathrm{Pr_w})^{0.11}$$

For gases in the range $0.5 < T/T_w < 2$ is:

$$\mathrm{Pr}/\mathrm{Pr_w} = 1$$

The *properties* of the media have to be determined at the average pressure p and the average temperature ϑ_m of the fluid:

$$\vartheta_m = \frac{\vartheta_e + \vartheta_a}{2}$$

The *characteristic length* on the calculation of Re and Nu is the hydraulic diameter of the circular gap:

$$d_h = d_a - d_i \,.$$

Definitions:

$$\mathrm{Nu} = \frac{\alpha \cdot d_h}{\lambda} \quad \text{Nusselt number}$$

$$\mathrm{Re} = \frac{w \cdot d_h}{v} = \frac{\dot{m} \cdot d_h}{\varrho \cdot A \cdot v} = \frac{\dot{V} \cdot d_h}{A \cdot v} \quad \text{Reynolds number}$$

$$\mathrm{Pr} = \frac{\eta}{\lambda/c_p} \quad \text{Prandtl number}$$

$\mathrm{Pr_w}$ Prandtl number at the temperature of the tube wall

A in m^2 cross-sectional area of the circular gap $= \pi(d_a^2 - d_i^2)/4$

c_p in J/(kgK)   specific heat capacity

d_a in m   inner diameter of the outer tube

d_h in m   hydraulic diameter of the circular gap

d_i in m   outer diameter of the inner tube

L in m   length of the tube

$\dot{m}$ in kg/s   mass flow

p in bar   average pressure

ϑ_a in °C   temperature at the exit of the circular gap

ϑ_e in °C   temperature at the entrance of the circular gap

ϑ_w in °C   temperature of the tube wall

T in K   average temperature of the fluid

T_w in K   temperature of the tube wall

$\dot{V}$ in m^3/s   volume flow

w in m/s   average flow velocity

α in W/(m^2 K)   heat transfer coefficient

λ in W/(mK)   thermal conductivity

v in m^2/s   kinematic viscosity

ϱ in kg/m^3 density

η in kg/(ms)   dynamic viscosity

Example: Air

average temperature of the air	$\vartheta_m = 0\,°\mathrm{C}$
pressure	$p = 1\,\mathrm{bar}$
inner diameter of the outer tube	$d_a = 0.02\,\mathrm{m}$
outer diameter of the inner tube	$d_i = 0.01\,\mathrm{m}$
length of the tube	$L = 1\,\mathrm{m}$
average flow velocity	$w = 2.7\,\mathrm{m/s}$
temperature of the wall of the outer tube	$\vartheta_w = 20\,°\mathrm{C}$

$$d_h = d_a - d_i = 0.01\,\mathrm{m}$$
$$d_h/L = 0.01$$
$$d_i/d_a = 0.5$$

Solution:

1. Determination of the Reynolds number from the values of ϑ_m, d_h, p, w: $\mathrm{Re} = 2000$ (sheet No 3.1.1.1).

   remark: If the mass flow is known, the average flow velocity has to be calculated from

   $$w = \frac{\dot{m}}{A \cdot \varrho}$$

   The density ϱ has to be determined by the average temperature of the air ϑ_m and the pressure p from nomograph No 3.1.1.2.

2. Determination of the Prandtl number by the values of ϑ_m, p: $\mathrm{Pr} = 0.7$ (sheet No 3.1.1.2).

3. Examination of the range of validity of the law of heat transfer by the values of Re, Pr, d_i/d_a, T/T_w (sheet No 3.2.2.2).

4. Evaluation of the law of heat transfer: $\mathrm{Nu} = 5.8$.

   remark: $\mathrm{Pr}/\mathrm{Pr_w} = 1$ (sheet No 3.2.2.2).

5. Determination of the heat transfer coefficient by the values of Nu, d_h, ϑ_m, p: $\alpha = 14\,\mathrm{W/(m^2\,K)}$ (sheet No 3.1.1.3).

Bibliography

[1] VDI-Wärmeatlas, 4th edition 1984. VDI-Verlag Düsseldorf

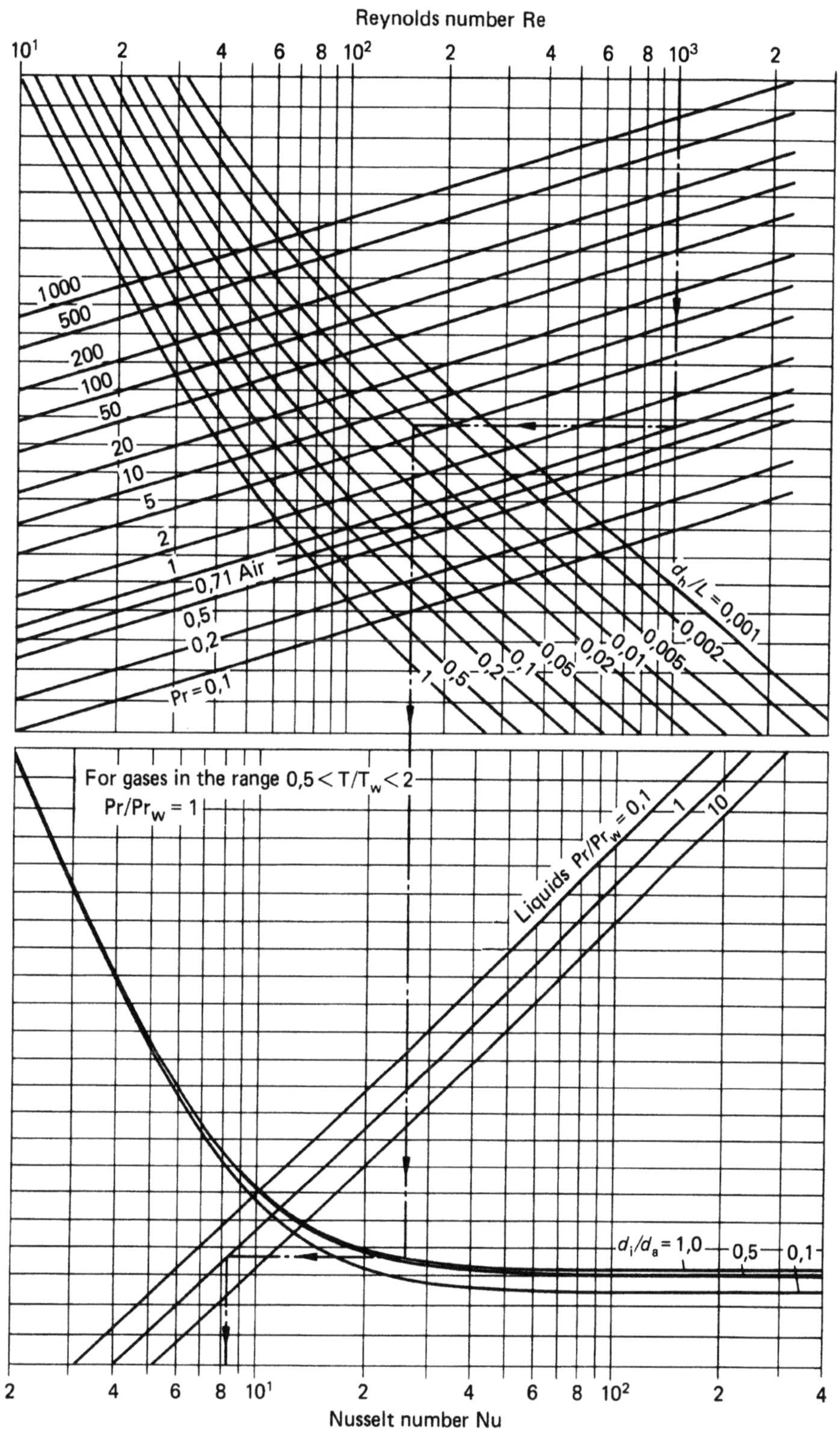

U. Renz

$$Nu = Nu_o \cdot K$$

$$Nu_o = Nu_\infty + f\left(\frac{d_i}{d_a}\right) \frac{0.19\left(RePr\,\dfrac{d_h}{L}\right)^{0.8}}{1 + 0.117\left(RePr\,\dfrac{d_h}{L}\right)^{0.467}}$$

$$Nu_\infty = 3.66 + \left[4 - \frac{0.102}{(d_i/d_a) + 0.02}\right]\left(\frac{d_i}{d_a}\right)^{0.04}$$

$$f\left(\frac{d_i}{d_a}\right) = 1 + 0.14\left(\frac{d_i}{d_a}\right)^{0.1} \quad [1]$$

Range of validity: concentric tubes

$$Re < 2300$$

$$0.1 < Pr < 1000$$

$$0 < \frac{d_i}{d_a} < 1$$

Correction factor K:

The correction factor K considers the influence of the direction of heat flow (heating or cooling).

For liquids this factor is:

$$K = (Pr/Pr_w)^{0.11}$$

For gases in the range $0.5 < T/T_w < 2$ is:

$$Pr/Pr_w = 1$$

The *properties* of the media have to be determined at the average pressure p and the average temperature ϑ_m of the fluid:

$$\vartheta_m = \frac{\vartheta_e + \vartheta_a}{2}$$

The *characteristic length* on the calculation of Re and Nu is the hydraulic diameter of the circular gap:

$$d_h = d_a - d_i\,.$$

Definitions:

$$Nu = \frac{\alpha \cdot d_h}{\lambda} \quad \text{Nusselt number}$$

$$Re = \frac{w \cdot d_h}{v} = \frac{\dot{m} \cdot d_h}{\varrho \cdot A \cdot v} = \frac{\dot{V} \cdot d_h}{A \cdot v} \quad \text{Reynolds number}$$

$$Pr = \frac{\eta}{\lambda/c_p} \quad \text{Prandtl number}$$

Pr_w Prandtl number at the temperature of the tube wall

A in m^2 cross-sectional area of the circular gap $= \pi(d_a^2 - d_i^2)/4$

c_p in $J/(kgK)$ specific heat capacity

d_a in m   inner diameter of the outer tube

d_h in m   hydraulic diameter of the circular gap

d_i in m   outer diameter of the inner tube

L in m   length of the tube

$\dot{m}$ in kg/s   mass flow

p in bar   average pressure

ϑ_a in °C   temperature at the exit of the circular gap

ϑ_e in °C   temperature at the entrance of the circular gap

ϑ_w in °C   temperature of the tube wall

T in K   average temperature of the fluid

T_w in K   temperature of the tube wall

$\dot{V}$ in m^3/s volume flow

w in m/s   average flow velocity

α in $W/(m^2\,K)$ heat transfer coefficient

λ in $W/(mK)$ thermal conductivity

v in m^2/s kinematic viscosity

ϱ in kg/m^3 density

η in $kg/(ms)$ dynamic viscosity

Example: Water (Steam)

average temperature of the steam	$\vartheta_m = 200\,°C$	
pressure	$p = 1\,bar$	
inner diameter of the outer tube	$d_a = 0.01\,m$	$d_h = d_a - d_i = 0.02\,m$
outer diameter of the inner tube	$d_i = 0.08\,m$	$d_h/L = 0.005$
length of the tube	$L = 4\,m$	$d_i/d_a = 0.8$
average flow velocity	$w = 1.8\,m/s$	
temperature of the wall of the inner tube	$\vartheta_w = 280\,°C$	

Solution:

1. Determination of the Reynolds number from the values of ϑ_m, d_h, p, w: $Re = 1000$ (sheet No 3.1.2.1).

   remark: If the mass flow is known, the average flow velocity has to be calculated from

   $$w = \frac{\dot{m}}{A \cdot \varrho}$$

   The density ϱ has to be determined with the average temperature of water ϑ_m and the pressure p from nomograph No 3.1.2.2.

2. Determination of the Prandtl number at the average temperature of the fluid (Pr), as well as at the temperature of the wall of the inner tube (Pr_w): $Pr = 1$, $Pr_w = 1$ (sheet No 3.1.2.2).

3. Examination of the range of validity of the law of heat transfer by the values of Re, Pr, d_i/d_a, T/T_w (sheet No 3.2.2.3).

4. Evaluation of the law of heat transfer: $Nu = 8.18$.

   remark: $Pr/Pr_w = 1$ (sheet No 3.2.2.3).

5. Determination of the heat transfer coefficient by the values of Nu, d_h, ϑ_m, p: $\alpha = 13.5\,W/(m^2\,K)$ (sheet No 3.1.2.3).

Bibliography

[1] VDI-Wärmeatlas, 4th edition 1984. VDI-Verlag Düsseldorf

Reynolds number Re
$d_h/L = 1$
0,2
0,001
Pr = 0,5
0,71
1,0
2,0
5,0
10
20
50
100
200
500
1000
Nusselt number Nu for liquids
Nusselt number Nu for gases
Gases $T/T_w \geq 1,0$
0,75
0,5
Liquids $Pr/Pr_w = 0,1$
1,0
10

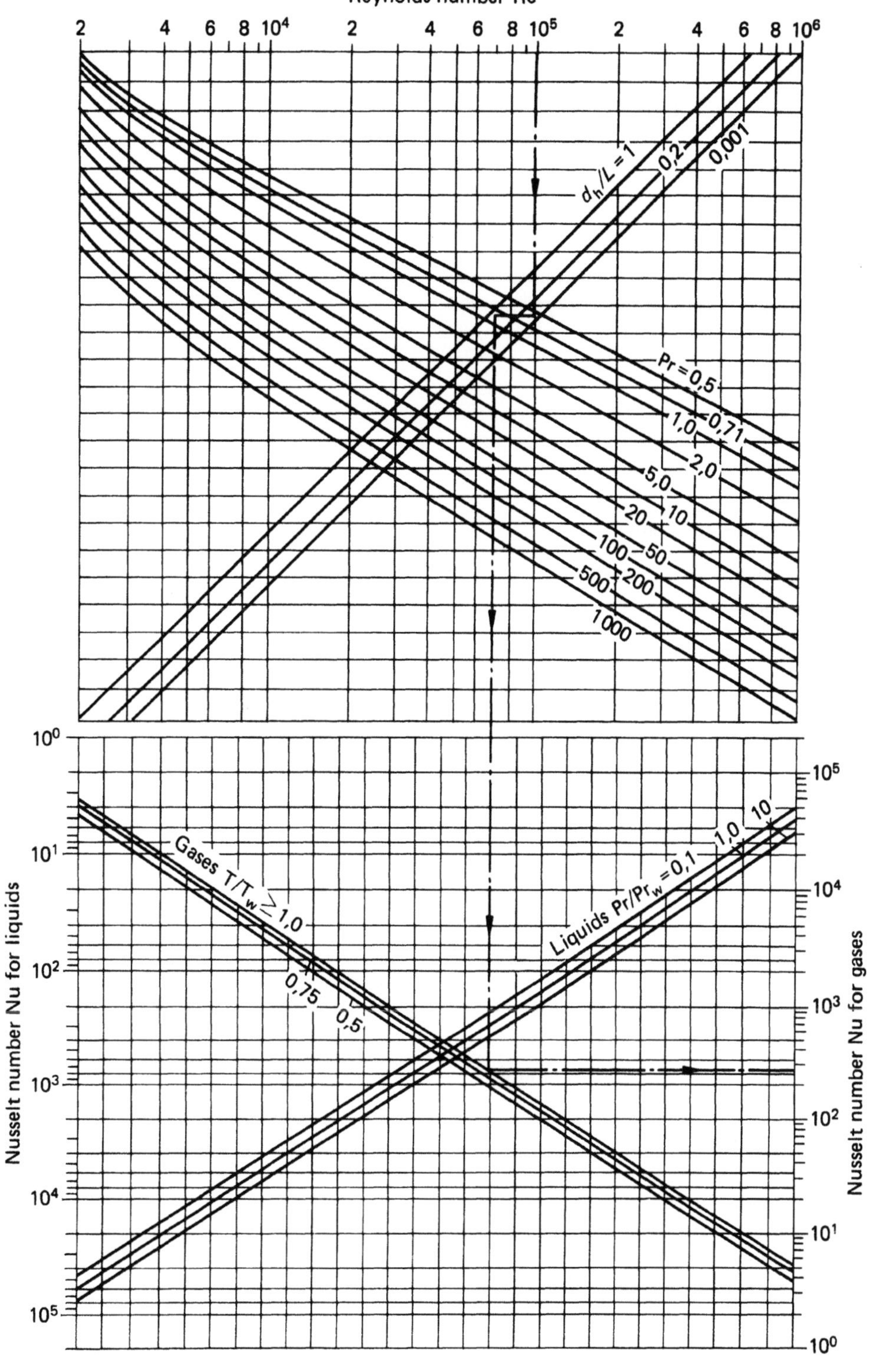

U. Renz

$$\mathrm{Nu} = f\left(\frac{d_i}{d_a}\right) \cdot \mathrm{Nu}_{\text{tube}}$$

Range of validity: concentric tubes

$$2300 < \mathrm{Re} < 10^6 \qquad 0.6 < \mathrm{Pr} < 1000$$

$$0 < \frac{d_h}{L} < 1 \qquad 0 < \frac{d_i}{d_a} < 1$$

$\mathrm{Nu}_{\text{tube}}$:

This is the Nusselt number, calculated by the equations of the turbulent tube flow (3.2.1.2). Thus, the necessary reference numbers Re, Nu, d_h/L have to be calculated by the hydraulic diameter of the circular gap.

$$d_h = d_a - d_i$$

The influence of the direction of heat flow is equivalent to that of the turbulent tube flow (3.2.1.2).

The *properties* of the media have to be determined at the average pressure p and the average temperature ϑ_m of the fluid:

$$\vartheta_m = \frac{\vartheta_e + \vartheta_a}{2}$$

$f\left(\dfrac{d_i}{d_a}\right)$: according to the boundary conditions.

Heat transfer to the inner tube only

$$f\left(\frac{d_i}{d_a}\right) = 0.86\left(\frac{d_i}{d_a}\right)^{-0.16}$$

Heat transfer to the outer tube only

$$f\left(\frac{d_i}{d_a}\right) = 1 - 0.14\left(\frac{d_i}{d_a}\right)^{0.6}$$

Heat transfer to the inner and outer tube

$$f\left(\frac{d_i}{d_a}\right) = \frac{0.86\left(\dfrac{d_i}{d_a}\right)^{0.84} + \left(1 - 0.14\left(\dfrac{d_i}{d_a}\right)^{0.8}\right)}{1 + \dfrac{d_i}{d_a}} \quad [1]$$

Definitions:

$$\mathrm{Nu} = \frac{\alpha \cdot d_h}{\lambda} \quad \text{Nusselt number}$$

$$\mathrm{Re} = \frac{w \cdot d_h}{v} = \frac{\dot{m} \cdot d_h}{\varrho \cdot A \cdot v} = \frac{\dot{V} \cdot d_h}{A \cdot v} \quad \text{Reynolds number}$$

$$\mathrm{Pr} = \frac{\eta}{\lambda/c_p} \quad \text{Prandtl number}$$

Pr_w	Prandtl number at the temperature of the tube wall
A in m^2	cross-sectional area of the circular gap $= \pi(d_a^2 - d_i^2)/4$
c_p in J/(kgK)	specific heat capacity
d_a in m	inner diameter of the outer tube
d_h in m	hydraulic diameter of the circular gap
d_i in m	outer diameter of the inner tube
L in m	length of the tube
$\dot{m}$ in kg/s	mass flow
p in bar	average pressure
ϑ_a in °C	temperature at the exit of the circular gap
ϑ_e in °C	temperature at the entrance of the circular gap
ϑ_w in °C	temperature of the tube wall
T in K	average temperature of the fluid
T_w in K	temperature of the tube wall
$\dot{V}$ in m^3/s	volume flow
w in m/s	average flow velocity
α in W/(m^2 K)	heat transfer coefficient
λ in W/(mK)	thermal conductivity
v in m^2/s	kinematic viscosity
ϱ in kg/m^3	density
η in kg/(ms)	dynamic viscosity

Bibliography

[1] VDI-Wärmeatlas, 4th edition 1984. VDI-Verlag Düsseldorf

Example: Cooling of steam (heat transfer to the inner tube only)

average temperature of the steam	$\vartheta_m = 300\,°C$
pressure	$p = 50\,bar$
average flow velocity	$w = 1\,m/s$

length of the tube $\quad L = 10\,m \quad$ $d_h = d_a - d_i = 0.1\,m$

inner diameter of the outer tube $\quad d_a = 0.25\,m \quad$ $d_i/d_a = 0.6$

outer diameter of the inner tube $\quad d_i = 0.15\,m \quad$ $d_h/L = 0.01$

Solution:

1. Determination of the Reynolds number from the values of ϑ_m, p, $\dot{m}$, d_h, d_i, d_a. Here, d_h is the characteristic length: d_h: $Re = 1.1 \cdot 10^5$ (sheet No 3.1.2.1).

   remark: If the mass flow is known, the average flow velocity has to be calculated from

   $$w = \frac{\dot{m}}{A \cdot \varrho}$$

   The density ϱ has to be determined with the average temperature ϑ_m and the pressure p from nomograph No 3.1.2.2.

2. Determination of the Prandtl number from the values of ϑ_m, p: $Pr = 1.2$ (sheet No 3.1.2.2).

3. Examination of the range of validity of the law of heat transfer by the values of Re, Pr, d_i/d_a, d_h/L (sheet No 3.2.2.4).

4. Calculation of Nu_{tube} by the equations for the turbulent tube flow: $Nu_{tube} = 282$ (sheet No 3.2.2.4).

   remark: $K = 1$

5. Determination of $f(d_i/d_a)$: $f(d_i/d_a) = 0.93$ and thus $Nu = f(d_i/d_a) \cdot Nu_{tube} = 0.93 \cdot 282 = 262.3$ (sheet No 3.2.2.4).

6. Determination of the heat transfer coefficient by the values of Nu, d_h, p, ϑ_m: $\alpha = 140\,W/(m^2\,K)$ (sheet No 3.1.2.3).

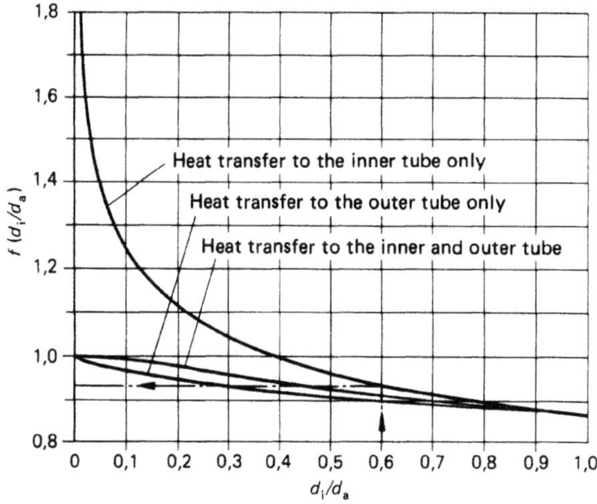

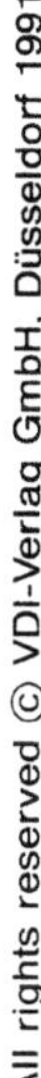

Nusselt Number at Flow past Flat Plates and along Single Bodies

$\mathrm{Nu_l} = \mathrm{Nu_{l,o}} \cdot K$

$\mathrm{Nu_{l,o}} = [\mathrm{Nu^2_{o,lam}} + \mathrm{Nu^2_{o,turb}}]^{1/2}$ [1]

Range of validity:

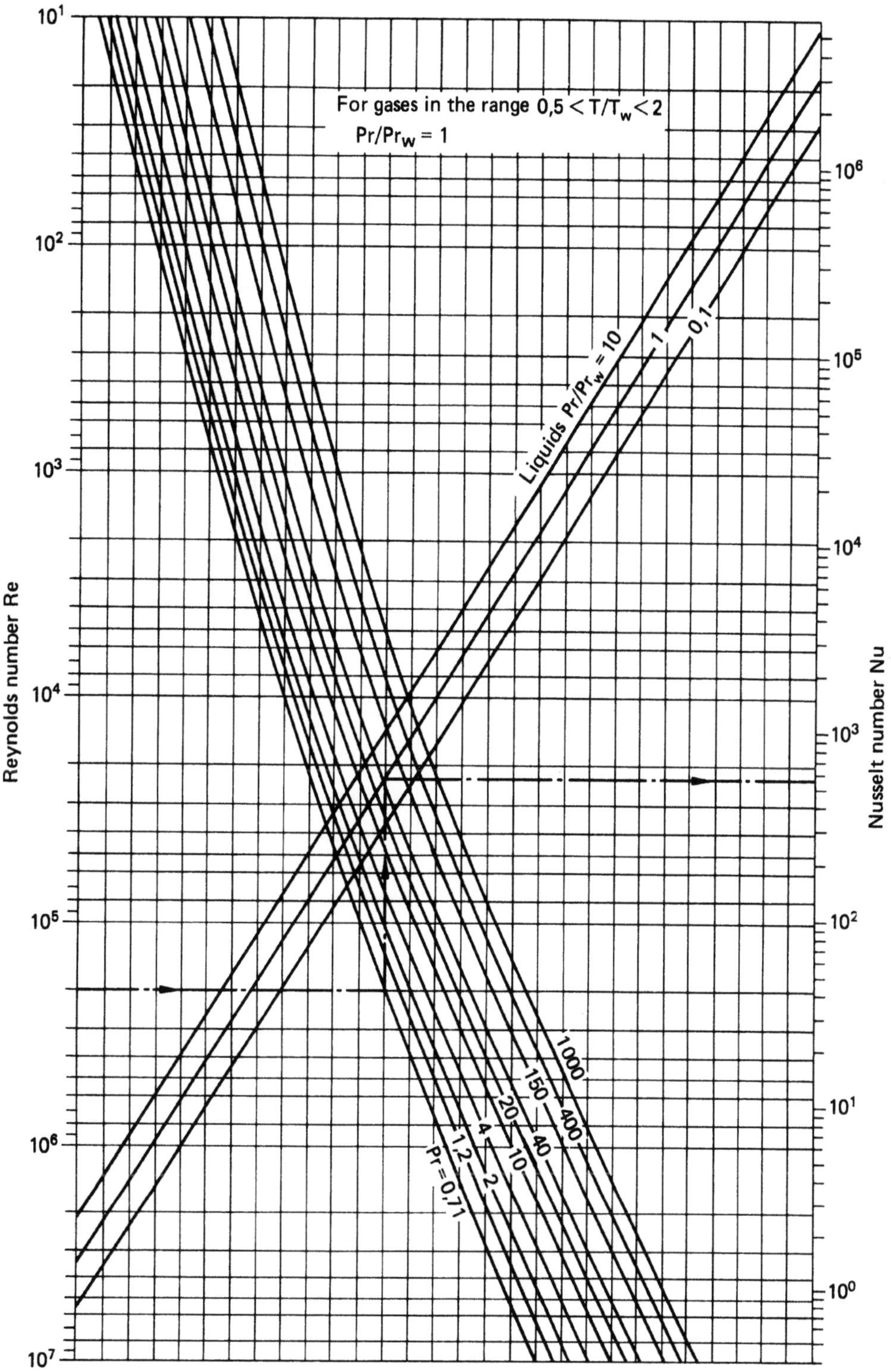

The equation for $\mathrm{Nu_{l,o}}$ is a summary of the equations for laminar and turbulent flow past a flat plate.

Flow with laminar boundary layer:

$\mathrm{Nu_{o,lam}} = 0.664(\mathrm{Re_l})^{0.5}\,(\mathrm{Pr})^{1/3}$

Range of validity:

$\mathrm{Re_l} < 10^5$

$0.6 < \mathrm{Pr} < 2000$

U. Renz

Flow with turbulent boundary layer:

$$\mathrm{Nu}_{\mathrm{o,turb}} = \frac{0.037\,\mathrm{Re}_l^{0.8} \cdot \mathrm{Pr}}{1 + 2.443\,\mathrm{Re}_l^{-0.1}\,(\mathrm{Pr}^{2/3} - 1)}$$

Range of validity:

$$5 \cdot 10^5 < \mathrm{Re}_l < 10^7$$
$$0.6 < \mathrm{Pr} < 2000$$

Correction factor K:

The correction factor K considers the influence of the direction of heat flow (heating or cooling).

For liquids this factor is:

$$K = (\mathrm{Pr}/\mathrm{Pr}_w)^{0.25}$$

For gases in the range $0.5 < T/T_w < 2$ is:

$$\mathrm{Pr}/\mathrm{Pr}_w = 1$$

The *properties* of the media have to be determined at the average pressure p and the average temperature ϑ_m of the fluid:

$$\vartheta_m = \frac{\vartheta_e + \vartheta_a}{2}$$

The *characteristic length* on the calculation of Re_l and Nu_l is the length of flow l.

Length of flow past a flat plate: length of the plate l

length of flow l
of tubes:

$$l = \frac{\pi}{2} \cdot D$$

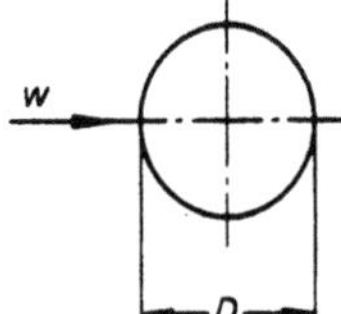

length of flow l
of channels:

$$l = a + b$$

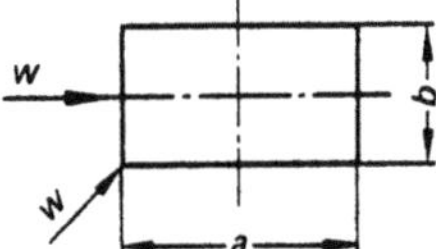

Definitions:

$$\mathrm{Nu}_l = \frac{d \cdot l}{\lambda} \quad \text{Nusselt number}$$

$$\mathrm{Re}_l = \frac{w \cdot l}{v} = \frac{\varrho \cdot w \cdot l}{\eta} \quad \text{Reynolds number}$$

$$\mathrm{Pr} = \frac{\eta}{\lambda/c_p} \quad \text{Prandtl number}$$

Pr_w Prandtl number at wall temperature ϑ_w
c_p in J/(kgK)   specific heat capacity
l in m   length of flow
p in bar   average pressure
ϑ_a in °C   temperature at the exit
ϑ_e in °C   temperature at the entrance
ϑ_m in °C   average temperature of the fluid
T in K   average temperature of the fluid
T_w in K   temperature of the wall
w in m/s   flow velocity
α in W/(m^2 K)   heat transfer coefficient
λ in W/(mK)   thermal conductivity
v in m^2/s   kinematic viscosity
ϱ in kg/m^3 density
η in kg/(ms)   dynamic viscosity

Example 1: Air flow past a flat plate

pressure	$p = 1$ bar
average temperature of the air	$\vartheta_m = 50\,°C$
velocity of the air	$w = 3.5$ m/s
length of the tube	$l = 1$ m

Solution:

1. Determination of the Reynolds number from the values of ϑ_m, p, w, l: $\mathrm{Re} = 2 \cdot 10^5$ (sheet No 3.1.1.1).

2. Determination of the Prandtl number by the values of ϑ_m, p: $\mathrm{Pr} = 0.7$ (sheet No 3.1.1.2).

3. Examination of the range of validity of the law of heat transfer by the values of Re_l, Pr, (sheet No 3.2.3).

4. Evaluation of the law of heat transfer by the values of Re_l, Pr: $\mathrm{Nu}_l = 581$ (sheet No 3.2.3).

5. Determination of the heat transfer coefficient by the values of ϑ_m, p, l, Nu_l: $\alpha = 16.4$ W/(m^2 K) (sheet No 3.1.1.3).

Example 2: Cross flow of air at a single lateral tube

average temperature of the air	$\vartheta_m = 20\,°C$
outer diameter of the tube	$D = 0.1$ m
temperature of the tube wall	$\vartheta_w = 100\,°C$
pressure of the air	$p = 1$ bar
flow velocity	$w = 20$ m/s

Solution:

1. Determination of the Reynolds number from the values of ϑ_m, p, l, w: $\mathrm{Re} = 2 \cdot 10^5$.
   remark: The length of flow l, surrounding a single lateral tube, is $l = \pi/2 \cdot D$ (sheet No 3.1.1.1).

2. Determination of the Prandtl number by the values of ϑ_m and p: $\mathrm{Pr} = 0.7$ (sheet No 3.1.1.2).

3. Examination of the range of validity of the law of heat transfer by the values of Re_l, Pr, T/T_w (sheet No 3.2.3).

4. Evaluation of the law of heat transfer by the values of Pr, Re_l: $\mathrm{Nu}_l = 581$ (sheet No 3.2.3).

5. Determination of the heat transfer coefficient by the values of ϑ_m, p, l, Nu_l: $\alpha = 96.3$ W/(m^2 K) (sheet No 3.1.1.3).

Bibliography

[1] VDI-Wärmeatlas, 4th edition 1984. VDI-Verlag Düsseldorf

a)

Ratio of vertical pitch a
4
3
2
1
For values of b > 1   b = 1
Ratio of horizontal pitch b = 0,6
0,7
0,8
0,9
1,0
0,2   0,3   0,4   0,5   0,6   0,7   0,8   0,9
Void fraction ψ

b)

Reynolds number $Re_{\psi,I}$
10^1 2   4   6 10^2 2   4   6 10^3 2   4   6 10^4 2   4   6 10^5 2   4   6 10^6
For gases
$Pr/Pr_w = 1$
Liquids $Pr/Pr_w = 0,1$ 1   10
Pr = 1000   400   150   40
Pr = 0,71   1,2   2   4   10   20
10^0 2   4   6 10^1 2   4   6 10^2 2   4   6 10^3 2   4   6 10^4 2   4   6 10^5
Nusselt number Nu

c)

Ratio of horizontal pitch b
1,0   1,5   2,0   2,5   3,0   3,5   4,0
Number of tube rows n = 2
4   6   10
1,05
1,1
1,2
a = 1,4
1,6
4   3,5   3   2,5   2   1,8
In line arrangement
0,8   1,0   1,2   1,4   1,6   1,8   2,0   2,2   2,4
Geometric factor F

Ratio of horizontal pitch b
0,5   1,0   1,5   2,0   2,5   3,0   3,5   4,0
2
Number of tube rows n = 4
6   10
Staggered arrangement
0,8   1,0   1,2   1,4   1,6   1,8   2,0   2,2   2,4
Geometric factor F

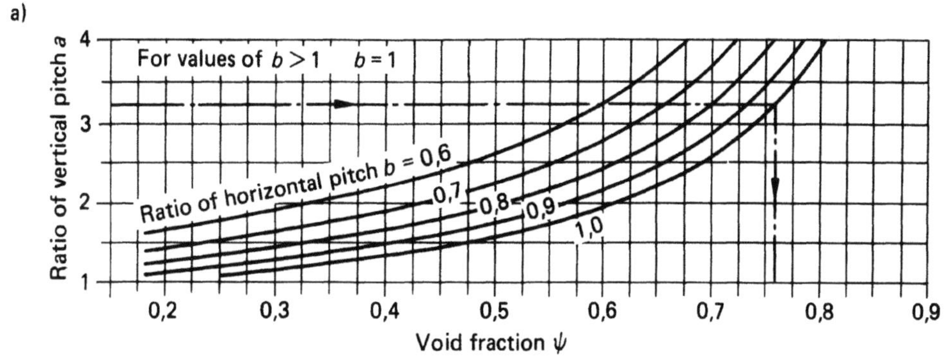

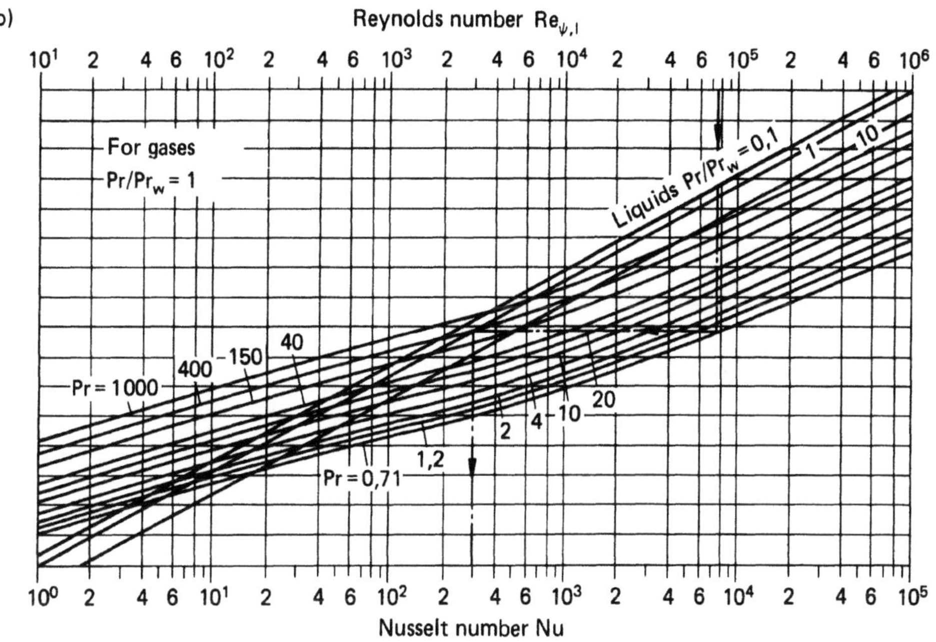

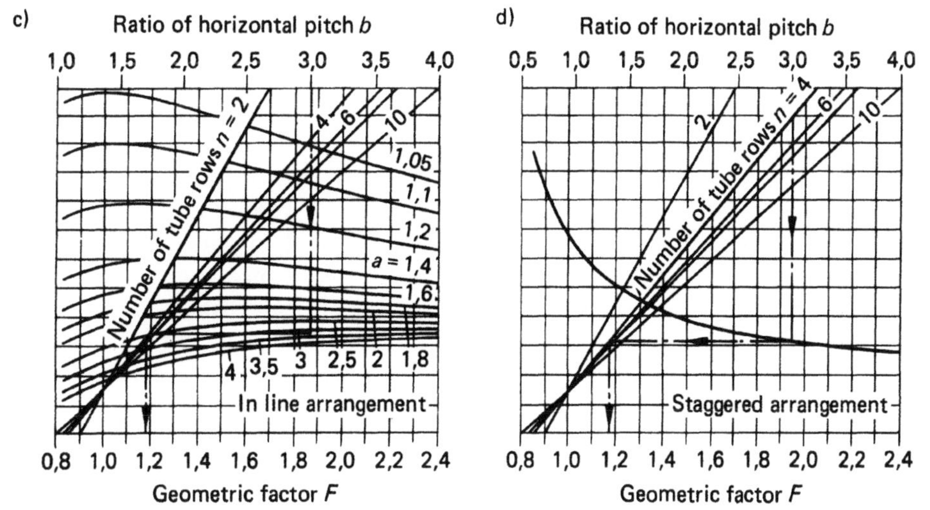

d)

U. Renz

$$\mathrm{Nu_{tube\ bank}} = F \cdot \mathrm{Nu}$$

with

$$\mathrm{Nu} = \mathrm{Nu_{1,o}} \cdot K$$

$$\mathrm{Nu_{1,o}} = 0.3 + \sqrt{\mathrm{Nu_{1,lam}^2} + \mathrm{Nu_{1,turb}^2}}$$

$$\mathrm{Nu_{1,lam}} = 0.664\,\mathrm{Re}_{\psi,1}^{1/2} \cdot \mathrm{Pr}^{1/3}$$

$$\mathrm{Nu_{1,turb}} = \frac{0.037\,\mathrm{Re}_{\psi,1}^{0.8} \cdot \mathrm{Pr}}{1 + 2.443 \cdot \mathrm{Re}_{\psi,1}^{-0.1}(\mathrm{Pr}^{2/3} - 1)} \quad [1]$$

Remarks:

The Reynolds number has to be determined by the previously calculated Reynolds number $\mathrm{Re_1}$ and by the void fraction ψ (sheet No 3.2.4a).

The Nusselt number Nu and the geometric factor F may each be determined by the nomographs 3.2.4b to d. The Nusselt number $\mathrm{Nu_{tube\ bank}}$ has to be determined by calculation.

Range of validity: $10 < \mathrm{Re}_{\psi,1} < 10^6$
$$0.6 < \mathrm{Pr} < 1000$$

Geometric factor F:

For tube banks with less than 10 tube rows is:

$$F = \frac{1 + (n-1) \cdot f_\mathrm{A}}{n}$$

For tube banks with 10 or more tube rows is:

$$F = f_\mathrm{A}$$

with n = number of tube rows
f_A = factor for tube arrangement

For in line tube arrangement is:

$$f_{\mathrm{A,fl}} = 1 + \frac{0.7\,(b/a - 0.3)}{\psi^{1.5}\,(b/a + 0.7)^2}$$

For staggered tube arrangement is:

$$f_{\mathrm{A,vers}} = 1 + \frac{2}{3b}$$

Correction factor K:

The correction factor K considers the influence of the direction of heat flow (heating or cooling).

For liquids this factor is:

$$K = (\mathrm{Pr}/\mathrm{Pr_w})^{0.25} \quad \text{for}\ \mathrm{Pr}/\mathrm{Pr_w} > 1$$
$$K = (\mathrm{Pr}/\mathrm{Pr_w})^{0.11} \quad \text{for}\ \mathrm{Pr}/\mathrm{Pr_w} < 1\,.$$

For gases is $\mathrm{Pr}/\mathrm{Pr_w} = 1$.

The *properties* of the media have to be determined at the average pressure p and the average temperature ϑ_m of the fluid:

$$\vartheta_\mathrm{m} = \frac{\vartheta_\mathrm{e} + \vartheta_\mathrm{a}}{2}\,.$$

The *characteristic length* for the calculation of the reference numbers $\mathrm{Nu_{tube}}$ and $\mathrm{Re_1}$ resp. $\mathrm{Re}_{\psi,1}$ is the length of flow surrounding a single tube (l):

$$l = \frac{\pi}{2} \cdot d_\mathrm{a}\,.$$

Definitions:

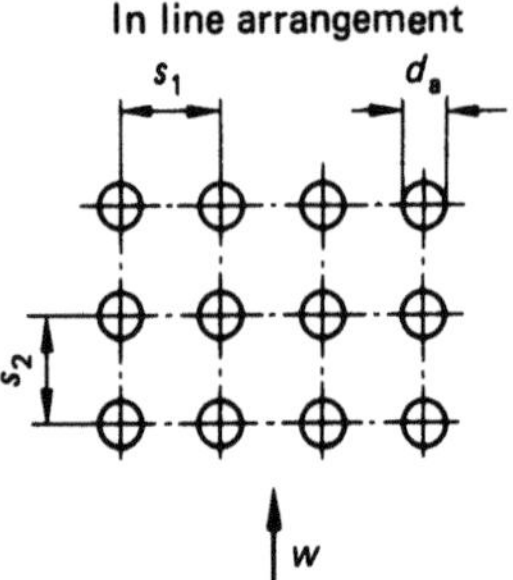

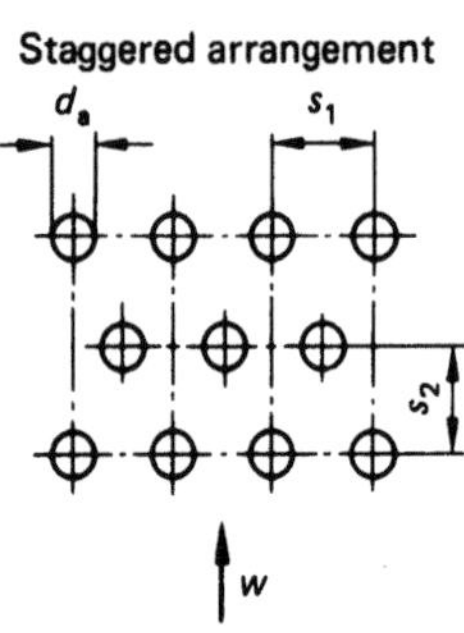

ratio of vertical pitch: $a = s_1/d_\mathrm{a}$

ratio of horizontal pitch: $b = s_2/d_\mathrm{a}$

void fraction: $\psi = 1 - \dfrac{\pi}{4 \cdot a} \quad \text{for}\ b \geq 1$

$\psi = 1 - \dfrac{\pi}{4 \cdot a \cdot b} \quad \text{for}\ b < 1$

$\mathrm{Nu_{tube\ bank}} = \dfrac{\alpha \cdot l}{\lambda}$ Nusselt number

$\mathrm{Nu} = \dfrac{\mathrm{Nu_{tube\ bank}}}{F}$ Nusselt number without geometric factor F

$\mathrm{Re_1} = \dfrac{w \cdot l}{v} = \dfrac{\dot{m} \cdot l}{A \cdot \varrho \cdot v} = \dfrac{\dot{V} \cdot l}{A \cdot v}$ Reynolds number without void fraction ψ

$\mathrm{Re}_{\psi,1} = \dfrac{\mathrm{Re_1}}{\psi}$ Reynolds number

$\mathrm{Pr} = \dfrac{\eta}{\lambda/c_\mathrm{p}}$ Prandtl number

A in m — free cross-sectional area in front of the tube bank
a — ratio of vertical pitch
b — ratio of horizontal pitch
c_p in J/(kgK) — specific heat capacity
d_a in m — outer diameter of a single tube
l in m — length of flow surrounding a single tube
$\dot{m}$ in kg/s — mass flow
n — number of tube rows
p in bar — average pressure
s_1 in m — vertical pitch
s_2 in m — horizontal pitch
ϑ_a in °C — temperature at the fluid at the exit of the tube bank
ϑ_e in °C — temperature of the fluid at the entrance of the tube bank
ϑ_m in °C — average temperature of the fluid
$\dot{V}$ in m³/s — volume flow
w in m/s — average flow velocity at the free cross-sectional area
α in W/(m² K) — heat transfer coefficient
λ in W/(mK) — thermal conductivity
v in m²/s — kinematic viscosity
ϱ in kg/m³ — density
η in kg/(ms) — dynamic viscosity
ψ — void fraction

Example: Air

average temperature of the air	$\vartheta_m = 50\,°C$	
pressure	$p = 5\,bar$	
number of tube rows	$n = 4$	
outer diameter of the tube	$d_a = 0.025\,m$	$l = \dfrac{\pi}{2} \cdot d_a = 0.04\,m$
vertical pitch	$S_1 = 0.080\,m$	$a = S_1/d_a = 3.2$
horizontal pitch	$S_2 = 0.075\,m$	$b = S_2/d_a = 3.0$
in line arrangement temperature of the tube wall	$\vartheta_w = 20\,°C$	
average velocity of the air at the free cross-sectional area	$w = 5\,m/s$	

Solution:

1. Determination of the Reynolds number Re_1 from the values of ϑ_m, w, p, l: $Re_1 = 56\,000$ (sheet No 3.1.1.1).

    remark: If the mass flow is known, the average flow velocity has to be calculated from

    $$w = \frac{\dot{m}}{A \cdot \varrho}.$$

    The density ϱ has to be determined with the average temperature of the air ϑ_m and the pressure p from nomograph No 3.1.1.2.

2. Determination of the void fraction by the values of a and b: $\psi = 0.76$ (diagram No 3.2.4a).

3. Calculation: $Re_{\psi,1} = Re_{1/\psi} = 74\,000$.

4. Determination of the Prandtl number from the values of ϑ_m, p: $Pr = 0.7$ (sheet No 3.1.1.2).

5. Examination of the range of validity of the law of heat transfer by the values of $Re_{\psi,1}$ and Pr (sheet No 3.2.4).

6. Evaluation of the law of heat transfer by the values of $Re_{\psi,1}$, Pr and Pr/Pr_w: $Nu = 293$.

    remark: $Pr/Pr_w = 1$ (diagram No 3.2.4b).

7. Determination of the geometric factor F by the values of a, b and n for the in line arrangement $F = 1.19$ (diagram No 3.2.4c).

    remark: In the case of staggered arrangement F would be 1.17 (diagram No 3.2.4d).

8. Calculation: $Nu_{tube\ bank} = F \cdot Nu = 348$.

9. Determination of the heat transfer coefficient by the values of $Nu_{tube\ bank}$, ϑ_m, p and l: $\alpha = 207\,W/(m^2\,K)$ (sheet No 3.1.1.3).

Bibliography

[1] VDI-Wärmeatlas, 4th edition 1984. VDI-Verlag Düsseldorf

Grashof number Gr
10¹²
6
4
2
10¹¹
6
4
2
10¹⁰
6
4
2
10⁹
6
4
2
10⁸
6
4
2
10⁷
6
4
2
10⁶
6
4
2
10⁵
6
4
2
10⁴
10⁰
2
4
6
8
10¹
2
4
Nusselt number Nu
0,1
Pr = 0,71
5
10
50
100
500
1000

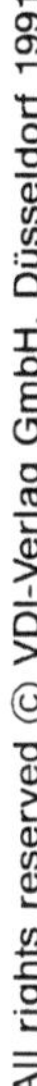

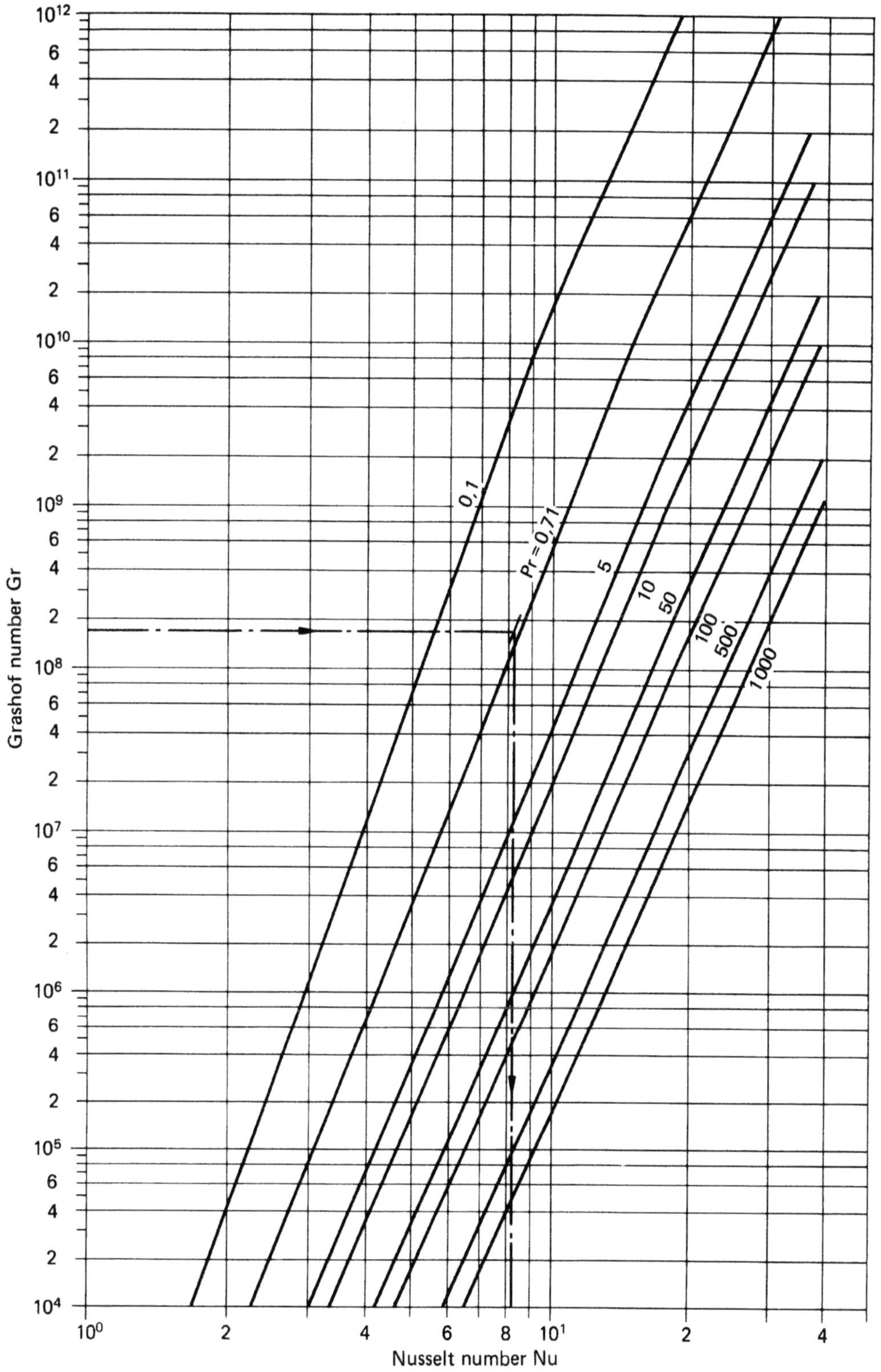

U. Renz

$$\text{Nu} = 0.825 + \frac{0.387\,(\text{Gr} \cdot \text{Pr})^{1/6}}{\left[1 + \left(\dfrac{0.492}{\text{Pr}}\right)^{9/16}\right]^{8/27}} \quad [1]$$

Range of validity:

vertical plates

$$0 < \text{Gr} \cdot \text{Pr} < 10^{12}$$
$$0 < \text{Pr} < \infty$$

horizontal cylinders

$$10^3 < \text{Gr} \cdot \text{Pr} < 10^{12}$$
$$0 < \text{Pr} < \infty$$

The *properties* of the media without index have to be determined at the average pressure p and at the average temperature ϑ_m of the flowing fluid:

$$\vartheta_m = \frac{\vartheta_w + \vartheta_\infty}{2}$$

The *flow length* l in the equations for the Nusselt- and the Grashof number is

for a vertical plate:

$$l = \text{height of the plate}$$

for a horizontal cylinder:

$$l = \pi/2 \cdot D$$

with the outer diameter of the cylinder D.

Definitions:

$$\text{Nu} = \frac{\alpha \cdot d_h}{\lambda} \quad \text{Nusselt number}$$

$$\text{Gr} = \frac{g \cdot l^3}{v^2}\,\frac{\varrho_\infty - \varrho_w}{\varrho_w} \quad \text{Grashof number}$$

$$\text{Pr} = \frac{\eta}{\lambda/c_p} \quad \text{Prandtl number}$$

c_p in J/(kgK)   specific heat capacity
g in m/s^2 acceleration due to gravity ($g = 9.81$ m/s^2)
l in m   length of flow
p in bar   average pressure
ϑ_m in °C   average temperature
ϑ_w in °C   temperature at the surface
α in W/(m^2 K)   heat transfer coefficient
λ in W/(mK)   thermal conductivity
v in m^2/s   kinematic viscosity
η in kg/(ms)   dynamic viscosity
ϱ_w in kg/m^3 density of the fluid at the temperature ϑ_w
ϱ_∞ in kg/m^3 density of the fluid at the temperature ϑ_∞

Example: Horizontal pipe line in unstirred air

length of the pipe line	$L = 5$ m
outer diameter of the pipe	$D = 0.19$ m
temperature of the pipe surface	$\vartheta_w = 80\,°\text{C}$
temperature of the air	$\vartheta_\infty = 20\,°\text{C}$
pressure of the air	$p = 1$ bar

$$\vartheta_m = \frac{\vartheta_w + \vartheta_\infty}{2} = 50\,°\text{C}\,.$$

Solution:

1. Determination of the density of the air at ambient temperature ϱ_∞ and at the temperature of the pipe surface ϱ_w by the values of ϑ_∞ resp. ϑ_w and p (sheet No 3.1.1.2).

2. Calculation of the value of $\dfrac{\varrho_\infty - \varrho_w}{\varrho_w} = 0.202$
   and of the stream length l.
   remark: For horizontal cylinders is

   $$l = \frac{\pi}{2} \cdot D = 0.3 \text{ m}\,.$$

3. Determination of the Grashof number by the values of ϑ_m, p, l, $\dfrac{\varrho_\infty - \varrho_w}{\varrho_w}$: $\text{Gr} = 1.7 \cdot 10^8$ (sheet No 3.1.1.4).

4. Determination of the Prandtl number by the values of p, ϑ_m: $\text{Pr} = 0.69$ (sheet No 3.1.1.2).

5. Examination of the range of validity of the law of heat transfer by the values of Pr, Gr (sheet No 3.3.1).

6. Evaluation of the law of heat transfer by the values of Gr, Pr: $\text{Nu} = 8$ (sheet No 3.3.1).

7. Determination of the heat transfer coefficient by the values of ϑ_m, l, Nu, p: $\alpha = 0.8$ W/(m^2 K) (sheet No 3.1.1.3).

Bibliography

[1] VDI-Wärmeatlas, 4th edition 1984. VDI-Verlag Düsseldorf

W/(m²K)
Heat transfer coefficient α
5000000
2000000
1000000
500000
200000
100000
50000
20000
10000
5000
2000
1000
500
200
q̇'' = 100 W/m2
Pressure p = 200 bar
180
150
120
100
70
40
20
10
5
2
1
0.5
0.2
0.1
Example 2
Example 1
Temperature difference T_w − T
K

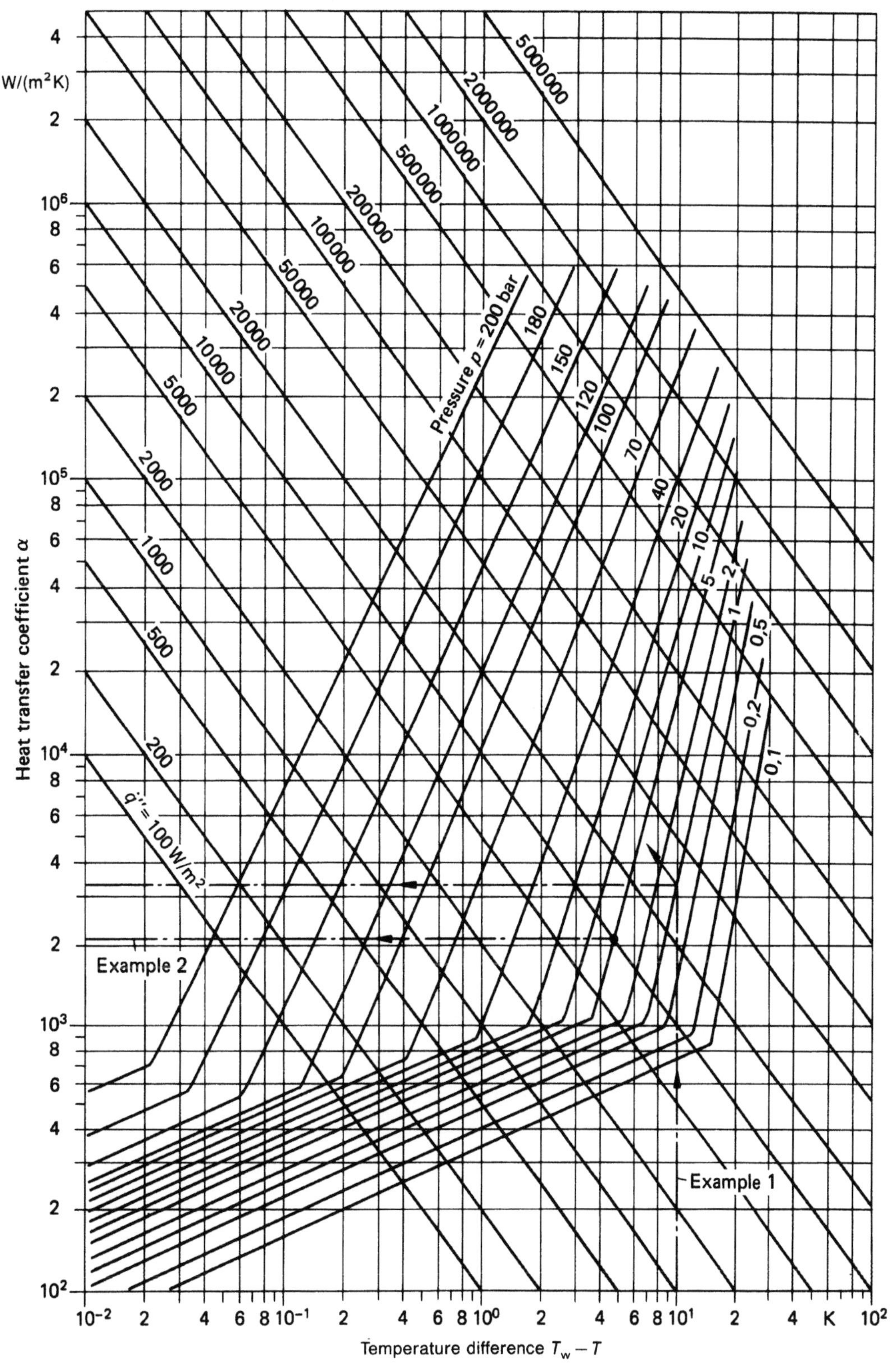

U. Renz

At *known difference of temperature* $T_w - T$ is

$$\alpha = \alpha_1 \cdot K_1$$

$$\dot{q}'' = \dot{q}_1'' \cdot K_1 = \alpha(T_w - T)$$

with the values for a roughness $R_{p_0} = 1\ \mu m$

$$\alpha_1 = (\alpha_0 \cdot f)^{1/(1-n)} \cdot [(T_w - T)/\dot{q}_o'']^{n/(1-n)}\ [1]$$

$$\dot{q}_1'' = \alpha_1(T_w - T),$$

These values can be found in the nomograph.

To the definition of f and n see the equations below.
At *known heat flow density* $\dot{q}''$ is

$$\alpha = \alpha_1 \cdot K_2,$$

where the heat transfer coefficient α_1 may be found in the nomograph. The difference of temperature can be calculated from

$$T_w - T = \dot{q}''/\alpha.$$

Range of validity: $0.02\ \text{bar} < p < 200\ \text{bar}$
The lower limit of this range is due to the heat transfer coefficient at free convection:

$$\alpha = 0.15 \cdot c^{1/3} \cdot (T_w - T)^{1/3}$$

with

$$c = g \cdot c_p \cdot \beta \cdot \lambda'^2 \cdot \varrho'^2/\eta'\ [1].$$

The upper limit is due to the maximum heat flux density $\dot{q}_{krit}''$:

$$\dot{q}_{krit}'' = 10^7 \cdot p^{*0.4} \cdot (1 - p^*)\ [1]$$

$$n = 0.9 - 0.3 \cdot p^{*0.15}$$

$$f = 2.55 \cdot p^{*0.27} + (9 + 1/(1 - p^{*2}))\, p^{*2}$$

$$p^* = p/p_c \text{ with } p_c = 221.2\ \text{bar}.$$

Remarks:
These limitations have been considered in the nomograph.

Correction factors K_1 and K_2:
The correction factors K_1 and K_2 consider the roughness of the wall and may be taken from the following tables.

Definitions:

c_p in J/(kgK)	specific heat capacity
g in m/s^2	acceleration due to gravity ($g = 9.81\ \text{m/s}^2$)
p in bar	pressure
R_p in μm	average roughness
R_{p_0} in μm	1 (reference roughness)
$\dot{q}''$ in W/m^2	heat flux density
$\dot{q}_o''$ in W/m^2	20 000 (reference heat flux density)
T in K	temperature of the water
T_w in K	temperature of the wall
α in W/(m^2 K)	heat transfer coefficient
α_0 in W/(m^2 K)	3800; heat transfer coefficient at $p = 0.03\ \text{bar}$ and $\dot{q}_o''$
α_1 in W/(m^2 K)	heat transfer coefficient at a roughness $R_{p_0} = 1\ \mu m$
β in K^{-1}	isobar expansion coefficient
λ' in W/(mK)	thermal conductivity of water
ϱ' in kg/m^3	density of water
η' in kg/(ms)	dynamic viscosity of water

Bibliography

[1] VDI-Wärmeatlas, 4th edition 1984. VDI-Verlag Düsseldorf

Correction factor K_1 at a known difference of temperature:						
Pressure	0.01 μm	0.2 μm	0.5 μm	1.0 μm	2.0 μm	5.0 μm	10.0 μm
0.1	0.2071	0.3327	0.6225	1.0000	1.6064	3.0059	4.8287
0.2	0.2243	0.3517	0.6376	1.0000	1.5684	2.8433	4.4593
0.5	0.2490	0.3784	0.6580	1.0000	1.5196	2.6424	4.0155
1.0	0.2694	0.3998	0.6738	1.0000	1.4842	2.5013	3.7124
2.0	0.2910	0.4220	0.6896	1.0000	1.4500	2.3697	3.4361
5.0	0.3216	0.4525	0.7107	1.0000	1.4071	2.2101	3.1098
10.0	0.3460	0.4762	0.7265	1.0000	1.3764	2.0998	2.8903
20.0	0.3714	0.5004	0.7422	1.0000	1.3474	1.9983	2.6924
40.0	0.3977	0.5249	0.7576	1.0000	1.3199	1.9051	2.5145
70.0	0.4194	0.5448	0.7698	1.0000	1.2990	1.8355	2.3843
100.0	0.4334	0.5575	0.7775	1.0000	1.2862	1.7938	2.3071
120.0	0.4407	0.5639	0.7814	1.0000	1.2798	1.7732	2.2693
150.0	0.4495	0.5719	0.7861	1.0000	1.2721	1.7487	2.2245
180.0	0.4568	0.5783	0.7899	1.0000	1.2660	1.7292	2.1891
200.0	0.4610	0.5820	0.7921	1.0000	1.2625	1.7181	2.1691
Correction factor K_2 at known heat flux density:						
	0.7362	0.8073	0.9119	1.0000	1.0966	1.2387	1.3583

Example 1:

Water boiling in a vessel with known difference of temperature

boiling pressure $\qquad p = 1\,\text{bar}$

difference of temperature $\qquad T_\text{w} - T = 10\,\text{K}$

roughness (shining surface) $\quad R_\text{p} = 0.1\,\mu\text{m}$

Solution:

1. Determination of the values for $R_{\text{p}_\text{o}} = 1\,\mu\text{m}$ from the nomograph (sheet No 3.4.1):

   $\alpha_1 = 3\,300\,\text{W/(m}^2\,\text{K)}$

   $\dot{q}_1'' = 33\,000\,\text{W/m}^2$.

2. Determination of the correction factor for $p = 1\,\text{bar}$ and $R_\text{p} = 0.1\,\mu\text{m}$: $K_1 = 0.2694$.

3. Results

   $\alpha = \alpha_1'' \cdot K_1 = 0.2694 \cdot 3\,300 = 889\,\text{W/(m}^2\,\text{K)}$

   $\dot{q}'' = \dot{q}_1'' \cdot K_1 = 0.2694 \cdot 33\,000 = 8\,890\,\text{W/m}^2$.

Example 2:

Water boiling in a vessel with known heat flux density

boiling pressure $\qquad p = 5\,\text{bar}$

heat flux density $\qquad \dot{q}'' = 10\,000\,\text{W/m}^2$

roughness (shining surface) $\quad R_\text{p} = 0.1\,\mu\text{m}$

Solution:

1. Determination of the heat transfer coefficient for $R_{\text{p}_\text{o}} = 1\,\mu\text{m}$ from the nomograph (sheet No 3.4.1):

   $\alpha_1 = 2\,050\,\text{W/m}^2$.

2. Determination of the correction factor for $R_\text{p} = 0.1\,\mu\text{m}$: $K_2 = 0.7362$.

3. Results

   $\alpha = \alpha_1 \cdot K_2 = 2\,050 \cdot 0.7362 = 1\,510\,\text{W/(m}^2\,\text{K)}$

   and

   $T_\text{w} - T = \dot{q}''/\alpha = 6.6\,\text{K}$.

O$_2$-Content $y_{O_2 T}$
%
$y_{CO_2 T} = 20\%$
18
16
14
12
10
CO$_2$-Content $y_{CO_2 T}$
Maximum CO$_2$-Content $\hat{y}_{CO_2 T}$
%
Solid fuels
Fuel oils
Natural gases
Net calorific value H_u
MJ/kg

F. Brandt

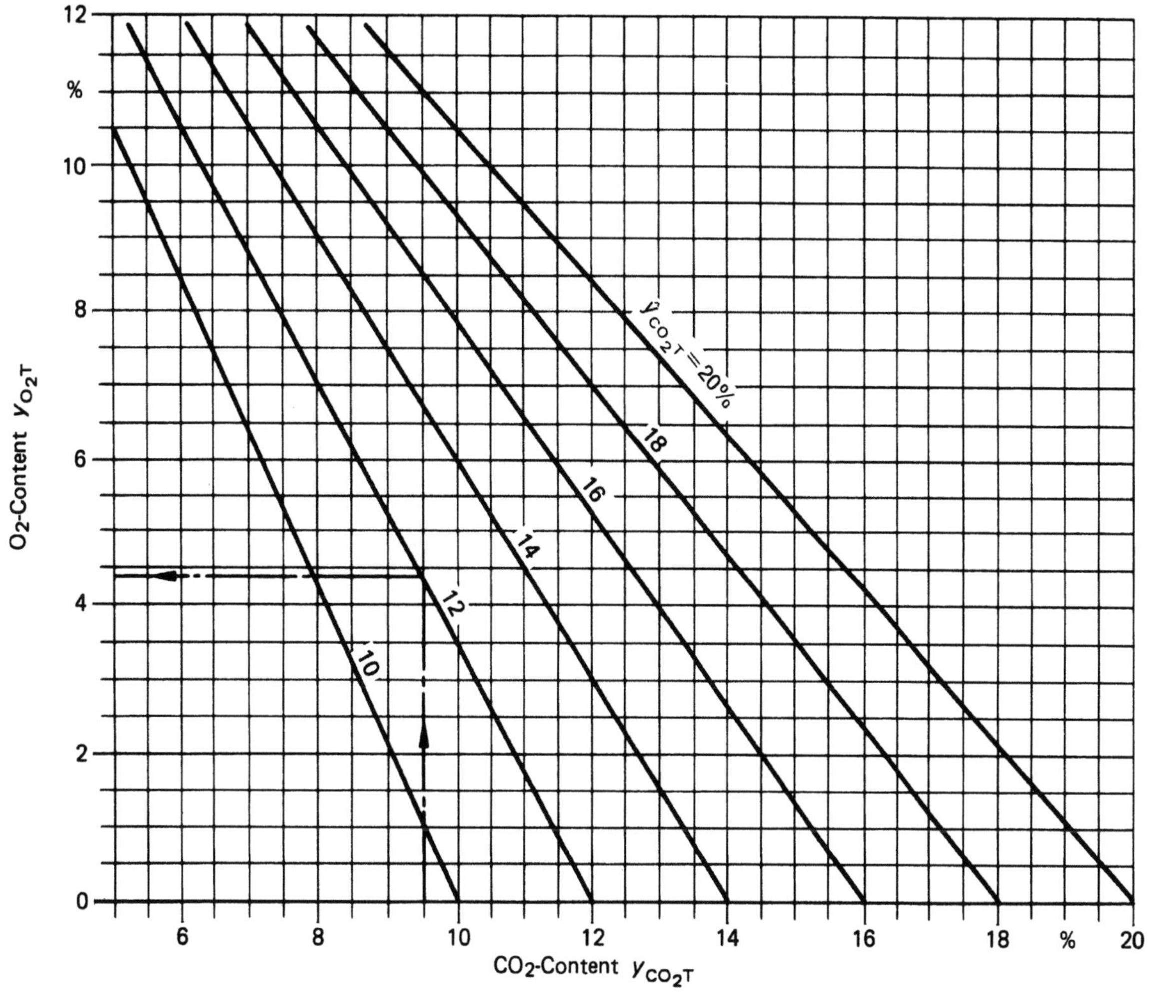

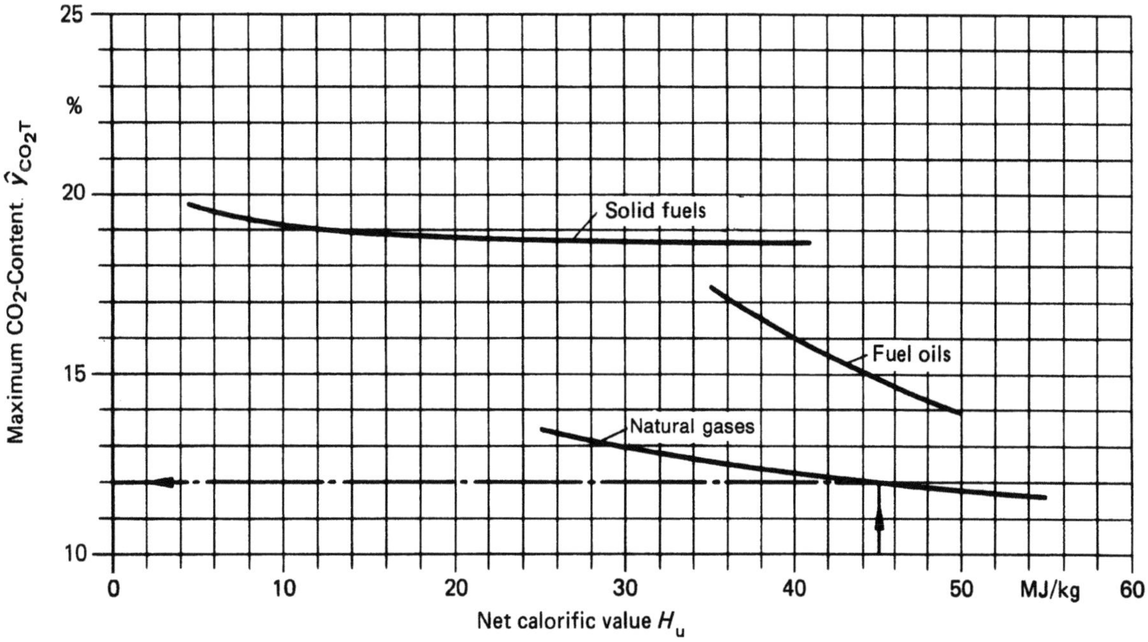

Explanation

With worksheet No 4.1 (lower diagram), one can determine the maximum (stoichiometric) CO_2-content in the dry flue gas $\hat{y}_{CO_2T}$ of solid fuels, fuel oil and natural gas dependent on the net calorific value H_u. The upper diagram shows the correlation between the CO_2-content y_{CO_2T} and the O_2-content of the dry flue gas y_{O_2T}.

Definitions:

y_{CO_2T} in %	Carbon dioxide content of dry flue gas
y_{O_2T} in %	Oxygen content of dry flue gas
$\hat{y}_{CO_2T}$ in %	Maximum carbon dioxide content of dry flue gas
V_{CO_2} in m^3/kg	Specific volume of CO_2
V_{GoT} in m^3/kg	Specific volume of dry flue gas at stoichiometric combustion
H_u in MJ/kg	Net calorific value of the fuel

Formulae for calculation:

$$y_{O_2T} = 0.21 \left(1 - \frac{y_{CO_2T}}{\hat{y}_{CO_2T}} \right) \qquad (1)$$

$$\hat{y}_{CO_2T} = \frac{V_{CO_2}}{V_{GoT}} \qquad (2)$$

Solid fuels:

$$\hat{y}_{CO_2T} = \frac{0.10162 + 0.04399\,H_u}{0.44971 + 0.23825\,H_u} \qquad (3)$$

Fuel oils:

$$\hat{y}_{CO_2T} = \frac{1.26601 + 0.007557\,H_u}{1.76427 + 0.20048\,H_u} \qquad (4)$$

Natural gases:

$$\hat{y}_{CO_2T} = \frac{0.27900 + 0.022575\,H_u}{0.64975 + 0.22538\,H_u} \qquad (5)$$

Example:

Known values: natural gas, $H_u = 45$ MJ/kg; $y_{CO_2T} = 9.5\%$

Results: $\hat{y}_{CO_2T} = 12\%$ from the lower diagram

$y_{O_2T} = 4.4\%$ (with $y_{CO_2T} = 9.5\%$ and $\hat{y}_{CO_2T} = 12\%$ from the upper diagram)

Engineering Reference Book on Energy and Heat	Specific Dry Air Mass of Solid Fuels	4.2.1

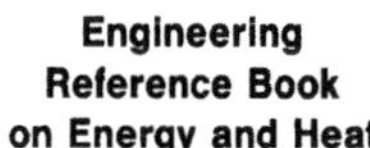

F. Brandt

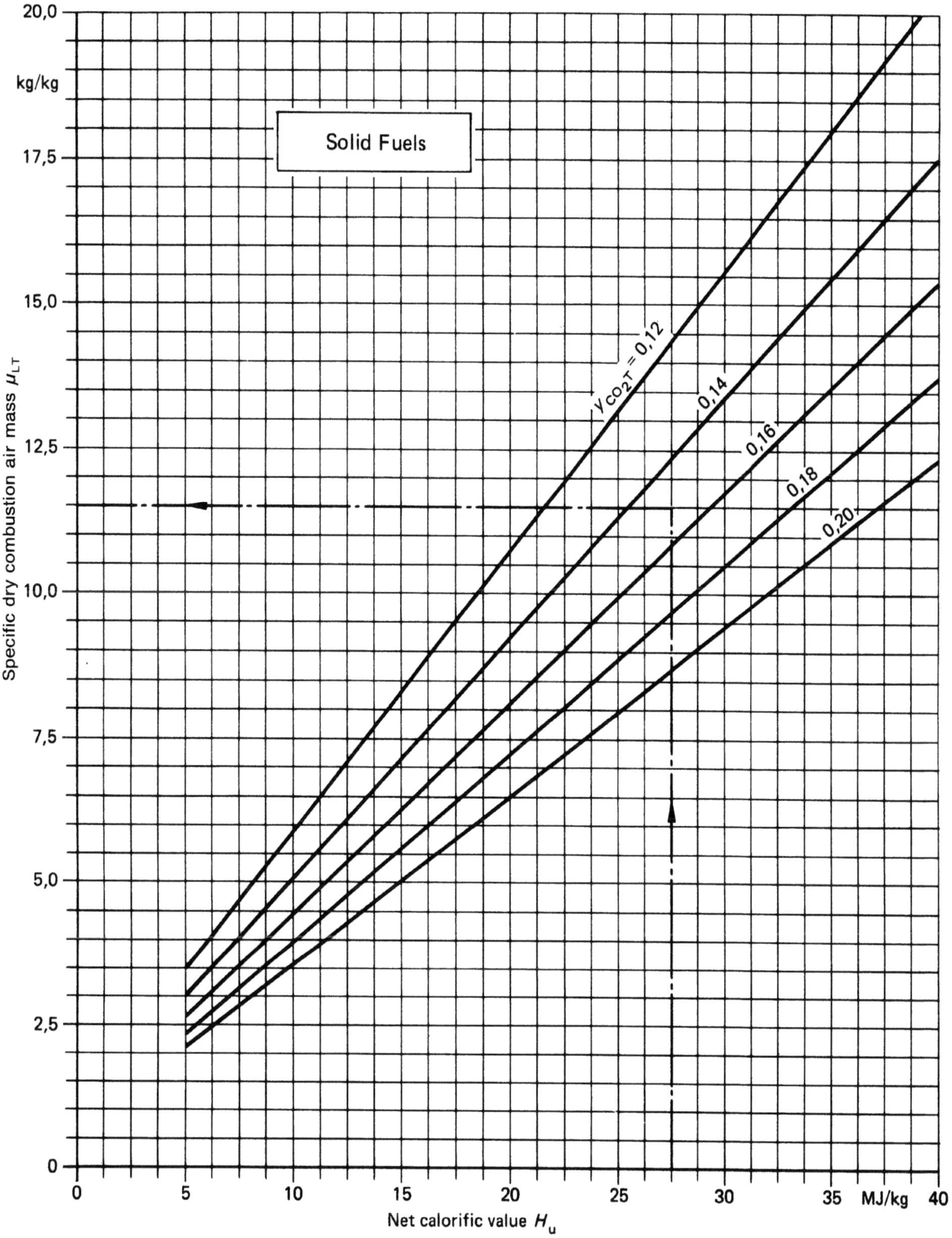

Explanation

With worksheet No 4.2.1, one can determine the specific dry combustion air mass μ_{LT} of solid fuels dependent on the net calorific value H_u and the measured CO_2-content of the dry flue gas y_{CO_2T}.

Definitions:

H_u in MJ/kg	Net calorific value of the fuel
y_{CO_2T} in %	Measured carbon dioxide content of dry flue gas
x_{H_2OL} in kg/kg	Absolute humidity of the combustion air
p in bar	Absolute pressure
p_s in bar	Saturation pressure of steam at air temperature
φ in %	Relative humidity of air
μ_{LT} in kg air/kg fuel	Specific dry combustion air mass
μ_L in kg air/kg fuel	Specific combustion air mass
V_{LT} in m^3/kg	Specific volume of dry combustion air

Formulae for calculation:

$$\mu_{LT} = (-0.0139 + 0.0089\,H_u) + (0.1314 + 0.0569\,H_u)\frac{1}{y_{CO_2T}} \tag{1}$$

$$\mu_L = \mu_{LT}(1 + x_{H_2OL}) \tag{2}$$

$$V_{LT} = \mu_{LT}/1.2930 \tag{3}$$

$$x_{H_2OL} = 0.622\,\frac{\varphi\,p_s}{p - \varphi\,p_s} \tag{4}$$

Example:

Known values: H_u = 27.5 MJ/kg; y_{CO_2T} = 0.15; x_{H_2OL} = 0.0095 kg/kg

Results: μ_{LT} = 11.54 kg/kg

$\qquad\qquad\quad \mu_L$ = 11.65 kg/kg

$\qquad\qquad\quad V_{LT}$ = 8.92 m^3/kg

Calculation of the combustion air required:

The equation of reaction for the complete combustion of the fuel is:

$$C_m H_n S_k O_j + x\,O_2 = m\,CO_2 + n/2\,H_2O + k\,SO_2$$

The molar oxygen demand at stoichiometric combustion results from the law of conservation of O-atoms:

$$2x + j = 2m + n/2 + k$$
$$x = m + n/4 + k/2 - j/2$$

Fuel Oils
Specific dry combustion air mass μ_{LT}
30,0
kg/kg
27,5
25,0
22,5
20,0
17,5
15,0
12,5
10,0
$y_{CO_2T} = 0,10$
0,12
0,13
0,14
0,16
0,18
0,20
34
36
38
40
42
44
46
48 MJ/kg
50
Net calorific value H_u

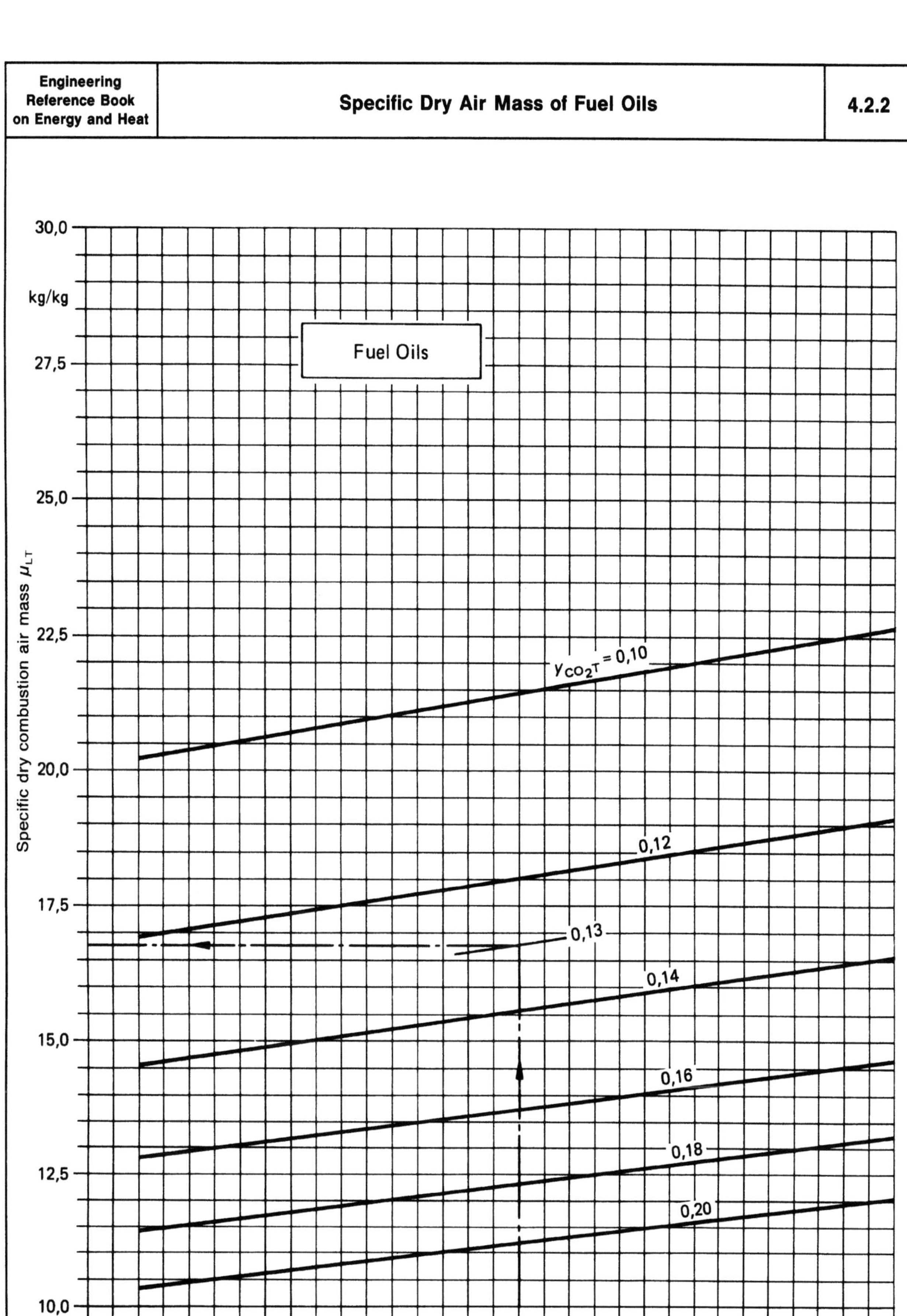

F. Brandt

Explanation

With worksheet No 4.2.2, one can determine the specific dry air mass μ_{LT} of fuel oils dependent on the net calorific value H_u and the measured CO_2-content of the dry flue gas y_{CO_2T}.

Definitions:

H_u in MJ/kg	Net calorific value of the fuel
y_{CO_2T} in %	Measured carbon dioxide content of dry flue gas
x_{H_2OL} in kg/kg	Absolute humidity of the combustion air
p in bar	Absolute pressure
p_s in bar	Saturation pressure of steam at air temperature
φ in %	Relative humidity of air
μ_{LT} in kg air/kg fuel	Specific dry air mass for combustion
μ_L in kg air/kg fuel	Specific air mass for combustion
V_{LT} in m³/kg	Specific volume of dry air for combustion

Formulae for calculation:

$$\mu_{LT} = (-1.8415 + 0.0649\,H_u) + (1.6370 + 0.0098\,H_u)\,\frac{1}{y_{CO_2T}} \tag{1}$$

$$\mu_L = \mu_{LT}(1 + x_{H_2OL}) \tag{2}$$

$$V_{LT} = \mu_{LT}/1.2930 \tag{3}$$

$$x_{H_2OL} = 0.622\,\frac{\varphi\,p_s}{p - \varphi\,p_s} \tag{4}$$

Example:

Known values: $H_u = 42.5$ MJ/kg; $y_{CO_2T} = 0.13$; $x_{H_2OL} = 0.0095$ kg/kg

Results:
$\mu_{LT} = 16.71$ kg/kg
$\mu_L = 16.87$ kg/kg
$V_{LT} = 12.93$ m³/kg

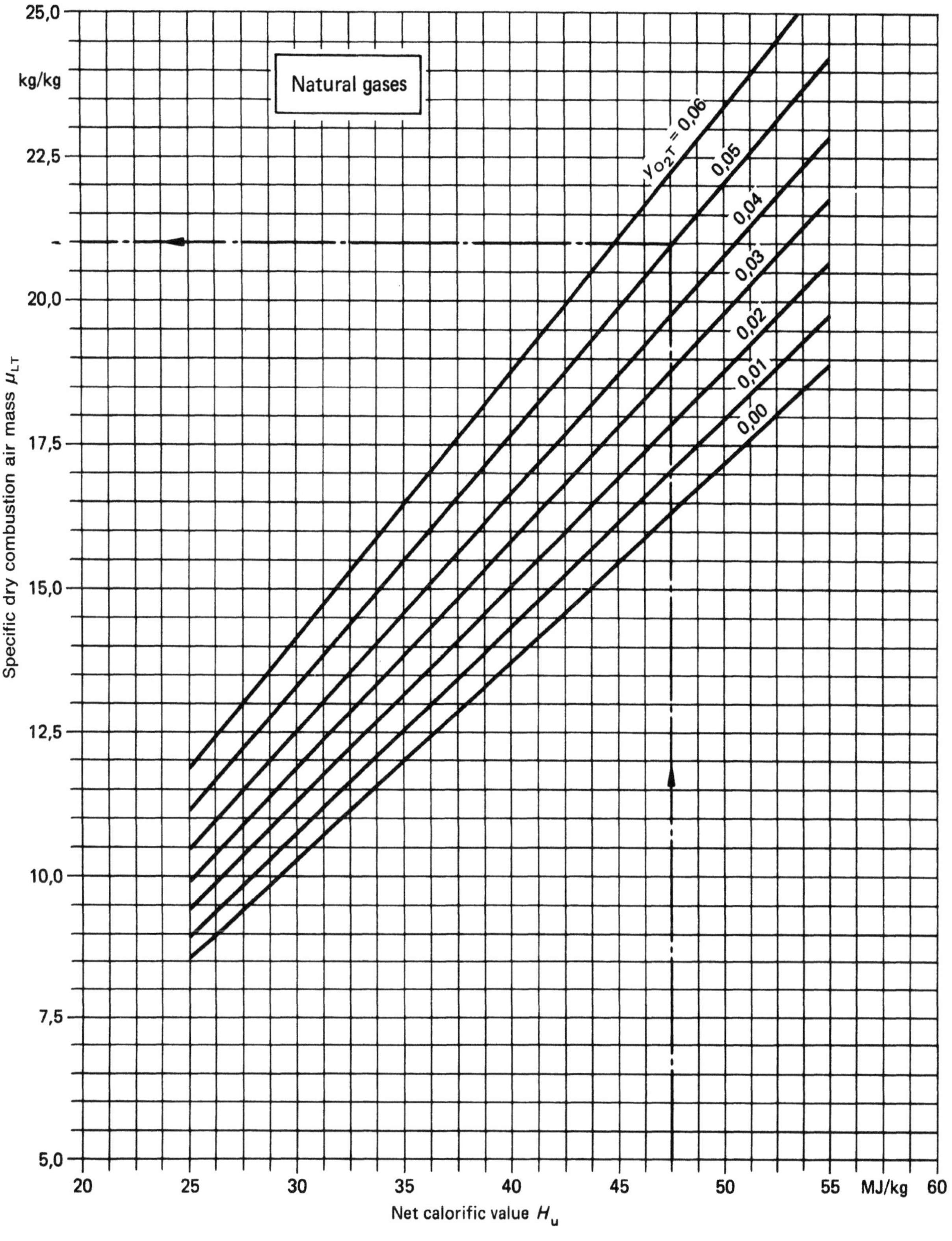

Specific Dry Air Mass of Natural Gas in kg/kg

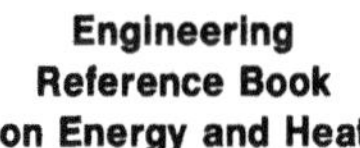

F. Brandt

Explanation

With worksheet No 4.2.3, one can determine the specific dry air mass μ_{LT} of natural gas dependent on the net calorific value H_u and the measured CO_2-content of the dry flue gas y_{O_2T}.

Definitions:

H_u in MJ/kg	Net calorific value of the fuel
y_{O_2T} in %	Measured carbon dioxide content of dry flue gas
x_{H_2OL} in kg/kg	Absolute humidity of the combustion air
p in bar	Absolute pressure
p_s in bar	Saturation pressure of steam at air temperature
φ in %	Relative humidity of air
μ_{LT} in kg air/kg fuel	Specific dry air mass for combustion
μ_L in kg air/kg fuel	Specific air mass for combustion
V_{LT} in m³/kg	Specific volume of dry air for combustion

Formulae for calculation:

$$\mu_{LT} = (-0.0630 + 0.3450\,H_u) + (0.8401 + 0.2914\,H_u)\frac{y_{O_2T}}{0.21 - y_{O_2T}} \qquad (1)$$

$$\mu_L = \mu_{LT}(1 + x_{H_2OL}) \qquad (2)$$

$$V_{LT} = \mu_{LT}/1.2930 \qquad (3)$$

$$x_{H_2OL} = 0.622\,\frac{\varphi\,p_s}{p - \varphi\,p_s} \qquad (4)$$

Example:

Known values: $H_u = 47.5$ MJ/kg; $y_{O_2T} = 0.05$; $x_{H_2OL} = 0.0095$ kg/kg

Results:
$\mu_{LT} = 20.91$ kg/kg
$\mu_L = 21.11$ kg/kg
$V_{LT} = 16.27$ m³/kg

kg/m³
Natural gases
Specific dry combustion air mass μ_{LTn}
25,0
22,5
20,0
17,5
15,0
12,5
10,0
7,5
5,0
$y_{O_2T} = 0,06$
0,05
0,04
0,03
0,02
0,01
0,00
20
25
30
35
40
45
50
55
MJ/m³
60
Gross calorific value H_{on}

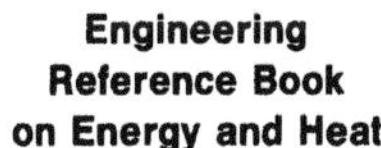

F. Brandt

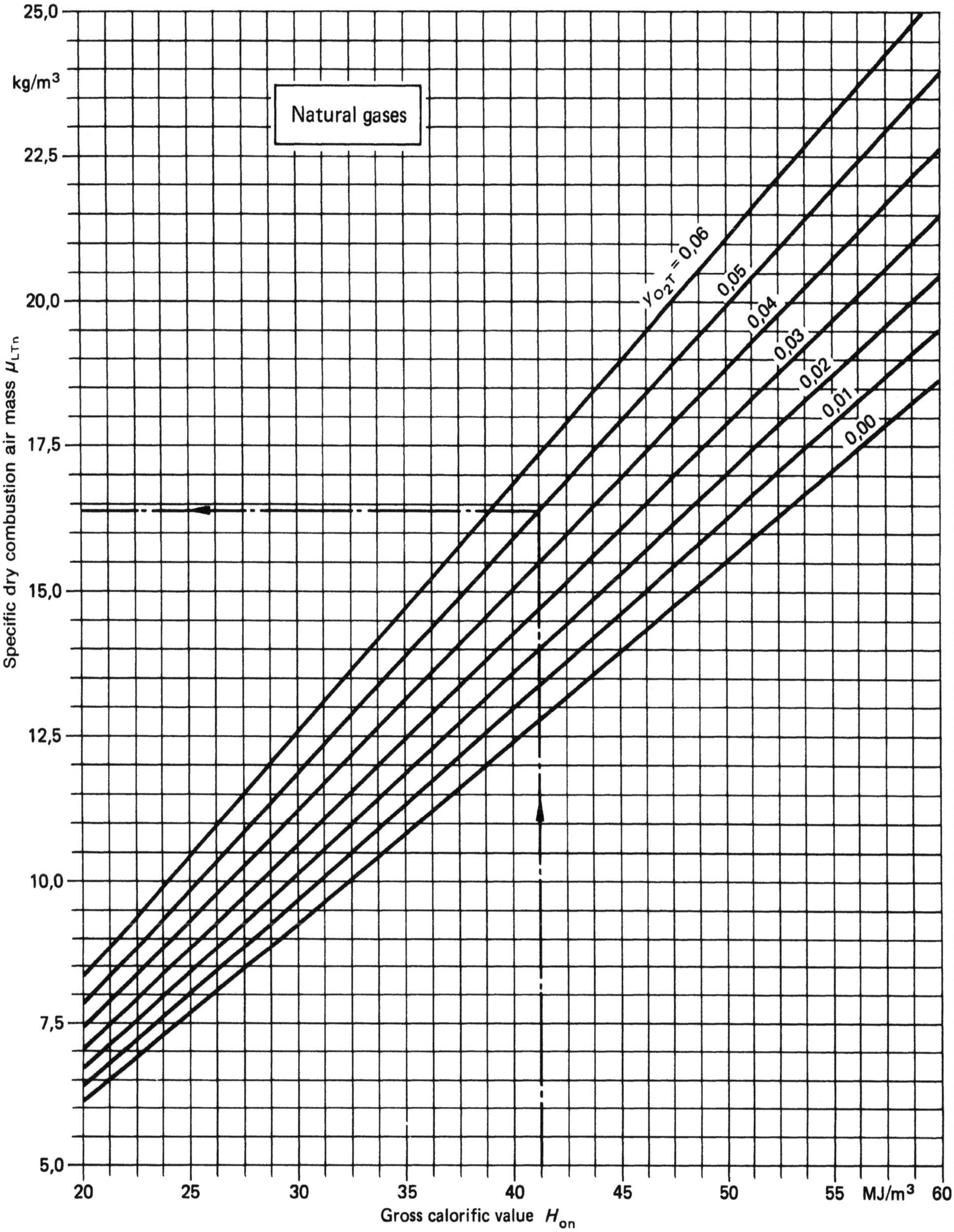

Explanation

With worksheet No 4.2.4, one can determine the specific dry air mass μ_{LTn} of natural gas dependent on the gross calorific value H_{on} and the measured O_2-content of the dry flue gas $y_{O_2\mathrm{T}}$.

Definitions:

H_{on} in MJ/m^3	Gross calorific value at standard conditions
H_{on}^* in kWh/m^3	Gross calorific value at standard conditions
$y_{O_2\mathrm{T}}$ in %	Measured oxygen content of dry flue gas
$x_{\mathrm{H_2OL}}$ in kg/kg	Absolute humidity of the combustion air
p in bar	Absolute pressure
p_{s} in bar	Saturation pressure of steam at air temperature
φ in %	Relative humidity of air
μ_{LTn} in kg air/m^3 fuel	Specific dry air mass for combustion at standard conditions
μ_{Ln} in kg air/m^3 fuel	Specific air mass for combustion at standard conditions
V_{LTn} in m^3/m^3	Specific volume of dry air for combustion

Formulae for calculation:

$$\mu_{\mathrm{LTn}} = (-0.1596 + 0.3140\,H_{\mathrm{on}}) + (0.082458 + 0.2786\,H_{\mathrm{on}})\,\frac{y_{O_2\mathrm{T}}}{0.21 - y_{O_2\mathrm{T}}} \tag{1}$$

$$\mu_{\mathrm{Ln}} = \mu_{\mathrm{LTn}}(1 + x_{\mathrm{H_2OL}}) \tag{2}$$

$$V_{\mathrm{LTn}} = \mu_{\mathrm{LTn}}/1.2930 \tag{3}$$

$$x_{\mathrm{H_2OL}} = 0.622\,\frac{\varphi\,p_{\mathrm{s}}}{p - \varphi\,p_{\mathrm{s}}} \tag{4}$$

$$H_{\mathrm{on}} = H_{\mathrm{on}}^*\,3.6 \tag{5}$$

Example:

Known values: $H_{\mathrm{on}} = 41.3$ MJ/m^3; $y_{O_2\mathrm{T}} = 0.05$; $x_{\mathrm{H_2OL}} = 0.0095$ kg/kg

Results:
$\mu_{\mathrm{LTn}} = 16.43$ kg/m^3

$\mu_{\mathrm{Ln}} = 16.59$ kg/m^3

$V_{\mathrm{LTn}} = 12.71$ m^3/kg

Solid Fuels
20,0
kg/kg
17,5
15,0
12,5
10,0
7,5
5,0
2,5
0
Specific flue gas mass μ_{GB}
y_{CO_2T} = 0,14
0,15
0,16
0,18
0,20
0
5
10
15
20
25
30
35
MJ/kg
40
Net calorific value H_u

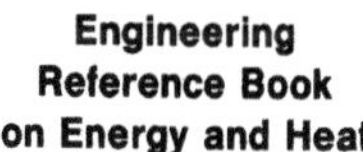

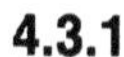

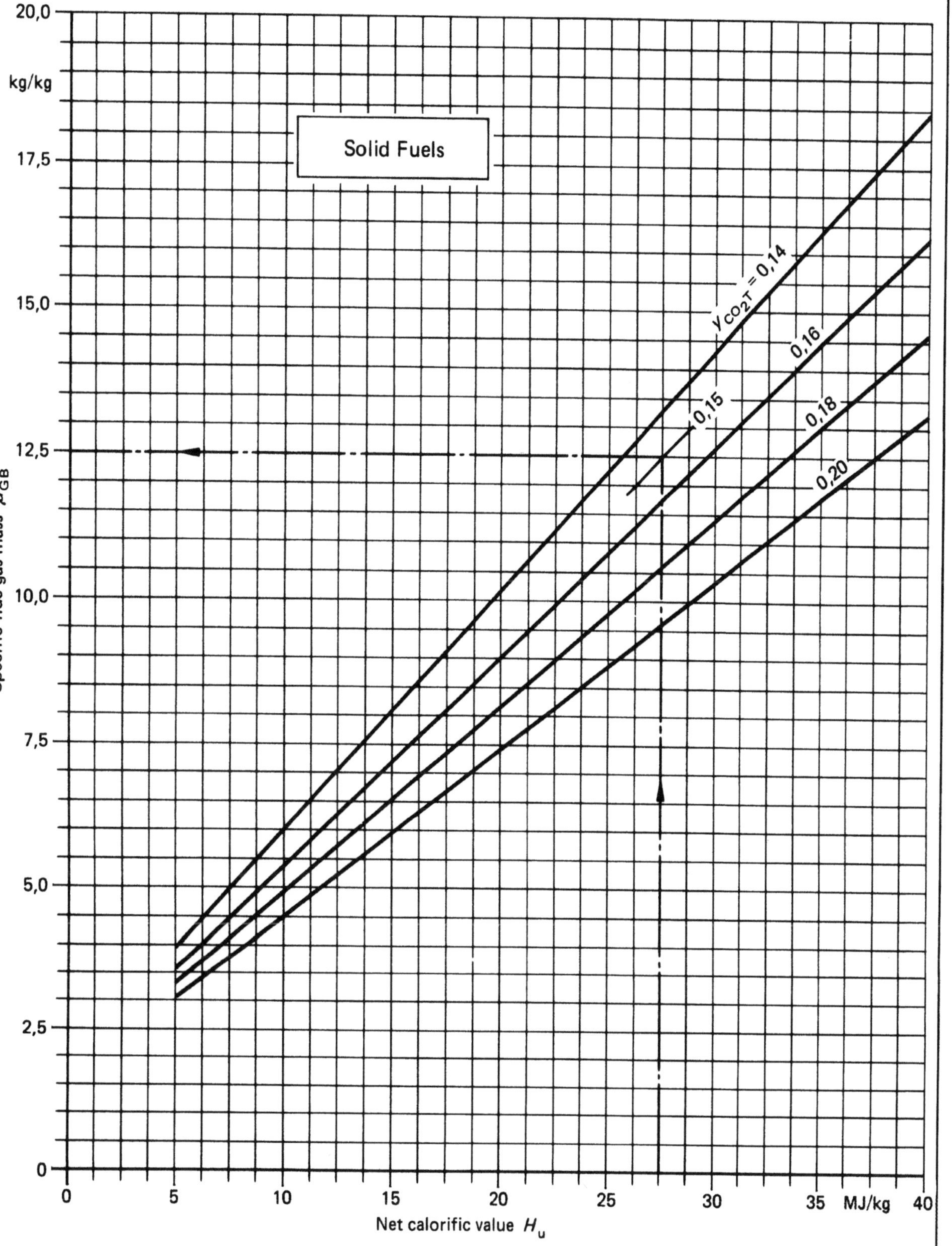

F. Brandt

Explanation

With worksheet No 4.3.1, one can determine the specific flue gas mass μ_{GB} of solid fuels dependent on the net calorific value H_u and the measured CO_2-content of the dry flue gas y_{CO_2T}.

Definitions:

H_u in MJ/kg	Net calorific value of the fuel
y_{CO_2T} in %	Measured carbon dioxide content of dry flue gas
x_{H_2OL} in kg/kg	Absolute humidity of the combustion air
μ_{LT} in kg air/kg fuel	Specific dry combustion air mass (from sheet No 4.2.1)
μ_{GB} in kg/kg	Specific flue gas mass (without air humidity)
μ_G in kg/kg	Specific flue gas mass

Formulae for calculation:

$$\mu_{GB} = (0.96569 + 0.00707\, H_u) + (1.13139 + 0.05688\, H_u)\,\frac{1}{y_{CO_2T}} \tag{1}$$

$$\mu_G = \mu_{GB} + \mu_{LT}\, x_{H_2OL} \tag{2}$$

Example:

Known values:
$H_u = 27.5\ \text{MJ/kg}$
$y_{CO_2T} = 0.15$
$x_{H_2OL} = 0.0095\ \text{kg/kg}$

Results:
$\mu_{LT} = 11.54\ \text{kg/kg from sheet No 4.2.1}$
$\mu_{GB} = 12.46\ \text{kg/kg}$
$\mu_G = 12.57\ \text{kg/kg}$

30,0
kg/kg
27,5
25,0
22,5
20,0
17,5
15,0
12,5
10,0
Specific flue gas mass μ_{GB}
Fuel Oils
y_{CO_2T} = 0,10
0,12
0,13
0,14
0,16
0,18
0,20
34
36
38
40
42
44
46
48
MJ/kg
50
Net calorific value H_u

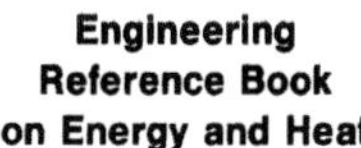

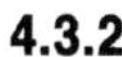

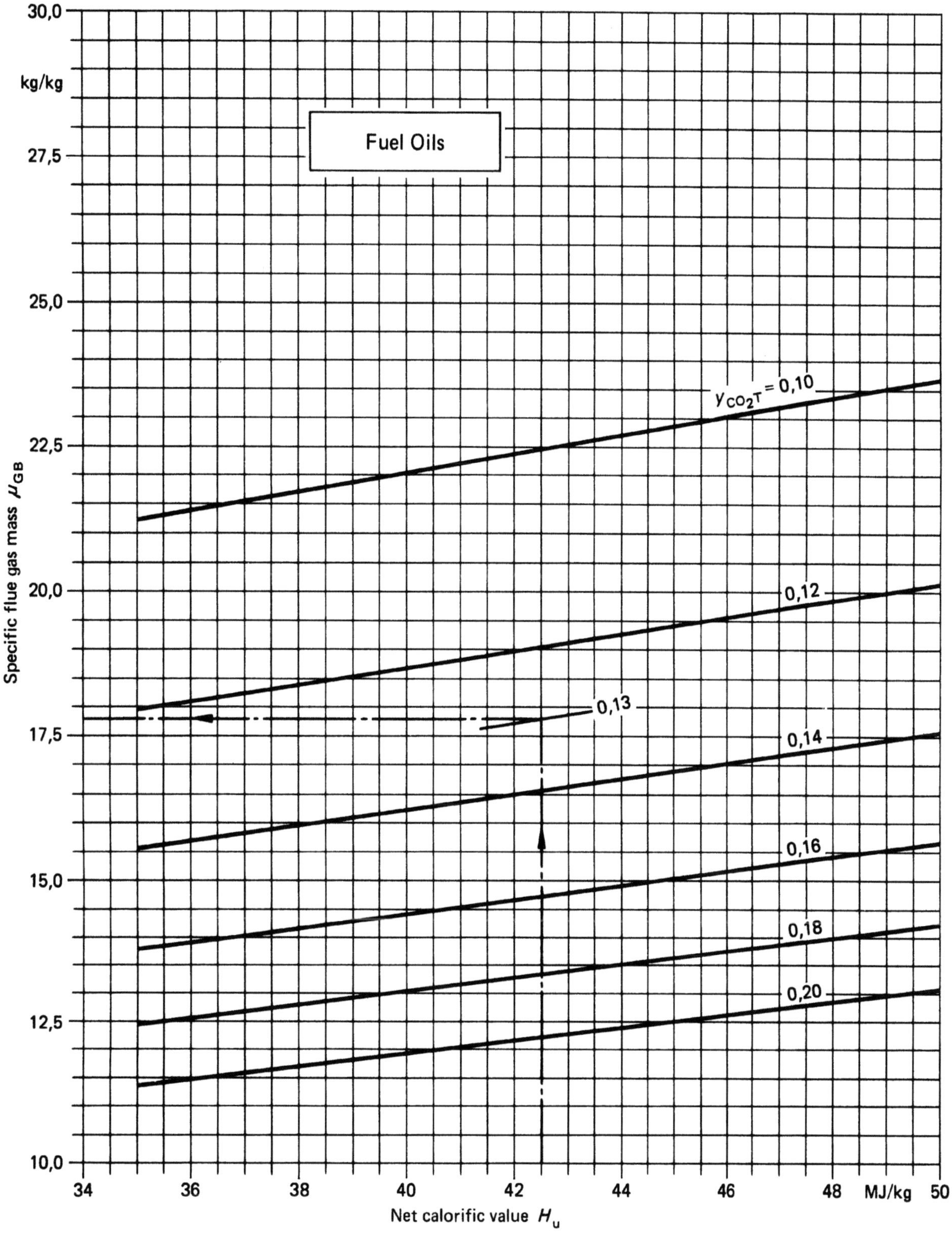

F. Brandt

Explanation

With worksheet No 4.3.2, one can determine the specific flue gas mass μ_{GB} of fuel oils dependent on the net calorific value H_u and the measured CO_2-content of the dry flue gas y_{CO_2T}.

Definitions:

H_u in MJ/kg	Net calorific value of the fuel
y_{CO_2T} in %	Measured carbon dioxide content of dry flue gas
x_{H_2OL} in kg/kg	Absolute humidity of the combustion air
μ_{LT} in kg air/kg fuel	Specific dry combustion air mass (from sheet No 4.2.2)
μ_{GB} in kg/kg	Specific flue gas mass (without air humidity)
μ_G in kg/kg	Specific flue gas mass

Formulae for calculation:

$$\mu_{GB} = -(0.84149 + 0.06491\,H_u) + (1.63696 + 0.009771\,H_u)\frac{1}{y_{CO_2T}} \tag{1}$$

$$\mu_G = \mu_{GB} + \mu_{LT}\,x_{H_2OL} \tag{2}$$

Example:

Known values:
$H_u = 42.5$ MJ/kg
$y_{CO_2T} = 0.13$
$x_{H_2OL} = 0.0095$ kg/kg

Results:
$\mu_{LT} = 16.71$ kg/kg from sheet No 4.2.2
$\mu_{GB} = 17.71$ kg/kg $= \mu_{LT} + 1$
$\mu_G = 17.86$ kg/kg

Natural gases
kg/kg
Specific flue gas mass μ_{GB}
$y_{O_2T} = 0,05$
0,04
0,03
0,02
0,01
0,00
25,0
22,5
20,0
17,5
15,0
12,5
10,0
7,5
5,0
20
25
30
35
40
45
50
55
MJ/kg
60
Net calorific value H_u

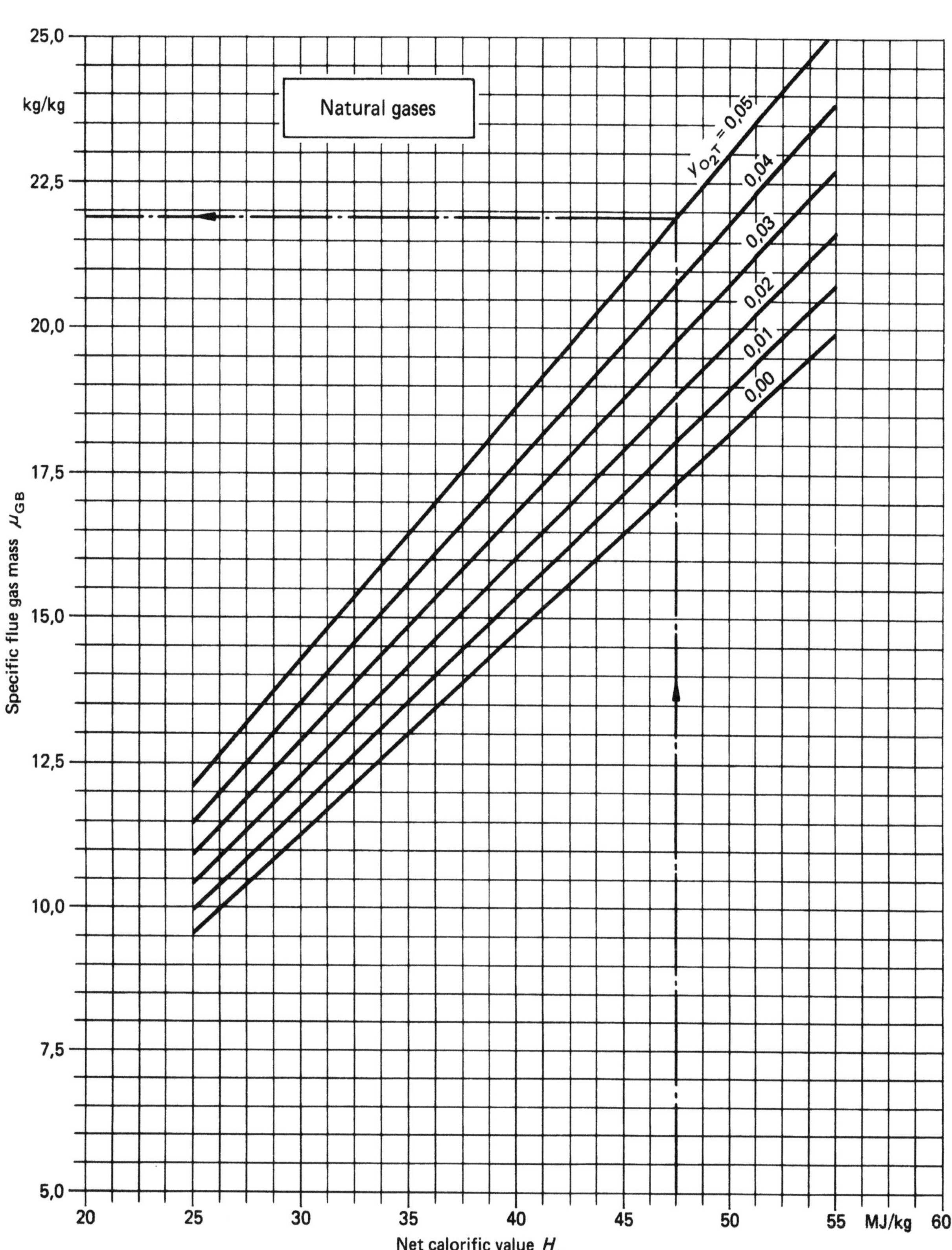

F. Brandt

Explanation

With worksheet No 4.3.3, one can determine the specific flue gas mass μ_{GB} of natural gas dependent on the net calorific value H_u and the measured O_2-content of the dry flue gas y_{O_2T}.

Definitions:

H_u in MJ/kg	Net calorific value of the fuel
y_{O_2T} in %	Measured carbon dioxide content of dry flue gas
x_{H_2OL} in kg/kg	Absolute humidity of the combustion air
μ_{LT} in kg air/kg fuel	Specific dry combustion air mass (from sheet No 4.2.3)
μ_{GB} in kg/kg	Specific flue gas mass (without air humidity)
μ_G in kg/kg	Specific flue gas mass

Formulae for calculation:

$$\mu_{GB} = (0.93697 + 0.34503\,H_u) + (0.84013 + 0.29142\,H_u)\,\frac{y_{O_2T}}{0.21 - y_{O_2T}} \tag{1}$$

$$\mu_G = \mu_{GB} + \mu_{LT}\,x_{H_2OL} \tag{2}$$

Example:

Known values: H_u = 47.5 MJ/kg

y_{O_2T} = 0.05

x_{H_2OL} = 0.0095 kg/kg

Results: μ_{LT} = 20.91 kg/kg from sheet No 4.2.3

μ_{GB} = 21.91 kg/kg = μ_{LT} + 1

μ_G = 22.11 kg/kg

Specific Flue Gas Mass (without Air Humidity) of Natural Gas in kg/m³

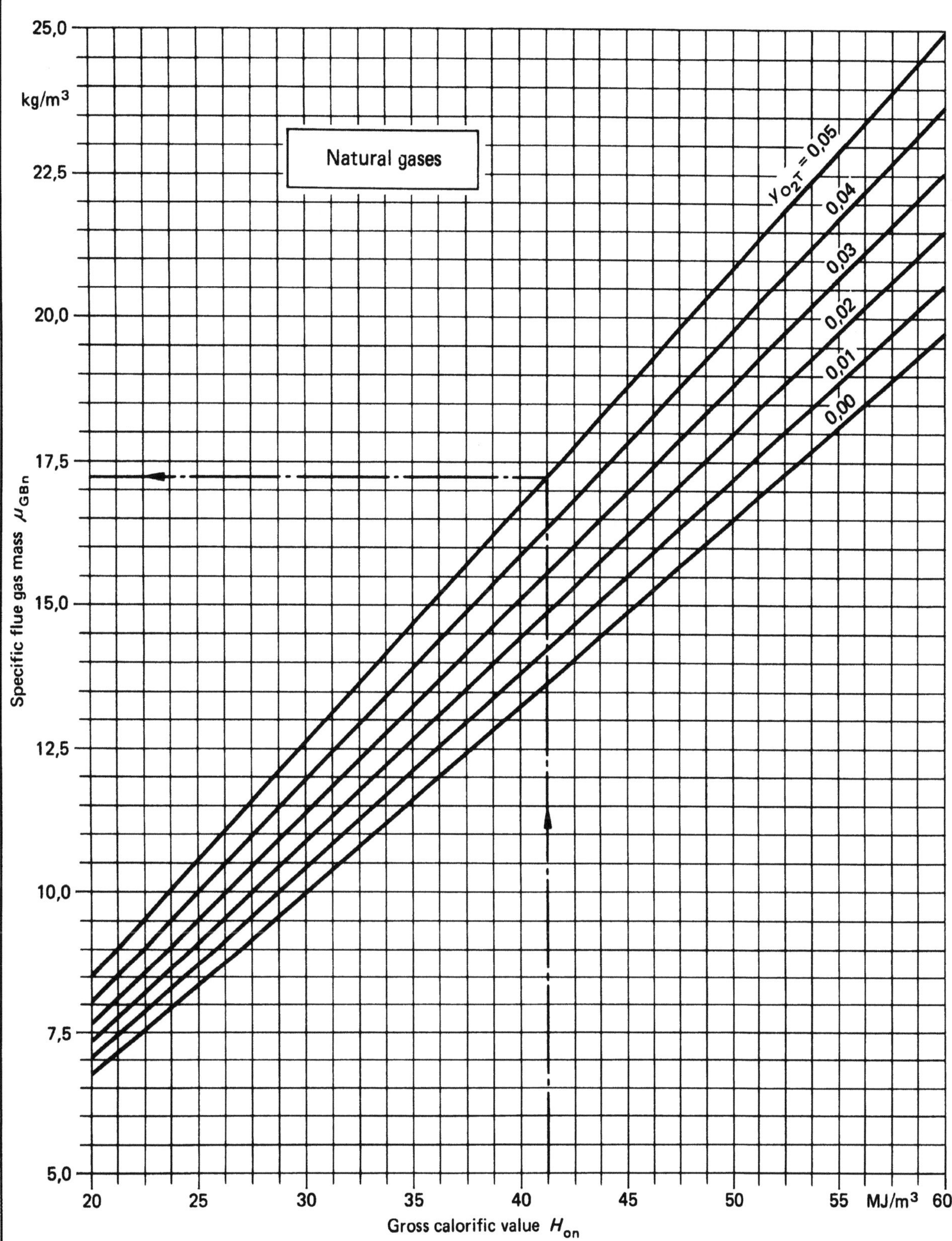

F. Brandt

Explanation

With worksheet No 4.3.4, one can determine the specific flue gas mass μ_{GBn} of natural gas dependent on the gross calorific value H_{on} and the measured O_2-content of the dry flue gas y_{O_2T}.

Definitions:

H_{on} in MJ/m^3	Gross calorific value of the fuel at standard conditions
H_{on}^* in kWh/m^3	Gross calorific value of the fuel at standard conditions
y_{O_2T} in %	Measured oxygen content of dry flue gas
x_{H_2OL} in kg/kg	Absolute humidity of the combustion air
μ_{LTn} in kg/m^3	Specific dry combustion air mass at standard conditions (from sheet No 4.2.4)
μ_{GBn} in kg/m^3	Specific flue gas mass (without air humidity) at standard conditions
μ_{Gn} in kg/m^3	Specific flue gas mass at standard conditions

Formulae for calculation:

$$\mu_{GBn} = (0.24737 + 0.32467\,H_{on}) + (0.08246 + 0.27860\,H_{on})\,\frac{y_{O_2T}}{0.21 - y_{O_2T}} \qquad (1)$$

$$H_{on} = H_{on}^*\,3.6 \qquad (2)$$

$$\mu_{Gn} = \mu_{GBn} + \mu_{LTn}\,x_{H_2OL} \qquad (3)$$

Example:

Known values:
$$H_{on} = 41.3\ \text{MJ/m}^3$$
$$y_{O_2T} = 0.05$$
$$x_{H_2OL} = 0.0095\ \text{kg/kg}$$

Results:
$$\mu_{LTn} = 16.43\ \text{kg/m}^3 \text{ from sheet No 4.2.4}$$
$$\mu_{GBn} = 17.28\ \text{kg/m}^3$$
$$\mu_{Gn} = 17.43\ \text{kg/m}^3$$

Engineering Reference Book on Energy and Heat	Specific Flue Gas Volume (without Air Humidity) of Solid Fuels	4.4.1

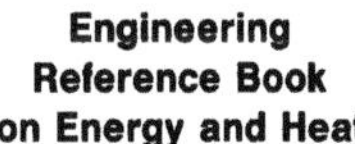

F. Brandt

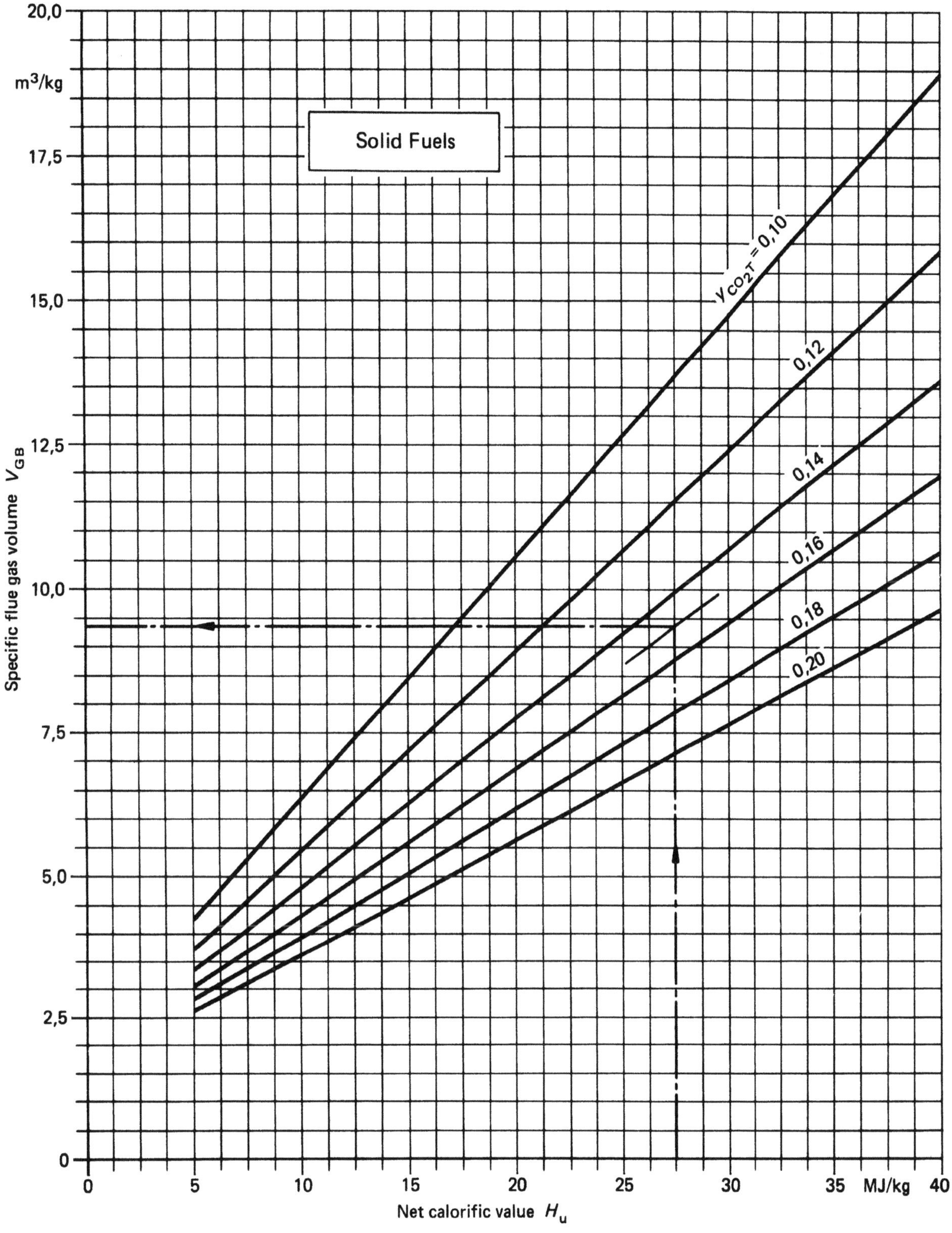

Explanation

With worksheet No 4.4.1, one can determine the specific flue gas volume V_{GB} of solid fuels dependent on the net calorific value H_u and the measured CO_2-content of the dry flue gas y_{CO_2T}.

Definitions:

H_u in MJ/kg	Net calorific value of the fuel
y_{CO_2T} in %	Measured carbon dioxide content of dry flue gas
y_{H_2OL} in m³/m³	Humidity of the combustion air
x_{H_2OL} in kg/kg	Absolute humidity of the combustion air
V_{LT} in m³/kg	Specific dry combustion air volume (from sheet No 4.2.1)
V_{GB} in m³/kg	Specific flue gas volume (without air humidity)
V_G in m³/kg	Specific flue gas volume

Formulae for calculation:

$$V_{GB} = (1.1296 + 0.02028\,H_u) + (0.10162 + 0.04399\,H_u)\,\frac{1}{y_{CO_2T}} \qquad (1)$$

$$V_G = V_{GB} + V_{LT}\,y_{H_2OL} \qquad (2)$$

$$y_{H_2OL} = 1.6086\,x_{H_2OL} \qquad (3)$$

Example:

Known values:
$$H_u = 27.5\ \text{MJ/kg}$$
$$y_{CO_2T} = 0.15$$
$$y_{H_2OL} = 0.0153\ \text{m}^3/\text{m}^3$$

Results:
$$V_{LT} = 8.92\ \text{m}^3/\text{kg from sheet No 4.2.1}$$
$$V_{GB} = 9.31\ \text{m}^3/\text{kg}$$
$$V_G = 9.45\ \text{m}^3/\text{kg}$$

Fuel Oils
25,0
m³/kg
22,5
20,0
17,5
15,0
12,5
10,0
7,5
5,0
Specific flue gas volume V_{GB}
y_{CO_2T} = 0,10
0,12
0,13
0,14
0,16
0,18
0,20
34
36
38
40
42
44
46
48
MJ/kg
50
Net calorific value H_u

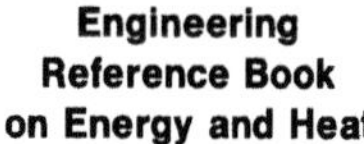

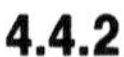

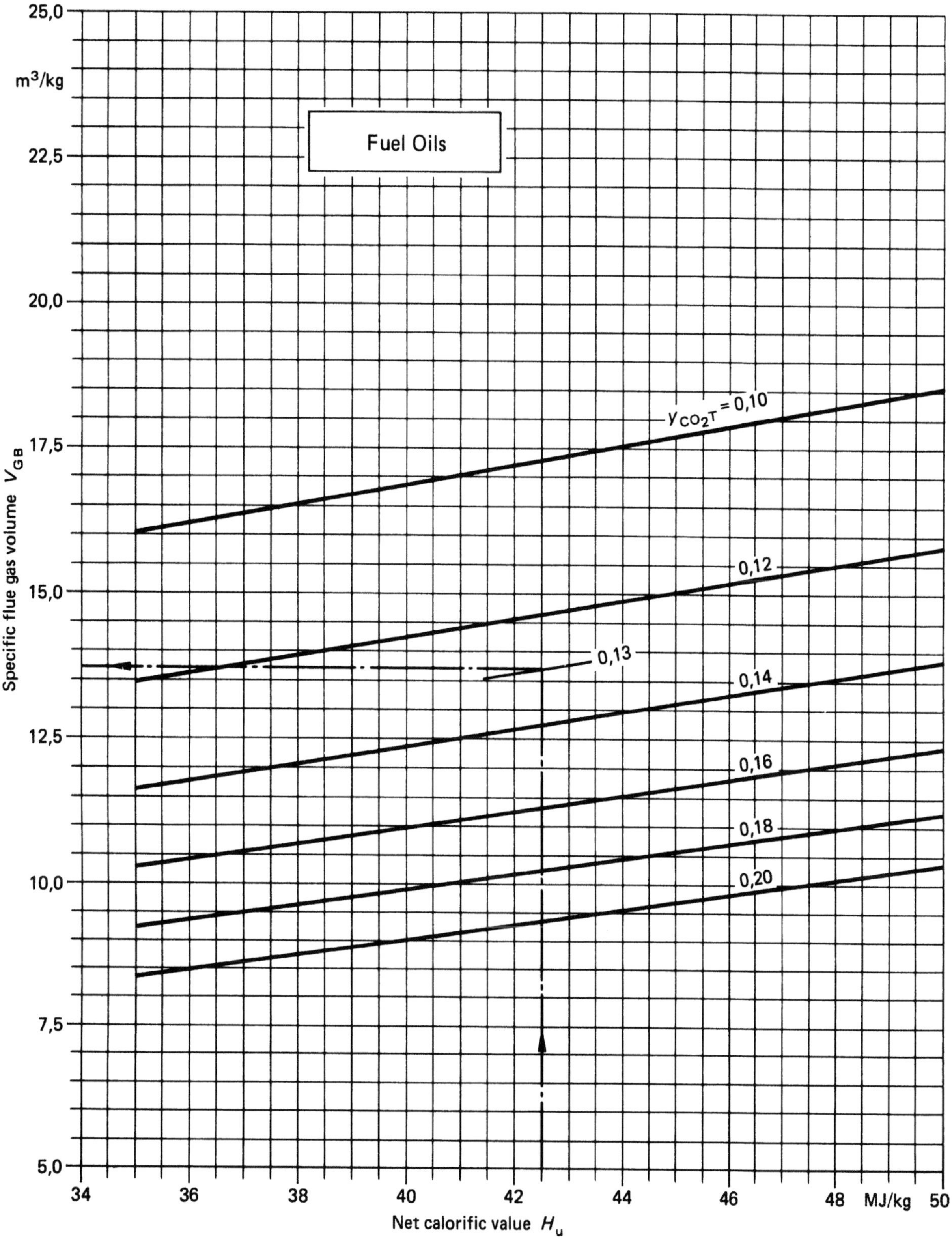

F. Brandt

Explanation

With worksheet No 4.4.2, one can determine the specific flue gas mass V_{GB} of fuel oils dependent on the net calorific value H_u and the measured CO_2-content of the dry flue gas y_{CO_2T}.

Definitions:

H_u in MJ/kg	Net calorific value of the fuel
y_{CO_2T} in %	Measured carbon dioxide content of dry flue gas
y_{H_2OL} in m³/m³	Humidity of the combustion air
x_{H_2OL} in kg/kg	Absolute humidity of the combustion air
V_{LT} in m³/kg	Specific dry combustion air volume (from sheet No 4.2.2)
V_{GB} in m³/kg	Specific flue gas volume (without air humidity)
V_G in m³/kg	Specific flue gas volume

Formulae for calculation:

$$V_{GB} = (-2.49319 + 0.09185\,H_u) + (1.26601 + 0.007557\,H_u)\,\frac{1}{y_{CO_2T}} \qquad (1)$$

$$V_G = V_{GB} + V_{LT}\,y_{H_2OL} \qquad (2)$$

$$y_{H_2OL} = 1.6086\,x_{H_2OL} \qquad (3)$$

Example:

Known values: $\quad H_u \quad = 42.5$ MJ/kg

$\qquad\qquad\qquad y_{CO_2T} = 0.13$

$\qquad\qquad\qquad y_{H_2OL} = 0.0153$ m³/m³

Results: $\qquad\quad V_{LT} \quad = 12.93$ m³/kg from sheet No 4.2.2

$\qquad\qquad\qquad V_{GB} \quad = 13.62$ m³/kg

$\qquad\qquad\qquad V_G \quad = 13.82$ m³/kg

Specific Flue Gas Volume (without Air Humidity) of Natural Gas in m³/kg

4.4.3

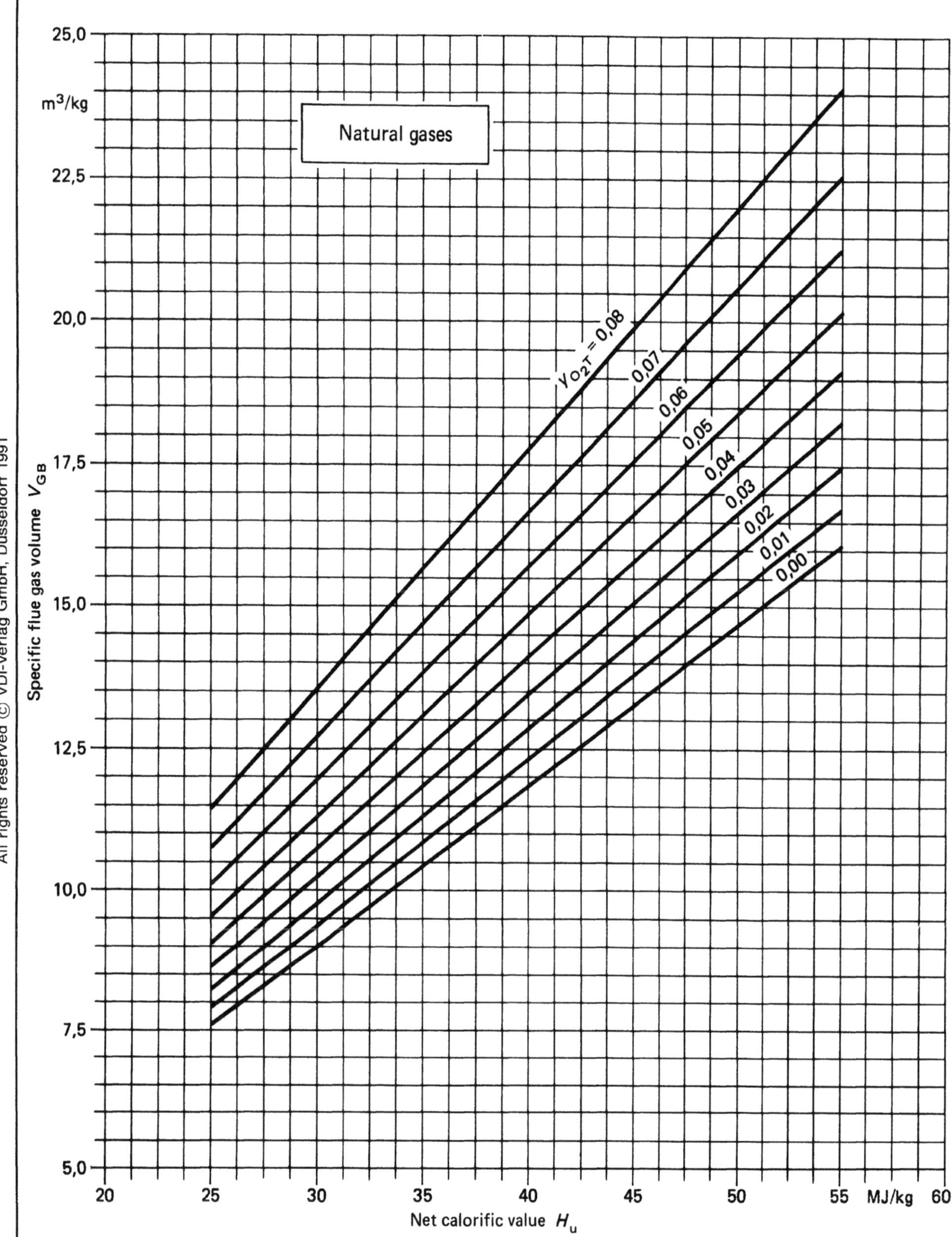

F. Brandt

Explanation

With worksheet No 4.4.3, one can determine the specific flue gas volume V_{GB} of natural gas dependent on the net calorific value H_u and the measured O_2-content of the dry flue gas y_{O_2T}.

Definitions:

H_u in MJ/kg	Net calorific value of the fuel
y_{O_2T} in %	Measured oxygen content of dry flue gas
y_{H_2OL} in m³/m³	Humidity of the combustion air
x_{H_2OL} in kg/kg	Absolute humidity of the combustion air
V_{LT} in m³/kg	Specific dry combustion air volume (from sheet No 4.2.3)
V_{GB} in m³/kg	Specific flue gas volume (without air humidity)
V_G in m³/kg	Specific flue gas volume

Formulae for calculation:

$$V_{GB} = (0.55281 + 0.28186\,H_u) + (0.64975 + 0.22538\,H_u)\,\frac{y_{O_2T}}{0.21 - y_{O_2T}} \tag{1}$$

$$V_G = V_{GB} + V_{LT}\,y_{H_2OL} \tag{2}$$

$$y_{H_2OL} = 1.6086\,x_{H_2OL} \tag{3}$$

Example:

Known values:	H_u	$= 47.5$ MJ/kg
	y_{O_2T}	$= 0.05$
	y_{H_2OL}	$= 0.0153$ m³/m³
Results:	V_{LT}	$= 16.27$ m³/kg from sheet No 4.2.3
	V_{GB}	$= 17.49$ m³/kg
	V_G	$= 17.74$ m³/kg

m³/m³
Natural gases
Specific flue gas volume V_{GBn}
$Y_{O_2T} = 0,09$
0,08
0,06
0,05
0,04
0,02
0,00
25,0
22,5
20,0
17,5
15,0
12,5
10,0
7,5
5,0
20
25
30
35
40
45
50
55
60
MJ/m³
Gross calorific value H_{on}

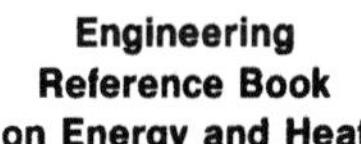

F. Brandt

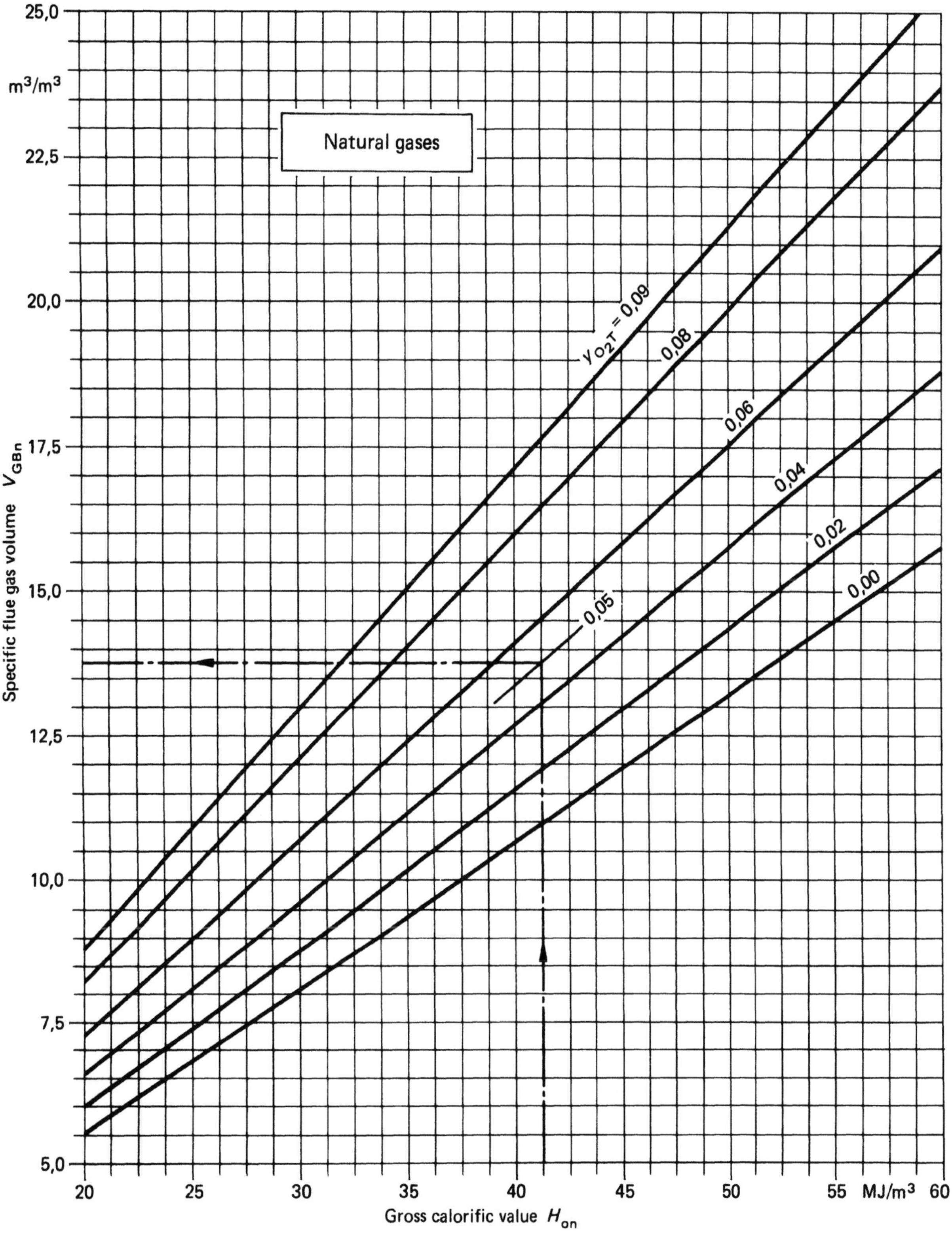

Explanation

With worksheet No 4.4.4, one can determine the specific flue gas volume V_{GBn} of natural gas dependent on the gross calorific value H_{on} and the measured O_2-content of the dry flue gas y_{O_2T}.

Definitions:

H_{on} in MJ/m^3	Gross calorific value at standard conditions
H_{on}^* in kWh/m^3	Gross calorific value at standard conditions
y_{O_2T} in %	Measured oxygen content of dry flue gas
y_{H_2OL} in m^3/m^3	Humidity of the combustion air
x_{H_2OL} in kg/kg	Absolute humidity of the combustion air
V_{LTn} in m^3/m^3	Specific dry combustion air volume at standard conditions (from sheet No 4.2.4)
V_{GBn} in m^3/m^3	Specific flue gas volume (without air humidity) at standard conditions
V_{Gn} in m^3/m^3	Specific flue gas volume at standard conditions

Formulae for calculation:

$$V_{GBn} = (0.42287 + 0.25549\,H_{on}) + (0.06378 + 0.21546\,H_{on})\,\frac{y_{O_2T}}{0.21 - y_{O_2T}} \qquad (1)$$

$$H_{on} = H_{on}^*\,3.6 \qquad (2)$$

$$V_{Gn} = V_{GBn} + V_{LTn}\,y_{H_2OL} \qquad (3)$$

$$y_{H_2OL} = 1.6086\,x_{H_2OL} \qquad (4)$$

Example:

Known values: $H_{on}\ = 41.3$ MJ/m^3

$\qquad\qquad\qquad y_{O_2T}\ = 0.05$

$\qquad\qquad\qquad y_{H_2OL} = 0.0153$ m^3/m^3

Results:$\qquad\quad V_{LTn}\ = 12.71$ m^3/m^3 from sheet No 4.2.4

$\qquad\qquad\qquad V_{GBn}\ = 13.78$ m^3/m^3

$\qquad\qquad\qquad V_{Gn}\ = 13.97$ m^3/m^3

Solid Fuels
Fuel Oils
Density ρ_{nGB}
kg/m³
$y_{CO_2T} =$
0,06
0,08
0,10
0,12
0,16
0,20
Net calorific value H_u
1,40
1,36
1,32
1,28
1,24
1,20
1,16
0
5
10
15
20
25
30
35
MJ/kg
40
1,38
1,34
1,30
1,26
0,20
0,15
0,10
0,05
$y_{CO_2T} =$
36
38
40
42
44
46
48
MJ/kg
50

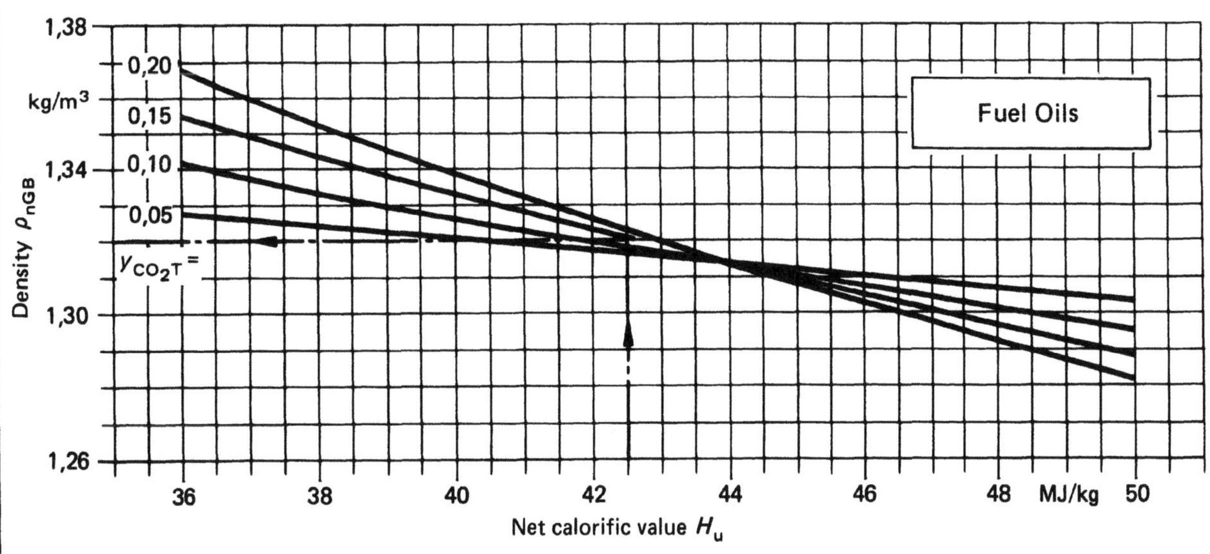

F. Brandt

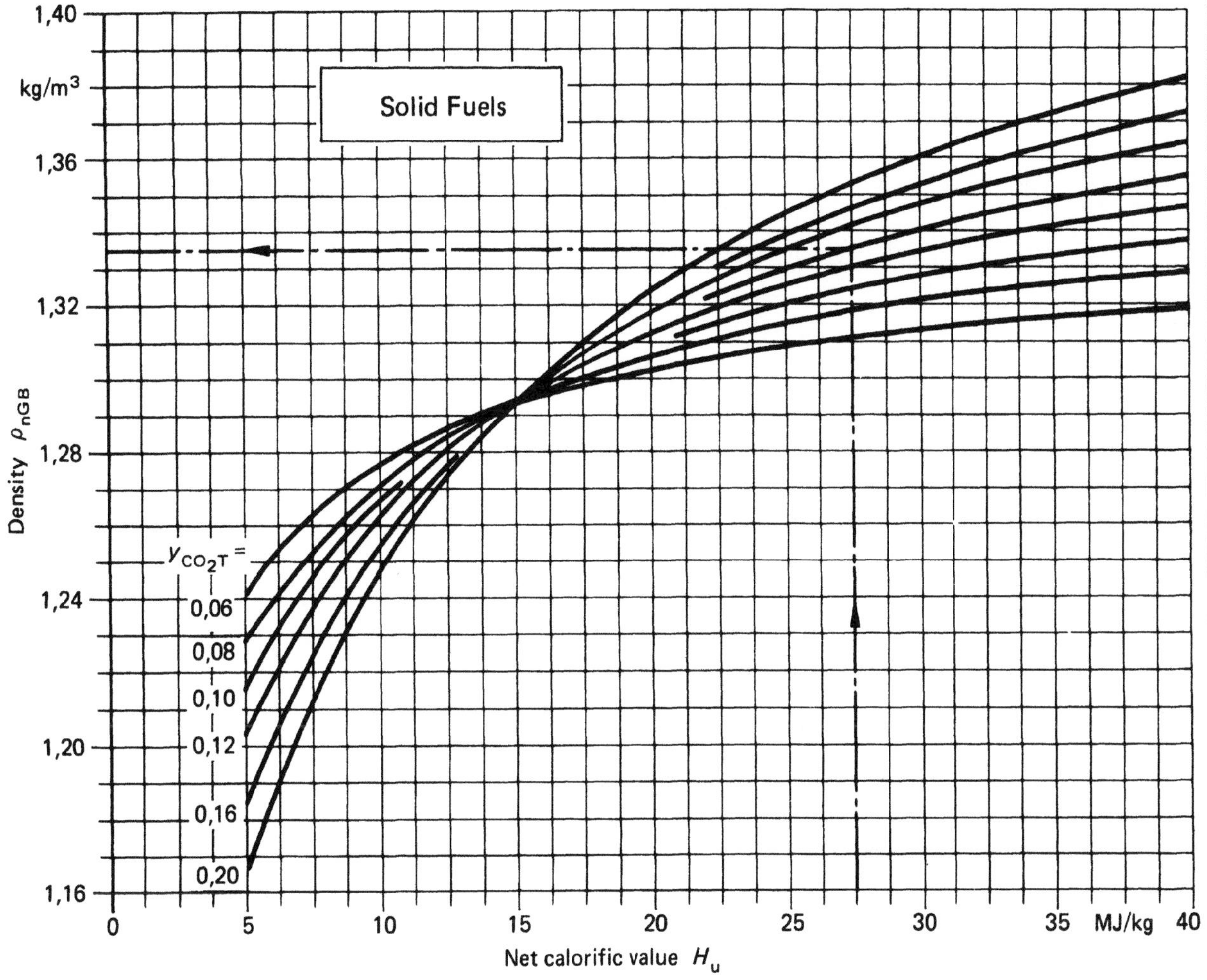

Explanation

With worksheet No 4.5.1, one can determine the flue gas density ϱ_{nGB} of solid fuels and fuel oils dependent on the net calorific value H_u and the measured CO_2-content of the dry flue gas y_{CO_2T}.

Definitions:

μ_{GB} in kg/kg   Specific flue gas mass (without air humidity) from sheet No 4.3.1 for solid fuels or sheet No 4.3.2 for fuel oils

V_{GB} in m³/kg   Specific flue gas volume (without air humidity) from sheet No 4.4.1 for solid fuels or sheet No 4.4.2 for fuel oils

ϱ_{nGB} in kg/m³   Density of flue gas volume at standard conditions

Formula for calculation:

$$\varrho_{nGB} = \mu_{GB}/V_{GB} \tag{1}$$

Example: solid fuels

Known values: $H_u = 27.5$ MJ/kg; $y_{CO_2T} = 0.14$; $x_{H_2OL} = 0$

Results: $\quad \mu_{GB} = 13.271$ kg/kg from sheet No 4.3.1

$\qquad\quad V_{GB} = 9.939$ m³/kg from sheet No 4.4.1

$\qquad\quad \varrho_{nGB} = 1.335$ kg/m³

Example: fuel oils

Known values: $H_u = 42.5$ MJ/kg; $y_{CO_2T} = 0.14$; $x_{H_2OL} = 0$

Results: $\quad \mu_{GB} = 17.71$ kg/kg from sheet No 4.3.2

$\qquad\quad V_{GB} = 13.62$ m³/kg from sheet No 4.4.2

$\qquad\quad \varrho_{nGB} = 1.300$ kg/m³

1,32
kg/m³
1,28
1,24
1,20
1,16
Density ρ_{nGB}
$y_{O_2T} = 0,12$
0,08
0,04
0,00
20 25 30 35 40 45 50 55 MJ/kg 60
Net calorific value H_u

1,32
kg/m³
1,28
1,24
1,20
1,16
Density ρ_{nGB}
$y_{O_2T} = 0,12$
0,08
0,04
0,00
20 25 30 35 40 45 50 55 MJ/m³ 60
Gross calorific value H_{on}

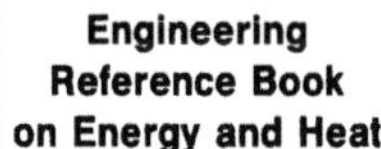

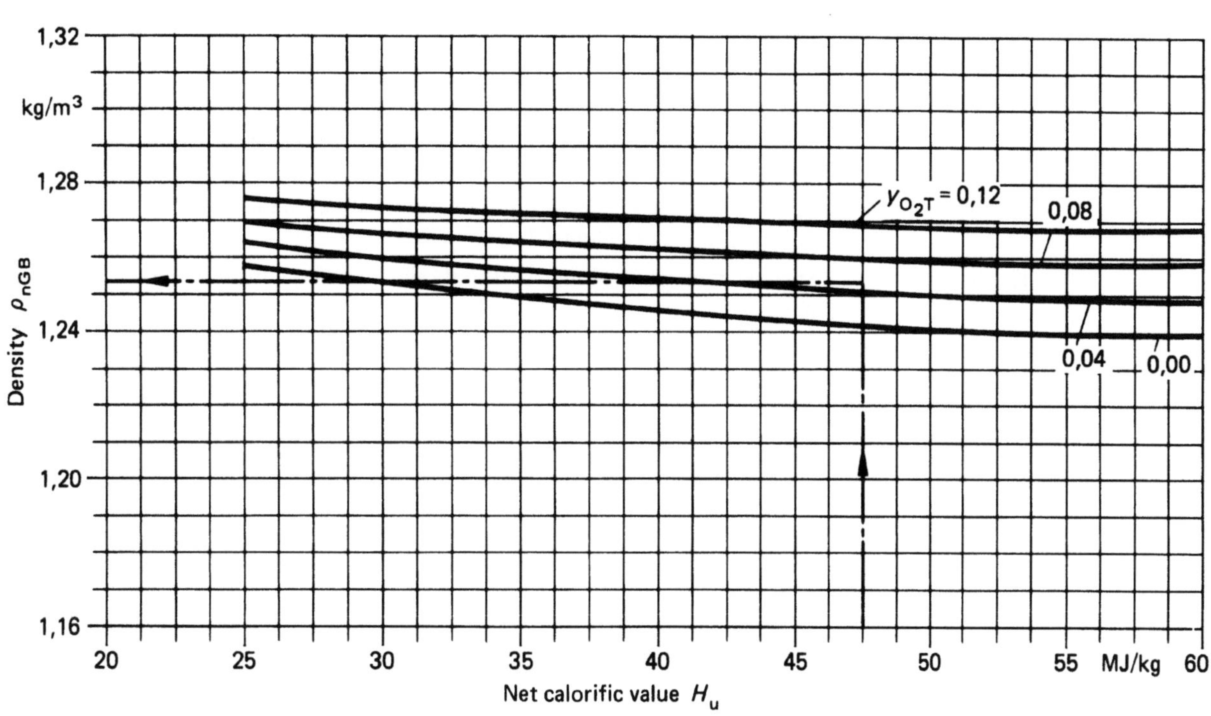

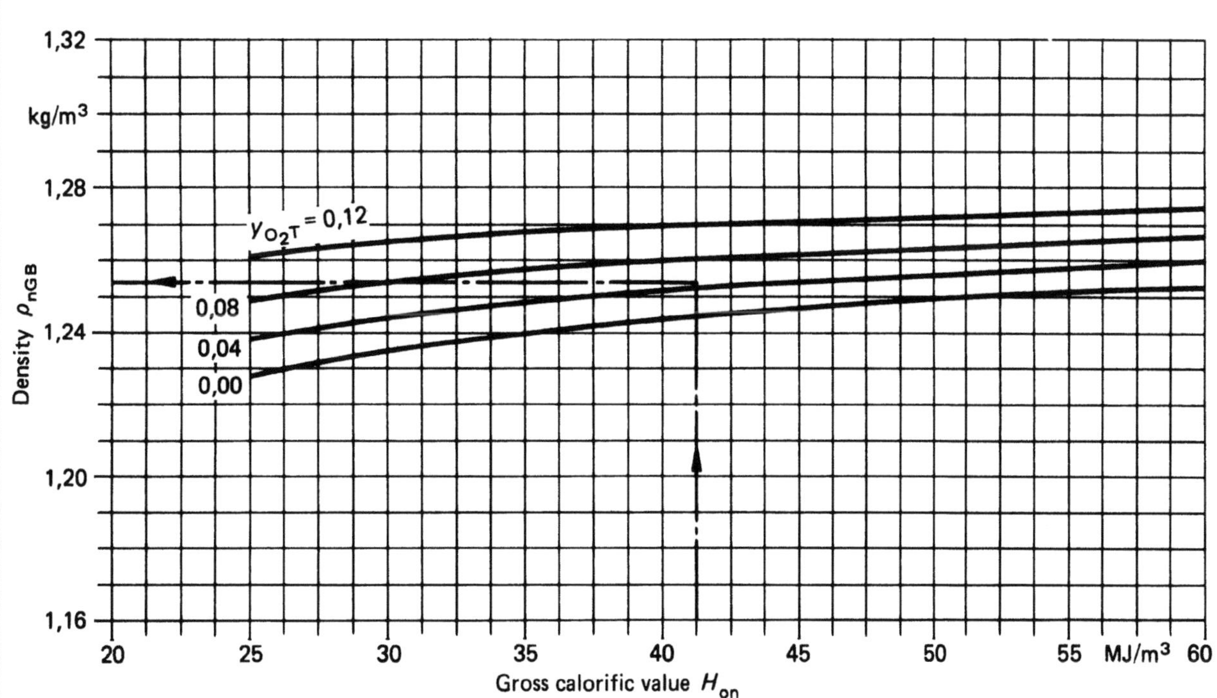

F. Brandt

Explanation

With worksheet No 4.5.2, one can determine the flue gas density ϱ_{nGB} of natural gas dependent on the net calorific value H_u or the gross calorific value H_{on} and the measured O_2-content of the dry flue gas y_{O_2T}.

Definitions:

μ_{GB} in kg/kg	Specific flue gas mass (without air humidity) from sheet No 4.3.3
μ_{GBn} in kg/m^3	Specific flue gas mass (without air humidity) at standard conditions from sheet No 4.3.4
V_{GB} in m^3/kg	Specific flue gas volume (without air humidity) from sheet No 4.4.3
V_{GBn} in m^3/m^3	Specific flue gas volume (without air humidity) at standard conditions from sheet No 4.4.4
ϱ_{nGB} in kg/m^3	Density of flue gas volume at standard conditions

Formulae for calculation:

$$\varrho_{nGB} = \mu_{GB}/V_{GB} = f(H_u) \tag{1}$$

$$\varrho_{nGB} = \mu_{GBn}/V_{GBn} = f(H_{on}) \tag{2}$$

Example 1: natural gas

Known values: $H_u = 47.5\ \text{MJ/kg}$; $y_{O_2T} = 0.05$; $x_{H_2OL} = 0$

Results:
$\mu_{GB} = 21.91\ \text{kg/kg}$ from sheet No 4.3.3
$V_{GB} = 17.49\ \text{m}^3/\text{kg}$ from sheet No 4.4.3
$\varrho_{nGB} = 1.253\ \text{kg/m}^3$

Example 2: natural gas

Known values: $H_{on} = 41.3\ \text{MJ/m}^3$; $y_{O_2T} = 0.05$; $x_{H_2OL} = 0$

Results:
$\mu_{GBn} = 17.28\ \text{kg/m}^3$ from sheet No 4.3.4
$V_{GBn} = 13.78\ \text{m}^3/\text{m}^3$ from sheet No 4.4.4
$\varrho_{nGB} = 1.254\ \text{kg/m}^3$

Table No. 1: Emission standards for plants fired with solid fuels

TA Luft No.   3.3.1.2.1   3.3.1.3.1

	Coal, coke coal briquettes	Wood, waste wood, peat, straw [1]	other solid fuels	Completions	
Field of application	1 to 50 MW		0.1 to 50 MW [2]		
Dust	$<$ 5 MW 150 mg/m^3 [3] $\geq$ 5 MW 50 mg/m^3 [3]		30 mg/m^3		
CO	250 mg/m^3 [4]		100 mg/m^3		
SO$_x$	best available technology; preferably 50% of sulphur emissions (s.e.); maximum 2 000 mg/m^3; fluidized bed combustors: 400 mg/m^3 or 25% s.e. [5]		100 mg/m^3 or 3% s.e. [6]		
NO$_x$	best available technology; maximum 500 mg/m^3 [3]; stationary fluidized bed combustors $>$ 20 MW and circulating fluidized bed combustors: 300 mg/m^3		500 mg/m^3 [7]		
Inorganic chlorides (HCL) and fluorides (HCF)	–	–	50 mg/m^3 or 0.25% Ch.e. [8] [9] 2 mg/m^3 or 0.25% Fl.e. [10]		
Other inorganic substances solid (3.1.4)		–	(independent of flow rates) ×		
vapour/gas (3.1.6)		–	halogenides only ×		
Organic substances (3.1.7)	×	50 mg/m^3	20 mg/m^3		

× section is valid
– section is not valid

1) No. 3.3.1.2.1 is valid for straw accordingly
2) See 4 Bundes-Immissionsschutz-Verordnung
3) Valid at cleaning of heating surface too
4) For single combustors $<$ 2.5 MW, limits are only valid at nominal load
5) For fluidized bed combustors 25% of sulphur emissions is only valid if the 400 mg/m^3 limit is not reasonable
6) Only valid, if 100 mg/m^3 is not reasonable
7) Compare 3.1.6
8) Ch.e. = chlorine emission; only valid, if 50 mg/m^3 is not reasonable
9) If the wood is plated by synthetic material 3.3.1.3.1 section 1 has to be considered
10) Fl.e. = fluorine emission; only valid if 2 mg/m^3 is not reasonable

Remark:
Changing standards for special circumstances have to be considered accordingly.

Table No. 2: Emission standards for plants fired with liquid fuels

TA Luft No.	3.3.1.2.2		3.3.1.3.2		
	Fuel oils of the first refinery, crude oils		other liquid fuels (containing PCB or PCP) [4]	Completions	
	DIN 51 603 part 1 (HEL)	DIN 51 603 part 2 (HS)			
Field of application	1 to 50 MW		0.1 to 50 MW		
Dust	soot number 1 [1]	<5 MW 80 mg/m^3 [2] ≥5 MW 50 mg/m^3 [2]	30 mg/m^3		
CO	170 mg/m^3		100 mg/m^3		
SO$_x$	best available technology maximum 1 700 mg/m^3 [3]		100 mg/m^3		
NO$_x$	250 mg/m^3	best available technology, maximum 450 mg/m^3	500 mg/m^3 [5]		
Inorganic chlorides (HCL) and fluorides (HCF)	according 3.1.6		50 mg/m^3 or 0.25% Ch.e. [6] 2 mg/m^3 or 0.25% Fl.e. [7]		
Other inorganic substances solid (3.1.4)	—	if of low ash content and if no dry ash removal facilities are necessary —	(independent of mass flow rates) ×		
vapour/gas (3.1.6)	×				
Organic substances (3.1.7)	×		20 mg/m^3		

× section is valid   — section is not valid

1) See 1 Bundes-Immissionsschutz-Verordnung Anlage 2
2) Valid at cleaning of heating surface too
3) If there is no desulphurisation, up to 5 MW thermal capacity only fuel oils with a sulphur content according to DIN 51 603 part 1 are permitted

4) Compare the defined limits in 3.3.1.3.2
5) Compare 3.1.6
6) 0.25% of Ch.e. = chlorine emission; only valid, if 50 mg/m^3 is not reasonable
7) 0.25% of Fl.e. = fluorine emission; only valid if 2 mg/m^3 is not reasonable

Table No. 3: Emission standards for plants fired with gaseous fuels

TA Luft No.	3.3.1.2.3		
Field of application	10 bis 100 MW	Completions	
Dust Blast furnace gas Industrial gases from steel mills Other gases	10 mg/m^3 50 mg/m^3 5 mg/m^3		
CO	100 mg/m^3		
SO$_x$ Coke oven gas, refinery gas LPG Producer gas from furnace and cokery Petroleum gas Other gases	100 mg/m^3 5 mg/m^3 200 up to 800 mg/m^3 1 700 mg/m^3 35 mg/m^3		
NO$_x$	200 mg/m^3 [1]		
Other inorganic substances solid (3.1.4)	×		
vapour/gas (3.1.6)	×		
Organic substances (3.1.7)	×		

× section is valid

1) Special regulations for process gases

Table No. 4: Emission standards for plants driven by combustion engines

TA Luft No.   3.3.1.4.1

Field of application	All plants from 1 MW up	Completions
Dust	130 mg/m^3 [1]	
CO	650 mg/m^3	
SO$_x$	limits similar to that of HEL	
NO$_x$ Four stroke motors Two stroke motors Diesel engine plants >3 MW Diesel engine plants <3 MW Plants for emergency service	500 mg/m^3 800 mg/m^3 2 000 mg/m^3 [2] 4 000 mg/m^3 [2] no limits	
Other inorganic substances solid (3.1.4)	×	
vapour/gas (3.1.6)	×	
Organic substances (3.1.7)	×	

× section is valid

---

1) Furthermore the installation of soot filters is recommended
2) Best available technology

Table No. 5: Emission standards for gas turbine plants (used as driving engine, open circle)

TA Luft No.   3.3.1.5.1

Field of application	All plants	Completions
Dust up to 60 000 m^3/h more than 60 000 m^3/h	soot number 4 soot number 2 (3)	
CO	100 mg/m^3	
SO$_x$	limits similar to that of HEL	
NO$_x$ up to 60 000 m^3/h more than 60 000 m^3/h	350 mg/m^3 [1] 300 mg/m^3 [1]	
Other inorganic substances solid (3.1.4)	×	
vapour/gas (3.1.6)	×	
Organic substances (3.1.7)	×	

If the efficiency surpasses 30% the emission values have to be increased accordingly.
× section is valid

---

1) Best available technology

Table No. 6: Collection of the substances with emission limits

	Limit of mass flow	Limit of concentration	TA Luft No.	Completions	
Total dust	up to 0.5 kg/h more than 0.5 kg/h	150 mg/m^3 50 mg/m^3	3.1.3		

Special inorganic substances [1]
– solid –
	class I:
	more than 0.001 kg/h
	class II:
	more than 0.005 kg/h
	class III:
	more than 0.025 kg/h
		0,2 mg/m^3
		
		1 mg/m^3
		
		5 mg/m^3
			3.1.4		

Special inorganic substances [1] – vapour/gas –	I arsine, hydrogen phosphide, chloric cyanide, phosgene: more than 0.01 kg/h II halogenides, HS, etc.: more than 0.05 kg/h III HCL: more than 0.3 kg/h IV SO_x, NO_x: more than 5 kg/h	 1 mg/m^3 5 mg/m^3 30 mg/m^3 0.50 g/m^3	3.1.6		

Organic substances [1]	class I:
	more than 0.1 kg/h
	class II:
	more than 2 kg/h
	class III:
	more than 35 kg/h
		20 mg/m^3
		
		0.10 g/m^3
		
		0.15 g/m^3
			3.1.7		

Odour intensive substances	—	—	3.1.9		

1) In this table the general requirements from TA Luft No. 3.1 have been listed. According to TA Luft No. 3.1.1 they are always valid, if there are no special regulations included in the tables No. 1 to 5

SO$_2$-Emission standards (solid fuels) [1]

	1 to 50 MW	50 to 100 MW	100 to 300 MW	>300 MW		Completions	
New plants	regulation not valid; TA Luft part 3 is valid	2 000 mg/m^3	40% S.e. *) 2 000 mg/m^3 maximum	15% S.e., 400 mg/m^3 maximum resp. 650 mg/m^3 [2]			
Old plants	regulation not valid; TA Luft part 3 and 4 are valid	up to the 1st April 1993					
		lifetime [h] ≤10 000 | >10 000	lifetime [h] ≤10 000 | >10 000	lifetime [h] ≤10 000 | >10 000 to 30 000	>30 000		
		according to per-mission | 2 500 mg/m^3	according to per-mission | 2 500 mg/m^3	according to per-mission | 2 500 mg/m^3	15% S.e. 400 mg/m^3 maximum resp. 650 mg/m^3 [2]		
		after the 1st April 1993					
		2 000 mg/m^3	40% S.e. 2 000 mg/m^3 maximum	15% S.e., 400 mg/m^3 maximum resp. 650 mg/m^3 [2]			

SO$_2$-Emission standards (liquid fuels) [1]

	1 to 50 MW	50 to 100 MW	100 to 300 MW	>300 MW		Completions	
New plants	regulation not valid; TA Luft part 3 is valid	1 700 mg/m^3	40% S.e. *) 1 700 mg/m^3 maximum	15% S.e., 400 mg/m^3 maximum resp. 650 mg/m^3 [2]			
Old plants	regulation not valid; TA Luft part 3 and 4 are valid	up to the 1st April 1993					
		lifetime [h] ≤10 000 | >10 000	lifetime [h] ≤10 000 | >10 000	lifetime [h] ≤10 000 | >10 000 to 30 000	>30 000		
		according to per-mission | 2 500 mg/m^3	according to per-mission | 2 500 mg/m^3	according to per-mission | 2 500 mg/m^3	15% S.e. 400 mg/m^3 maximum resp. 650 mg/m^3 [2]		
		after the 1st April 1993					
		1 700 mg/m^3	40% S.e. 1 700 mg/m^3 maximum	15% S.e., 400 mg/m^3 maximum resp. 650 mg/m^3 [2]			

*) S.e. = sulphur emission degree
1) Time limited exceptions are possible
2) 650 mg/m^3 i.v. with maximum filter load

NO$_x$-Emission standards [1]

	Completions	
New plants		

Solid fuels:
    general   800 mg/m^3 maximum
    hard coal, slag tap furnace 1 800 mg/m^3 maximum		

Liquid fuels 450 mg/m^3 maximum		
Gaseous fuels 350 mg/m^3 maximum		
Old plants		

Solid fuels:
    hard coal, dry bottom furnace   1 300 mg/m^3 maximum
    hard coal, slag tap furnace   2 000 mg/m^3 maximum
    other plants fired by solid fuels 1 000 mg/m^3 maximum		

Liquid fuels 700 mg/m^3 maximum		
Gaseous fuels 500 mg/m^3 maximum		

1) **Important:** According to §§ 5, 10 and 15 (new plants) and § 19 (old plants) the emission limits are connected with the following obligation: "All possibilities to lower the emissions by combustion technological measures ore other ones have to be achieved according to the state of technology."

Bibliography

Umweltrecht für Energieanlagen. Edited by: Vereinigung Industrieller Kraftwirtschaft e.V. Essen, Verl. Energieberatung GmbH, Essen

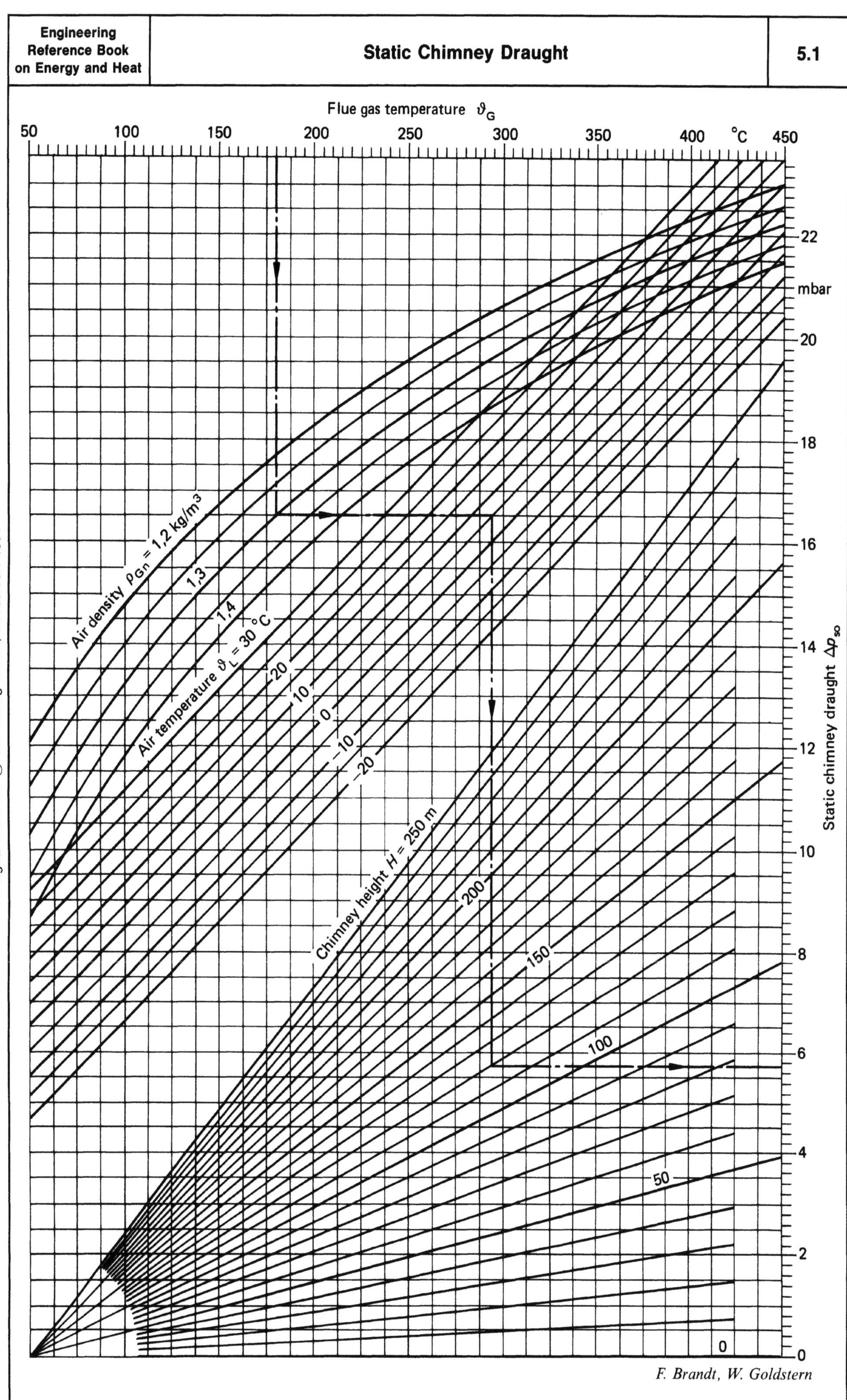

F. Brandt, W. Goldstern

Explanation

With worksheet No 5.1, one can determine the static chimney draught Δp_{so} dependent on the average flue gas temperature T_G, the flue gas density ϱ_{Gn} (from sheet No 4.5.1 or 4.5.2), the temperature of the air T_L and the chimney height H.

Definitions:

ϱ_L in kg/m^3	Air density
ϱ_{Ln} in kg/m^3	Air density at standard conditions according to DIN 1343
ϱ_G in kg/m^3	Flue gas density
ϱ_{Gn} in kg/m^3	Flue gas density standard conditions according to DIN 1343
T_L in K	$= \vartheta_L + 273.15$ Air temperature
T_G in K	$= \vartheta_G + 273.15$ Flue gas temperature
g in m/s^2	Acceleration due to gravity
H in m	Chimney height
Δp_{so} in mbar	Static chimney draught
Δp_s in mbar	Static chimney draught at ground level
p^* in mbar	Barometric pressure measured at ground level
p_o in mbar	$= p_G = 1013$ mbar (standard barometric pressure)

If indexed by o, the value is related to $p_o = p_G$.

Formulae for calculation:

$$\varrho_G = \varrho_{Gn}\,\frac{273}{T_G} \tag{1}$$

$$\varrho_L = \varrho_{Ln}\,\frac{273}{T_L} \tag{2}$$

$$\Delta p_{so} = gH(\varrho_L - \varrho_G) \tag{3}$$

$$\Delta p_{so} = H\left(\frac{34.6}{T_L} - \varrho_{Gn}\,\frac{26.8}{T_G}\right)\text{mbar} \tag{4}$$

$$\Delta p_s = \Delta p_{so}\,\frac{p^*}{p_o} \tag{5}$$

Example:

Known values: $T_G = 453\,$K; $\varrho_{Gn} = 1.3$ kg/m^3 from sheet No 4.5.1 or 4.5.2;
$\varrho_{Ln} = 1.293$ kg/m^3; $T_L = 278$ K; $g = 9.81$ m/s^2; $H = 120$ m

Result: $\Delta p_{so} = 5.75$ mbar

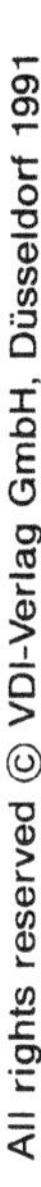

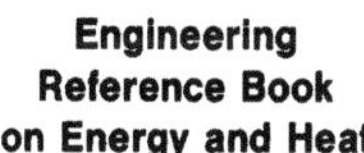

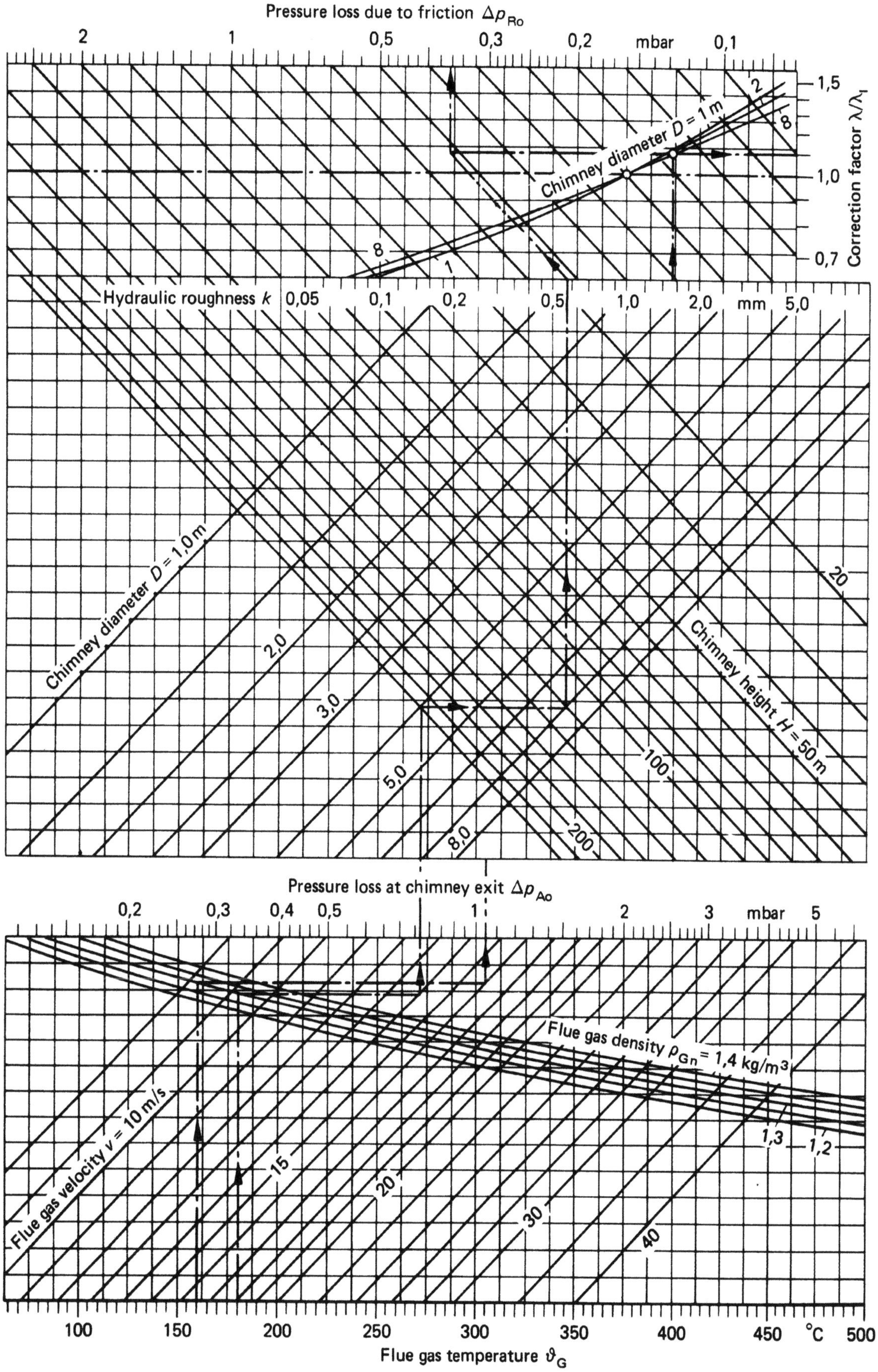

Pressure loss due to friction Δp_{Ro}

F. Brandt, W. Goldstern

Explanation

With worksheet No 5.2, one can determine the draught loss in chimneys (pressure loss at chimney exit Δp_{Ao} and pressure loss due to friction Δp_{Ro}) dependent on the average flue gas temperature $\vartheta_{G\,(m)}$, the average flue gas density ϱ_{an} (from sheet No 4.5.1 or 4.5.2), the average velocity at chimney exit v_m, the average chimney diameter D_m, the chimney height H, the hydraulic roughness k (from sheet No 9.1) and the correction factor λ/λ_1.

Definitions:

Δp_{Ao} in mbar	Pressure loss at chimney exit
Δp_A in mbar	Pressure loss at chimney exit at different barometric pressure p^*
Δp_{Ro} in mbar	Pressure loss due to friction
Δp_R in mbar	Pressure loss due to friction at different barometric pressure p^*
p_o in mbar	$= 1013$ mbar (standard barometric pressure)
p^* in mbar	Barometric pressure measured at ground level
v in m/s	Flue gas velocity at chimney exit
ϱ_G in kg/m³	Flue gas density
ϱ_{Gn} in kg/m³	Flue gas density at standard conditions according to DIN 1343
T_G in K	Flue gas temperature
D in m	Chimney diameter
H in m	Chimney height
k in mm	Hydraulic roughness
$\lambda\,(-)$	Friction factor
$\lambda_1\,(-)$	Friction factor at $k = 1$ mm

If indexed by o, the value is related to p_o.

The index m indicates an average value.

Formulae for calculation:

$$\Delta p_{Ao} = \frac{v^2}{2}\,\varrho_G = \frac{v^2}{2}\,\varrho_{Gn}\,\frac{273}{T_G} \tag{1}$$

$$\Delta p_A = \Delta p_{Ao}\,\frac{p_o}{p^*} \tag{2}$$

$$\Delta p_{Ro} = \frac{\lambda}{D}\,H\,\frac{v^2}{2}\,\varrho_G \tag{3}$$

$$\varrho_G = \varrho_{Gn}\,\frac{273}{T_G} \tag{4}$$

$$\Delta p_{Ro} = \frac{\lambda}{D}\,H\,\frac{v^2}{2}\,\varrho_{Gn}\,\frac{273}{T_G} \tag{5}$$

$$\Delta p_{Ro} = \frac{\lambda_1}{D}\,\Delta p_{Ao}\,\frac{\lambda}{\lambda_1}\,H \tag{6}$$

for hydraulically smooth tubes

$$\frac{1}{\sqrt{\lambda}} = 2\log\frac{D}{2k} + 1.74 \tag{7}$$

The correction factor λ/λ_1 thus is a function of D and k.

Within the assumed range of values of diameter and velocity, the values of the Reynolds number are between 10^5 and 10^7. If one considers the technical feasible range for the values of the relative hydraulic roughness between 10^{-2} and 10^{-7}, the pressure loss due to friction may be always calculated for smooth tubes, for high values of relative hydraulic roughness are always connected to low values of the Reynolds number.

$$\Delta p_r = \Delta p_{Ro}\,\frac{p_o}{p^*} \tag{8}$$

Example 1:

Known values:
$T_G \quad = 433$ K
$\varrho_{Gn} = 1.3$ kg/m³ from sheet No 4.5.1 or 4.5.2
$v \quad = 16$ m/s

Results: $\Delta p_{Ao} = 1.05$ mbar

Example 2:

Known values:
$T_{Gm} \quad = 453$ K
$\varrho_{Gn} = 1.3$ kg/m³ from sheet No 4.5.1 or 4.5.2
$v_m \quad = 14$ m/s
$D_m \quad = 4$ m
$H \quad = 120$ m
$k \quad = 1.5$ mm

Result:
$\lambda/\lambda_1 = 1.1$
$\Delta p_{Ro} = 0.365$ mbar

Bibliography

Eck, B.: Technische Strömungslehre. Berlin: Springer 1961; p. 135

Flue gas temperature at the exit ϑ_G
°C
400
350
300
250
200
150
120
100
80
60
40
20
0
Air temperature at the entrance ϑ_L and wall temperature ϑ_W
°C
400
350
300
250
200
150
120
100
80
60
40
20
0
∞ 20 10 5 2,0 1,0 0,7 0,5 0,4 0,3 0,2 0,1 0
$\alpha_G A_G / \alpha_L A_L$

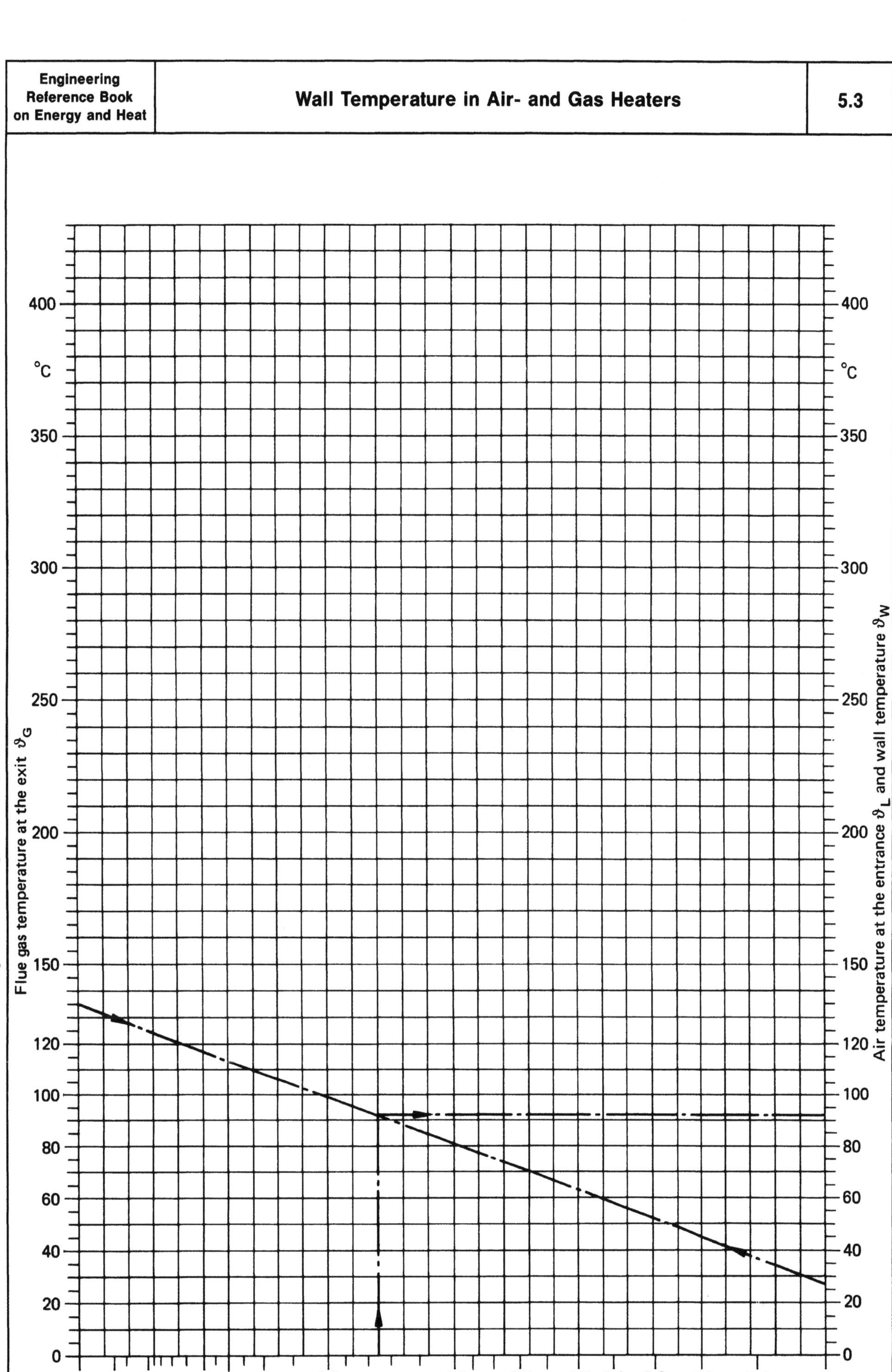

F. Brandt

Explanation

With worksheet No 5.3, one can determine the wall temperature ϑ_W in air- and gas heaters dependent on the entrance temperature of the air ϑ_L and the exit temperature of the flue gas ϑ_G as well as the flue gas- and air related heat transfer areas A_G, A_L, and the heat transfer coefficients α_G, α_L. To simplify the calculation, the heat transmission coefficient of the metal wall has been neglected.

For the calculation in the diagram the values of the flue gas temperature and the air temperature have to be connected by a straight line. The value for the wall temperature can be determined on the right axis at the point of intersection between the straight line and a vertical line to the value of $\alpha_G A_G/(\alpha_L \cdot A_L)$ on the bottom axis.

Definitions:

T_G in K	Flue gas temperature at the exit
T_L in K	Air temperature at the entrance
T_W in K	Average wall temperature
α_G in W/(m²K)	Heat transfer coefficient of the flue gas
α_L in W/(m²K)	Heat transfer coefficient of the air
A_G in m²	Heat transfer area of the flue gas
A_L in m²	Heat transfer area of the air

Formula for calculation:

$$T_W = T_L + \cfrac{1}{1 + \cfrac{\alpha_L\, A_L}{\alpha_G\, \alpha_G}}\ (T_G - T_L) \tag{1}$$

Example:

Known values: $T_G = 300$ K $T_L = 408$ K
               $\alpha_G = 77.5$ W/(m²K) $\alpha_L = 77.0$ W/(m²K)
               $A_G = 33\,400$ m² $A_L = 22\,450$ m²

Result: $T_W = 365$ K

Bibliography

Härtel, S.: Vermeidung von Taupunkt-Unterschreitungen an Kessellufterhitzern. BWK 7 (1955) p. 10/13

Flue Gas Loss of Solid Fuels
$(\vartheta_b = 25\,°C;\; x_{H_2OL} = 0.0062)$

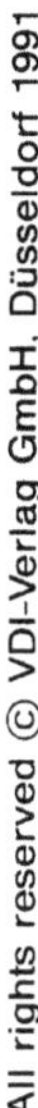

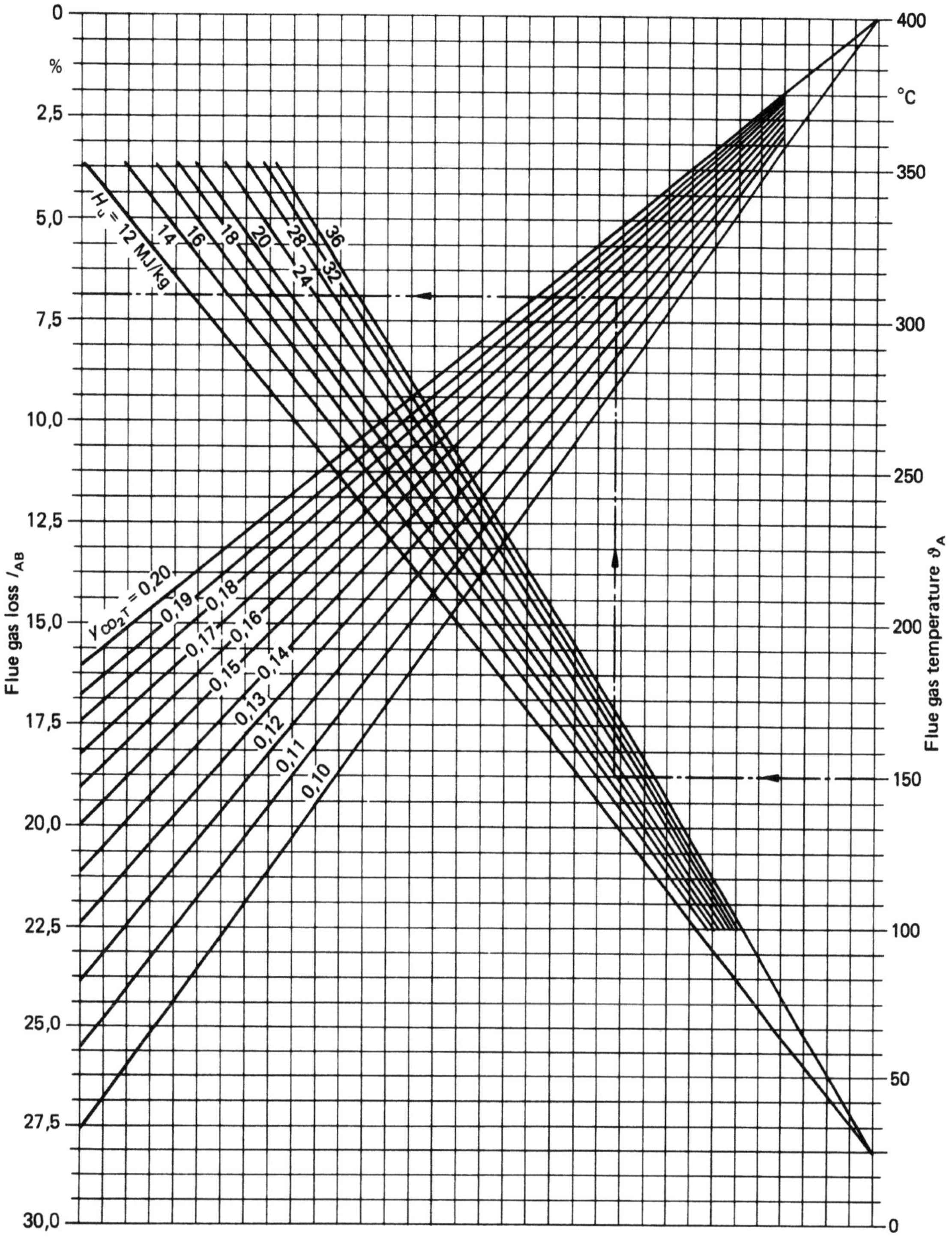

F. Brandt

Explanation

With worksheet No 5.4.1, one can determine the flue gas loss l_{AB} of solid fuels dependent on the flue gas temperature ϑ_A, the net calorific value H_u and the CO_2-content of the dry flue gas y_{CO_2T}.

Definitions:

l_{AB} in %	Flue gas loss of solid fuel
μ_G in kg/kg	Specific flue gas mass from sheet No 4.3
μ_{H_2O} in kg/kg	Specific steam content of flue gas
μ_{CO_2} in kg/kg	Specific CO_2-content of flue gas
μ_{GB} in kg/kg	Specific flue gas mass (without air humidity) from sheet No 4.3.1
μ_{LT} in kg/kg	Specific dry air mass for combustion from sheet No 4.3.1
μ_{H_2OB} in kg/kg	Specific steam content of flue gas (without air humidity)
H_u in MJ/kg	Net calorific value of the fuel
$\bar{c}_{pG}$ in kJ/(kgK)	Integral specific heat capacity of the flue gas
ϑ_A in °C	Flue gas temperature at the exit
ϑ_b in °C	= 25 °C (standard temperature)
ϑ in °C	Flue gas temperature
h_A in kJ/kg	Specific enthalpy of the flue gas at exit temperature
h_b in kJ/kg	Specific enthalpy of the flue gas at standard temperature
h in kJ/kg	Specific enthalpy of the flue gas
h_{LT} in kJ/kg	Specific enthalpy of the dry air
$h_{1,2}\,(-)$	Polynomials
x_{H_2O}	H_2O-content of the humid flue gas
x_{CO_2}	CO_2-content of the humid flue gas
x_{H_2OL} in kg/kg	= 0.0062 kg/kg absolute humidity of the combustion air ($t = 10$ °C, $\varphi = 80\%$)
y_{CO_2T} in %	Measured carbon dioxide content

Formulae for calculation:

$$l_{AB} = \frac{\mu_G}{H_u}(\bar{c}_{pG}\,\vartheta_A - \bar{c}_{pG}\,\vartheta_b) = \frac{\mu_G}{H_u}(h_A - h_b) \quad (1)$$

$$h = \bar{c}_{pG}\,\vartheta = h_{LT} + h_1\,x_{H_2O} + h_2\,x_{CO_2} \quad (2)$$

$$l_{AB} = \frac{1}{H_u}\left[(h_{LT}(\vartheta_A) - h_{LT}(\vartheta_b))\,\mu_G + (h_1(\vartheta_A) - h_1(\vartheta_b)) + \mu_{H_2O} + (h_2(\vartheta_A) - h_2(\vartheta_b))\,\mu_{CO_2}\right] \quad (3)$$

$$h_{LT} = \left(a\vartheta + \frac{b}{2}\vartheta^2 + \frac{c}{3}\vartheta^3 + \frac{d}{4}\vartheta^4 + \frac{e}{5}\vartheta^5 + \frac{f}{6}\vartheta^6\right) \quad (4)$$

$$h_1 = \left(a_1\vartheta + \frac{b_1}{2}\vartheta^2 + \frac{c_1}{3}\vartheta^3 + \frac{d_1}{4}\vartheta^4 + \frac{e_1}{5}\vartheta^5\right) \quad (5)$$

$$h_2 = \left(a_2\vartheta + \frac{b_2}{2}\vartheta^2 + \frac{c_2}{3}\vartheta^3 + \frac{d_2}{4}\vartheta^4 + \frac{e_2}{5}\vartheta^5\right) \quad (6)$$

$$\mu_{GB} = (0.96569 + 0.00707\,H_u) + (0.13139 + 0.05688\,H_u)\,\frac{1}{y_{CO_2T}} \quad (7)$$

$$\mu_{LT} = (-0.0139 + 0.0890\,H_u) + (0.13139 + 0.05688\,H_u)\,\frac{1}{y_{CO_2T}} \quad (8)$$

$$\mu_{H_2OB} = 0.90809 - 0.01630\,H_u \quad (9)$$

$$\mu_{CO_2} = 0.2009 + 0.08697\,H_u$$

with H_u in MJ/kg $\quad (10)$

$$\mu_G = \mu_{GB} + \mu_{LT}\,x_{H_2OL} \quad (11)$$

$$\mu_{H_2O} = \mu_{H_2OB} + \mu_{LT}\,x_{H_2OL} \quad (12)$$

Example:

Known values:

$\vartheta_A = 150$ °C; $H_u = 18$ MJ/kg; $y_{CO_2T} = 0.14$.

$\mu_{CO_2} = 1.766$ kg/kg,
$\mu_{H_2OB} = 0.6147$ kg/kg,
$\mu_{H_2O} = 0.6688$ kg/kg,
$\mu_{LT} = 8.398$ kg/kg from sheet No 4.3.1.

$\mu_{GB} = 9.345$ kg/kg $\qquad$ $\mu_G = 9.397$ kg/kg
from sheet No 4.3.1 $\qquad$ from sheet No 4.3.1

ϑ	150		25	(150 − 25)
h_{LT}	151.42		25.11	126.31
h_1	131.09		21.45	109.64
h_2	− 7.39		− 2.27	− 5.12

Result: $\quad l_{AB} = \dfrac{1\,251}{18\,000} = 0.0695$

Bibliography

Brandt, F.: Wärmeübertragung in Dampferzeugern und Wärmeaustauschern. FDBR Fachbuchreihe, vol. 2, Vulkan-Verlag, Essen, 1985

h_{LT}		h_1		h_2	
a	0.1004173 E + 01	a_1	0.8554535 E $\quad$ 00	a_2 −	0.1002311 E $\quad$ 00
b	0.1919210 E − 04	b_1	0.2036005 E − 03	b_2	0.7661864 E − 03
c	0.5883438 E − 06	c_1	0.4583082 E − 06	c_2 −	0.9259622 E − 06
d −	0.7011184 E − 09	d_1 −	0.2798080 E − 09	d_2	0.5293496 E − 09
e	0.3309525 E − 12	e_1	0.5634413 E − 13	e_2 −	0.1093573 E − 12
f −	0.5673876 E − 16				

0
%
2
4
6
8
10
12
14
16
18
20
22
24
Flue gas loss l_{AB}
$H_u = 42{,}76$ MJ/kg
$H_u = 39$ MJ/kg
$y_{CO_2T} = 0{,}16$
0,15
0,14
0,13
0,12
0,11
0,10
0,09
0,08
350
°C
300
250
200
150
100
50
0
Flue gas temperature ϑ_A

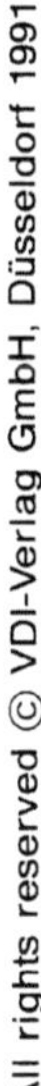

F. Brandt

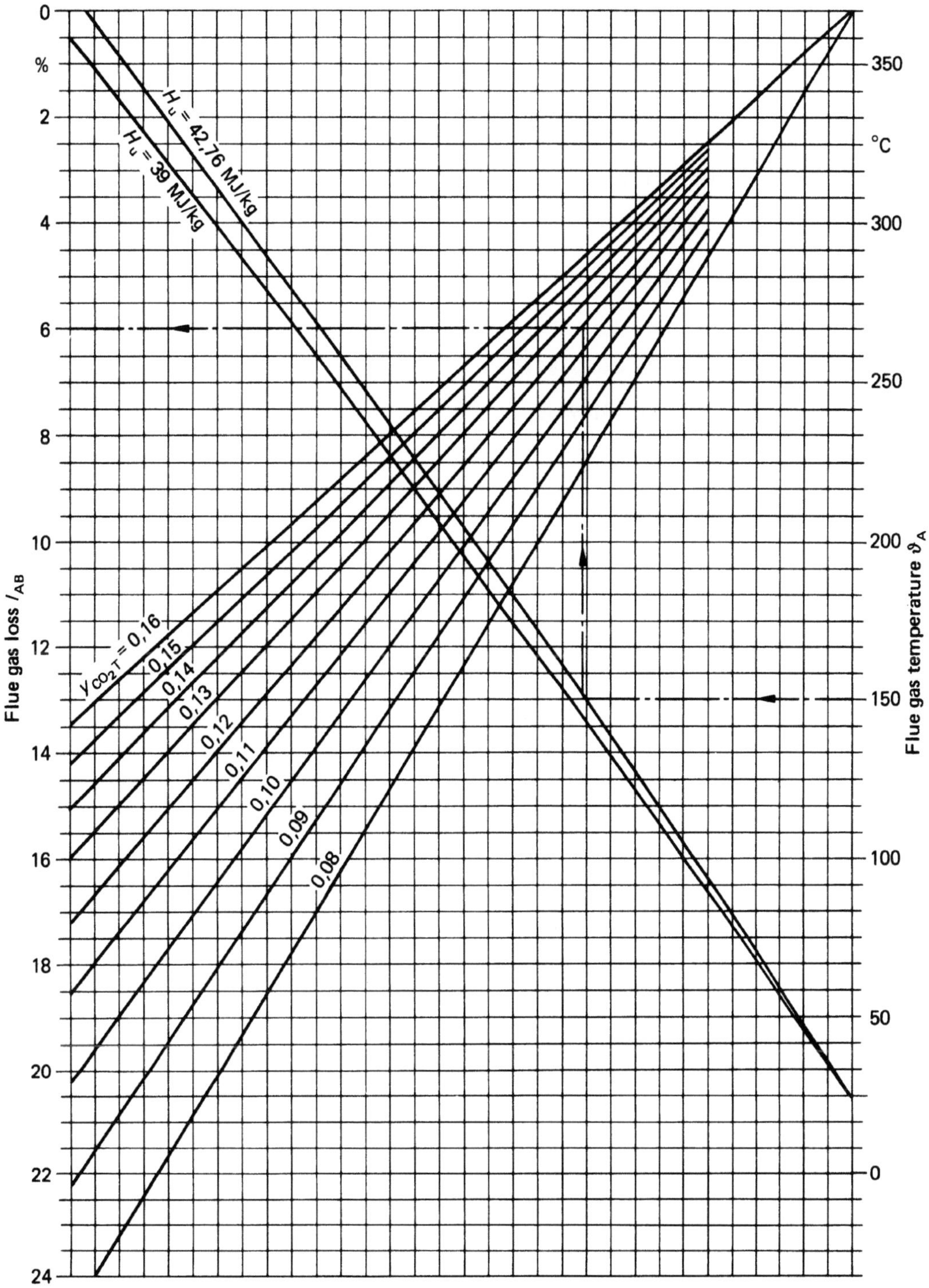

Explanation

With worksheet No 5.4.2, one can determine the flue gas loss l_{AB} of fuel oils dependent on the flue gas temperature ϑ_A, the net calorific value H_u and the CO_2-content of the dry flue gas y_{CO_2T}.

Definitions:

See worksheet No 5.4.1.

Formulae for calculation:

l_{AB} see worksheet No 5.4.1 $\hspace{3cm}$ (1) to (6)

$$\mu_{GB} = (-0.84159 + 0.06491\,H_u) + (1.63696 + 0.009771\,H_u)\,\frac{1}{y_{CO_2T}} \tag{7}$$

$$\mu_{LT} = \mu_{GB} - 1 \tag{8}$$

$$\mu_{H_2OB} = -2.00425 + 0.07384\,H_u \tag{9}$$

$$\mu_{CO_2} = 2.50291 + 0.01494\,H_u \tag{10}$$

$$\mu_G = \mu_{GB} + \mu_{LT}\,x_{H_2OL} \tag{11}$$

$$\mu_{H_2O} = \mu_{H_2OB} + \mu_{LT}\,x_{H_2OL} \tag{12}$$

Example:

Known values: $\quad\vartheta_A \quad = 150\,°C;\ H_u = 42.76\ MJ/kg;\ y_{CO_2T} = 0.14$

$\qquad\qquad\qquad\quad \mu_{CO_2} = 3.142\ kg/kg$

$\qquad\qquad\qquad\quad \mu_{H_2OB} = 1.153\ kg/kg;\ \mu_{H_2O} = 1.265\ kg/kg$

$\qquad\qquad\qquad\quad \mu_{LT} \quad = 18.06\ kg/kg$ from sheet No 4.3.2

$\qquad\qquad\qquad\quad \mu_{GB} \quad = 19.06\ kg/kg$ from sheet No 4.3.2

$\qquad\qquad\qquad\quad \mu_G \quad = 19.172\ kg/kg$ from sheet No 4.3.2

$\qquad\qquad\qquad\quad h_{LT}, h_1, h_2$ see worksheet No 5.4.1

Result: $\quad l_{AB} = \dfrac{2\,544}{42\,760} = 0.0595$

Natural gas H $H_u = 45{,}5$ MJ/kg
Natural gas L $H_u = 38{,}3$ MJ/kg
Flue gas loss l_{AB}
Flue gas temperature ϑ_A
%
°C
0,00
0,01
0,02
0,03
0,04
0,05
0,06
0,07
$y_{O_2T} = 0{,}08$
2
4
6
8
10
12
14
16
18
20
22
24
350
300
250
200
150
100
50
25
0

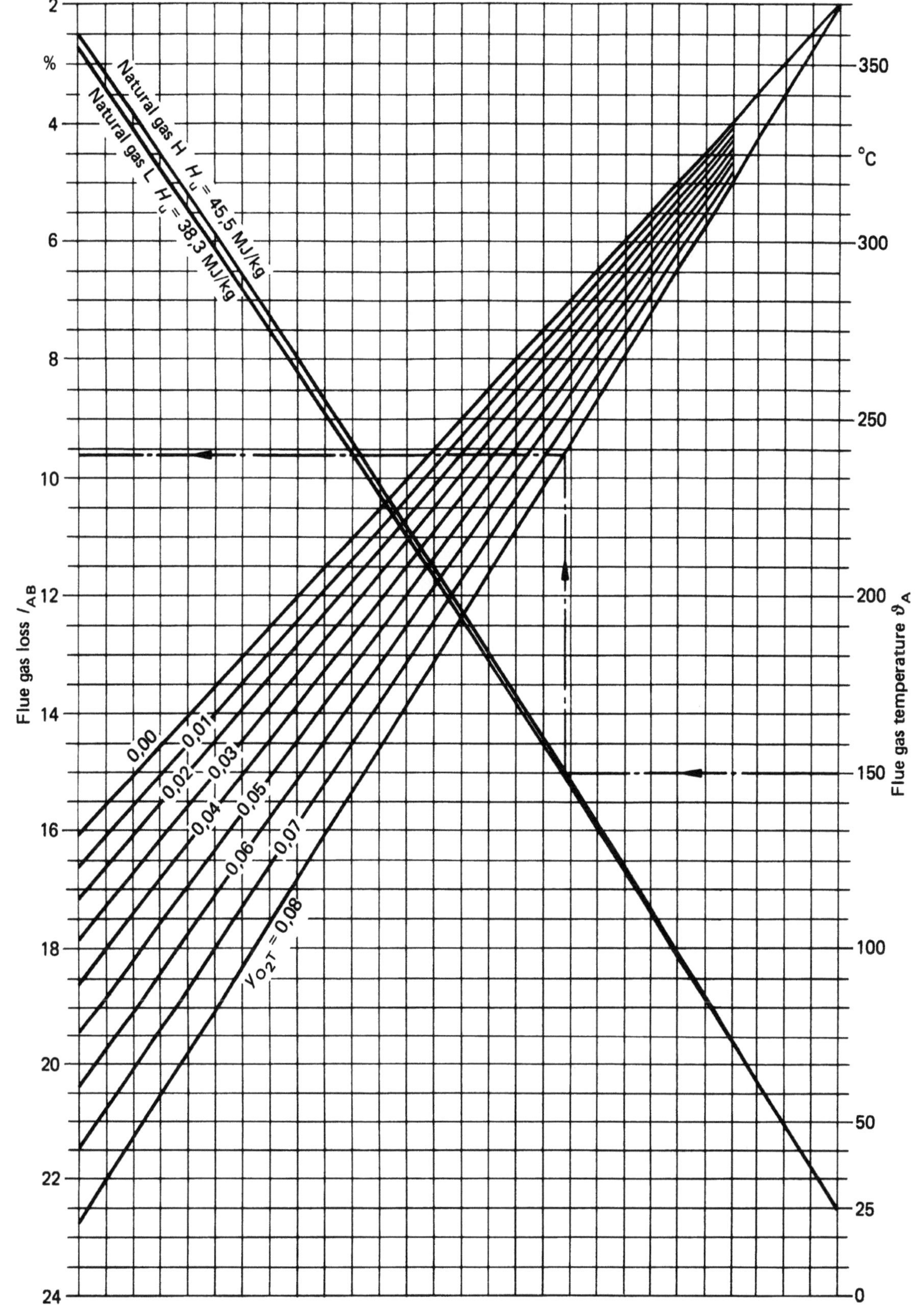

F. Brandt

Explanation

With worksheet No 5.4.3, one can determine the flue gas loss l_{AB} of natural gas dependent on the flue gas temperature ϑ_A, the net calorific value H_u and the O_2-content of the dry flue gas y_{O_2T}.

Definitions:

See worksheet No 5.4.1, where the CO_2-content has to be replaced by the equivalent O_2-content.

Formulae for calculation:

l_{AB} see worksheet No 5.4.1 $\qquad\qquad$ (1) to (6)

$$\mu_{GB} = (0.93697 + 0.34503\,H_u) + (0.84013 + 0.29142\,H_u)\frac{y_{O_2T}}{0.21 - y_{O_2T}} \qquad (7)$$

$$\mu_{LT} = \mu_{GB} - 1 \qquad (8)$$

$$\mu_{H_2OB} = -0.07793 + 0.0454\,H_u \qquad (9)$$

$$\mu_{O_2} = 0.55159 + 0.04463\,H_u \qquad (10)$$

$$\mu_G = \mu_{GB} + \mu_{LT}\,x_{H_2OL} \qquad (11)$$

$$\mu_{H_2O} = \mu_{H_2OB} + \mu_{LT}\,x_{H_2OL} \qquad (12)$$

Example natural gas H:

Known values: $\quad\vartheta_A \quad = 150\,°C;\ H_u = 45.5\ \text{MJ/kg};\ y_{O_2T} = 0.08$
$\qquad\qquad\qquad\ \ \mu_{O_2} \quad = 2.582\ \text{kg/kg}$
$\qquad\qquad\qquad\ \ \mu_{H_2OB} = 1.988\ \text{kg/kg};\ \mu_{H_2O} = 2.139\ \text{kg/kg}$
$\qquad\qquad\qquad\ \ \mu_{LT} \quad = 24.31\ \text{kg/kg from sheet No 4.3.3}$
$\qquad\qquad\qquad\ \ \mu_{GB} \quad = 25.31\ \text{kg/kg from sheet No 4.3.3}$
$\qquad\qquad\qquad\ \ \mu_G \quad = 25.461\ \text{kg/kg from sheet No 4.3.3}$
$\qquad\qquad\qquad\ \ h_{LT}, h_1, h_2$ see worksheet No 5.4.1

Result: $\qquad\qquad l_{AB} = \dfrac{3\,437}{45\,500} = 0.0755$

Loss due to incomplete combustion l_{CO}
%
H_u = 12 MJ/kg
14
16
20
24
28
36
y_{CO_2T}
CO_2-content of the dry flue gas y_{CO_2T}
y_{COT} = 0,00250
0,00225
0,00200
0,00175
0,00150
0,00125
0,00100
0,00075
0,00050
0,00025
2,0
1,8
1,6
1,4
1,2
1,0
0,8
0,6
0,4
0,2
0
0,20
0,19
0,18
0,17
0,16
0,15
0,14
0,13
0,12
0,11
0,10

F. Brandt

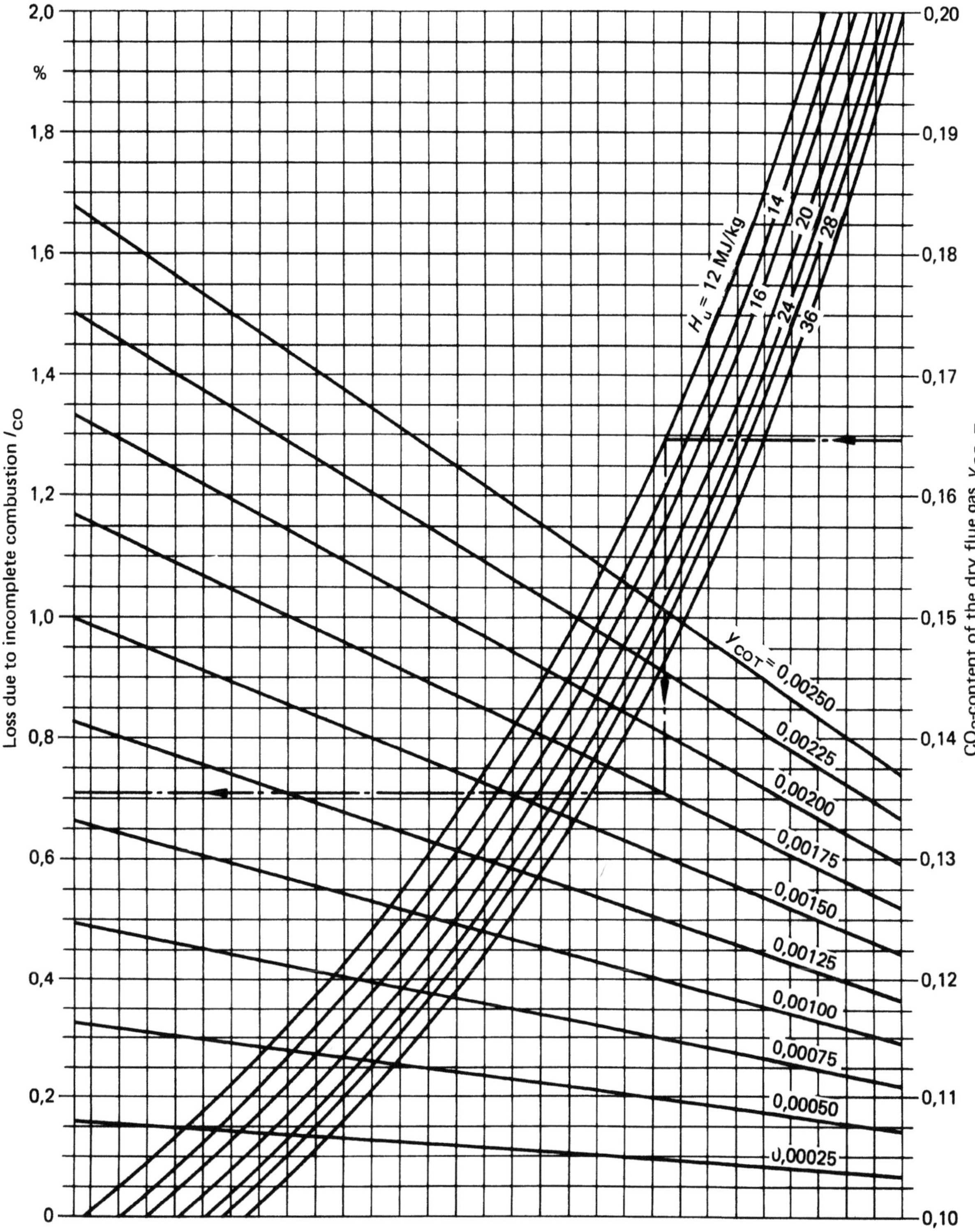

Explanation

With worksheet No 5.5.1, one can determine the loss due to incomplete combustion l_{CO} of solid fuels dependent on the CO_2-content of the dry flue gas y_{CO_2T}, the net calorific value H_u and the CO-content of the dry flue gas y_{COT}.

Definitions:

l_{CO} in %	Loss due to incomplete combustion
V_{GT} in m^3/kg	Specific dry flue gas volume
V_{CO_2} in m^3/kg	Specific CO_2 volume
y_{CO_2T} in %	Measured carbon dioxide content
y_{COT} in %	Measured carbon monoxide content
μ_{CO_2} in kg/kg	Specific CO_2-content of flue gas from sheet No 5.4.1
ϱ_{nCO_2} in kg/m^3	$= 1.977\ kg/m^3$; density of carbon monoxide at standard conditions
H_u in MJ/kg	Net calorific value
H_{unCO} in MJ/m^3	$= 12.633\ MJ/m^3$; net calorific value of carbon monoxide at standard conditions

Formulae for calculation:

$$V_{GT} = \frac{V_{CO_2}}{y_{CO_2T}} = \frac{\mu_{CO_2}}{y_{CO_2T}\,\varrho_{nCO_2}} \tag{1}$$

$$l_{CO} = \frac{1}{H_u}\left(V_{GT}\,y_{COT}\,H_{unCO}\right) = \frac{1}{H_u}\left(\frac{\mu_{CO_2}}{y_{CO_2T}\,\varrho_{nCO_2}}\,y_{COT}\,H_{unCO}\right) \tag{2}$$

Example:

Known values: $y_{COT} = 0.00175$; $y_{CO_2T} = 0.165$; $H_u = 12.0$ MJ/kg

$\mu_{CO_2} = 1.2445$ kg/kg from sheet No 5.4.1

$V_{GT} = 3.815\ m^3/kg$

Result: $l_{CO} = \dfrac{1}{12.0}\,(3.815 \cdot 0.00175 \cdot 12.633) = 0.00703 = 0.703\%$

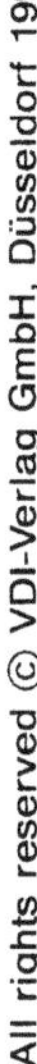

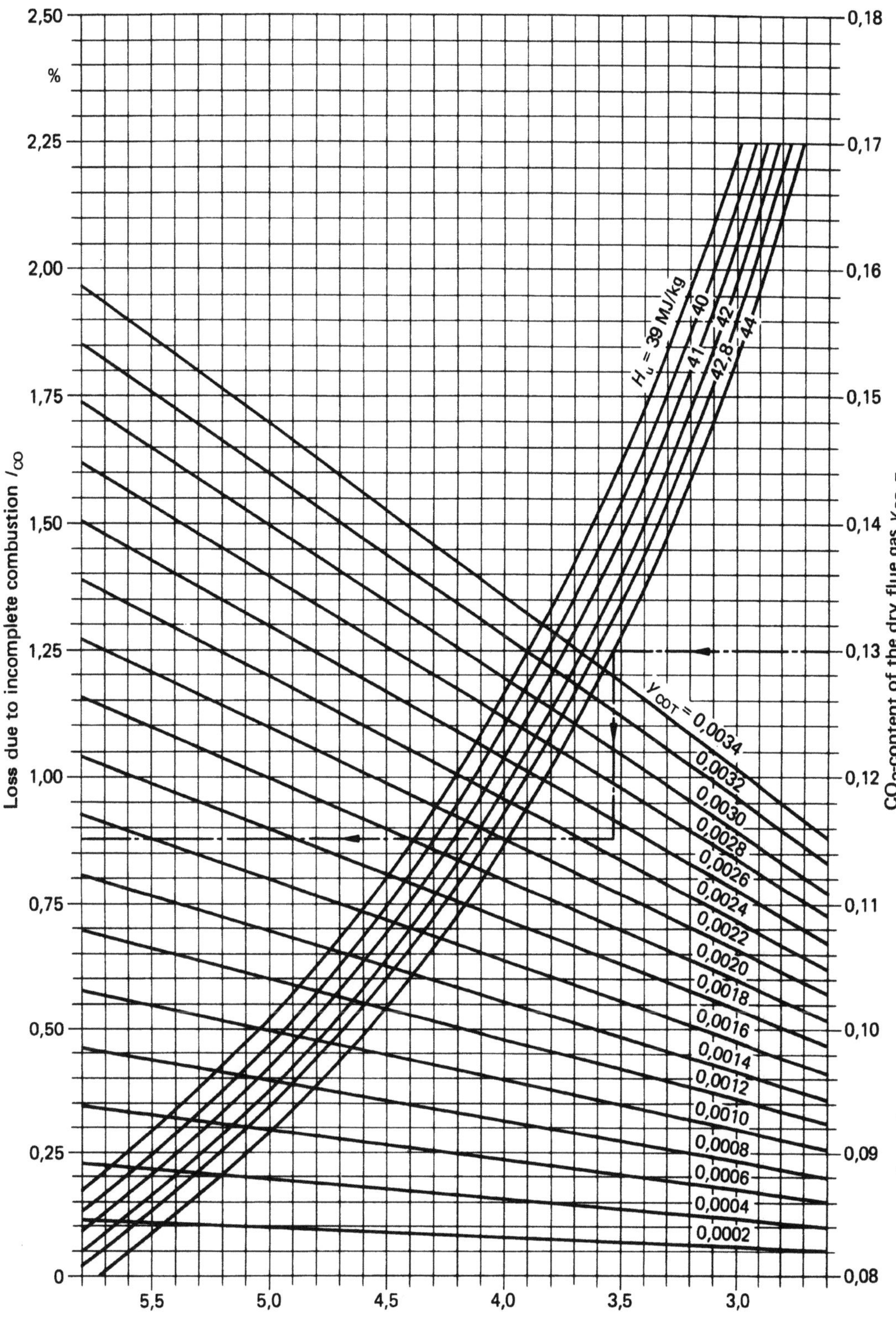

2,50
%
2,25
2,00
1,75
1,50
1,25
1,00
0,75
0,50
0,25
0
Loss due to incomplete combustion l_{CO}
0,18
0,17
0,16
0,15
0,14
0,13
0,12
0,11
0,10
0,09
0,08
CO_2-content of the dry flue gas y_{CO_2T}
5,5
5,0
4,5
4,0
3,5
3,0
H_u = 39 MJ/kg
40
41
42
42,8
44
y_{COT} = 0,0034
0,0032
0,0030
0,0028
0,0026
0,0024
0,0022
0,0020
0,0018
0,0016
0,0014
0,0012
0,0010
0,0008
0,0006
0,0004
0,0002

F. Brandt

Explanation

With worksheet No 5.5.2, one can determine the loss due to incomplete combustion l_{CO} of fuel oils dependent on the CO_2-content of the dry flue gas y_{CO_2T}, the net calorific value H_u and the CO-content of the dry flue gas y_{COT}.

Definitions:

l_{CO} in %	Loss due to incomplete combustion
V_{GT} in m³/kg	Specific dry flue gas volume
V_{CO_2} in m³/kg	Specific CO_2 volume
y_{CO_2T} in %	Measured carbon dioxide content
y_{COT} in %	Measured carbon monoxide content
μ_{CO_2} in kg/kg	Specific CO_2-content of flue gas from sheet No 5.4.1
ϱ_{nCO_2} in kg/m³	= 1.977 kg/m³; density of carbon monoxide at standard conditions
H_u in MJ/kg	Net calorific value
H_{unCO} in MJ/m³	= 12.633 MJ/m³; net calorific value of carbon monoxide at standard conditions

Formulae for calculation:

$$V_{GT} = \frac{V_{CO_2}}{y_{CO_2T}} = \frac{\mu_{CO_2}}{y_{CO_2T}\, \varrho_{nCO_2}} \tag{1}$$

$$l_{CO} = \frac{1}{H_u}\left(V_{GT}\, y_{COT}\, H_{unCO}\right) = \frac{1}{H_u}\left(\frac{\mu_{CO_2}}{y_{CO_2T}\, \varrho_{nCO_2}}\, y_{COT}\, H_{unCO}\right) \tag{2}$$

Example:

Known values: $y_{COT} = 0.0025$; $y_{CO_2T} = 0.13$; $H_u = 44.0$ MJ/kg

$\mu_{CO_2} = 3.160$ kg/kg from sheet No 5.4.2

$V_{GT} = 12.296$ m³/kg

Result: $l_{CO} = \dfrac{1}{44}\,(12.296 \cdot 0.0025 \cdot 12.633) = 0.00883 = 0.883\%$

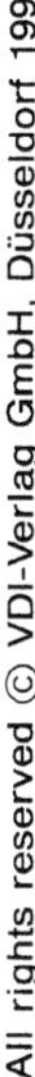

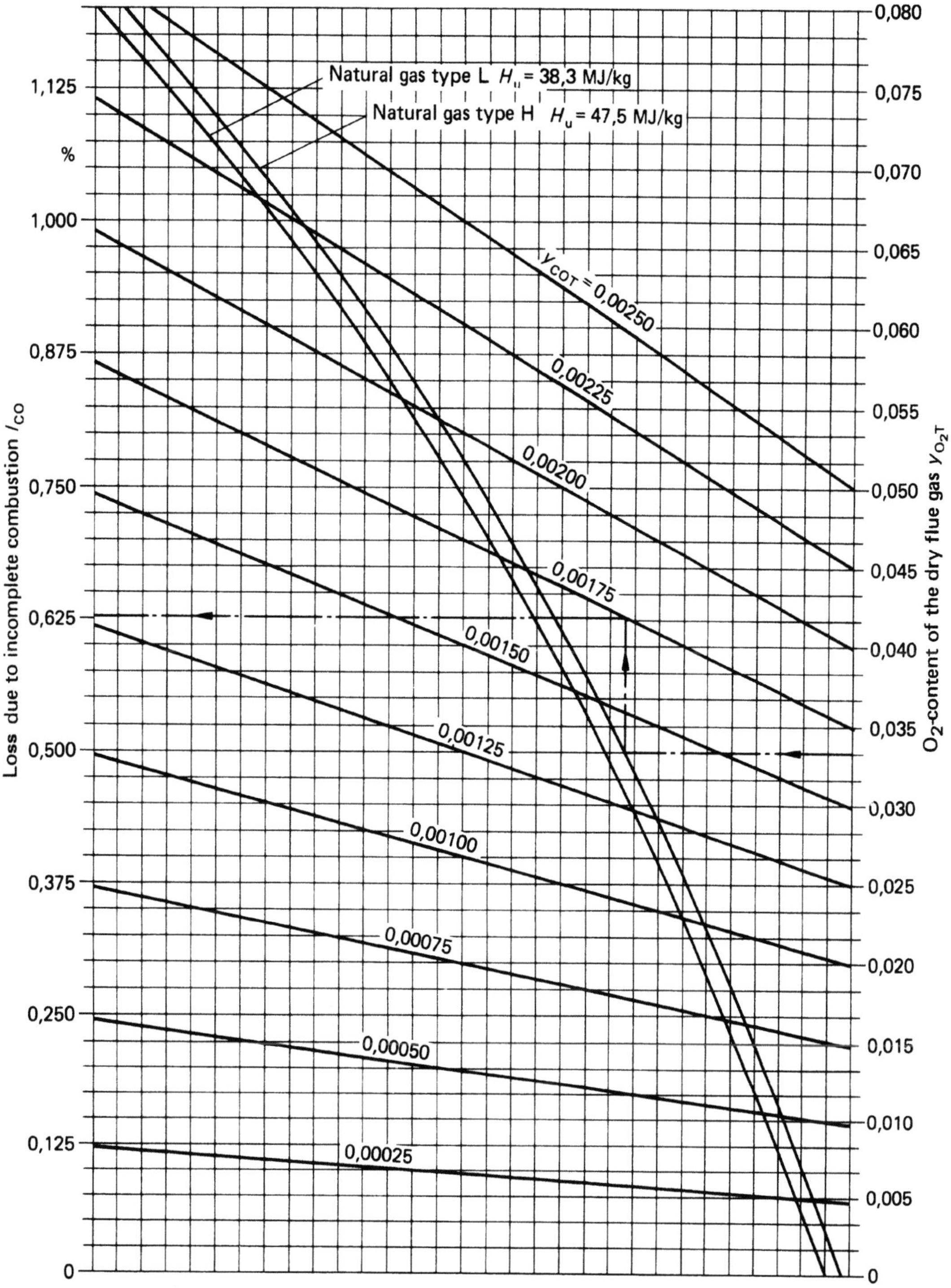

Natural gas type L H_u = 38,3 MJ/kg
Natural gas type H H_u = 47,5 MJ/kg
y_{COT} = 0,00250
0,00225
0,00200
0,00175
0,00150
0,00125
0,00100
0,00075
0,00050
0,00025
Loss due to incomplete combustion l_{CO}
%
1,125
1,000
0,875
0,750
0,625
0,500
0,375
0,250
0,125
0
O_2-content of the dry flue gas y_{O_2T}
0,080
0,075
0,070
0,065
0,060
0,055
0,050
0,045
0,040
0,035
0,030
0,025
0,020
0,015
0,010
0,005
0

F. Brandt

Explanation

With worksheet No 5.5.3, one can determine the loss due to incomplete combustion l_{CO} of natural gas dependent on the O_2-content of the dry flue gas y_{O_2T}, the net calorific value H_u and the CO-content of the dry flue gas y_{COT}.

Definitions:

l_{CO} in %	Loss due to incomplete combustion
V_{GT} in m³/kg	Specific dry flue gas volume
V_{GoT} in m³/kg	Specific dry flue gas volume at stoichiometric combustion
y_{O_2T} in %	Measured oxygen content
y_{COT} in %	Measured carbon monoxide content
H_u in MJ/kg	Net calorific value
H_{unCO} in MJ/m³	= 12.633 MJ/m³; net calorific value of carbon monoxide at standard conditions

Formulae for calculation:

$$V_{GT} = V_{GoT} = \frac{0.21}{0.21 - y_{O_2T}} \tag{1}$$

$$l_{CO} = \frac{1}{H_u}(V_{GT}\, y_{COT}\, H_{unCO}) = \frac{1}{H_u}\left(V_{GoT}\, \frac{0.21}{0.21 - y_{O_2T}}\, y_{COT}\, H_{unCO}\right) \tag{2}$$

$$V_{GoT} = 0.64975 + 0.22538\, H_u \tag{3}$$

Example:

Known values: $y_{COT} = 0.00175;\ y_{O_2T} = 0.0335;\ H_u = 47.5$ MJ/kg

$V_{GoT} = 11.355$ m³/kg

$V_{GT} = 13.511$ m³/kg

Result: $l_{CO} = \dfrac{1}{47.5}(13.511 \cdot 0.00175 \cdot 12.633) = 0.0063 = 0.63\%$

Partial load factor $\varphi = 0,1$ (10%)
0,2 (20%)
0,3 (30%)
0,4 (40%)
0,5 (50%)
0,6 (60%)
0,8 (80%)
1,0 (100%)
Brown coal, blast furnace gas
Hard coal
Fuel oil, natural gas
Nominal steam generator output $\dot{Q}_N$
MJ/s
1000
800
500
400
300
200
100
80
50
40
30
20
10
8
6
4
2
1
Radiation loss at partial load $l_{st\,\varphi}$
%
10
8
7
6
5
4
3
2
1
0,9
0,8
0,7
0,6
0,5
0,4
0,3
0,2
0,15
Radiation loss at nominal load l_{st}
0,15 0,2 0,3 0,4 0,5 0,6 0,7 0,8 1,0 1,5 % 2,0

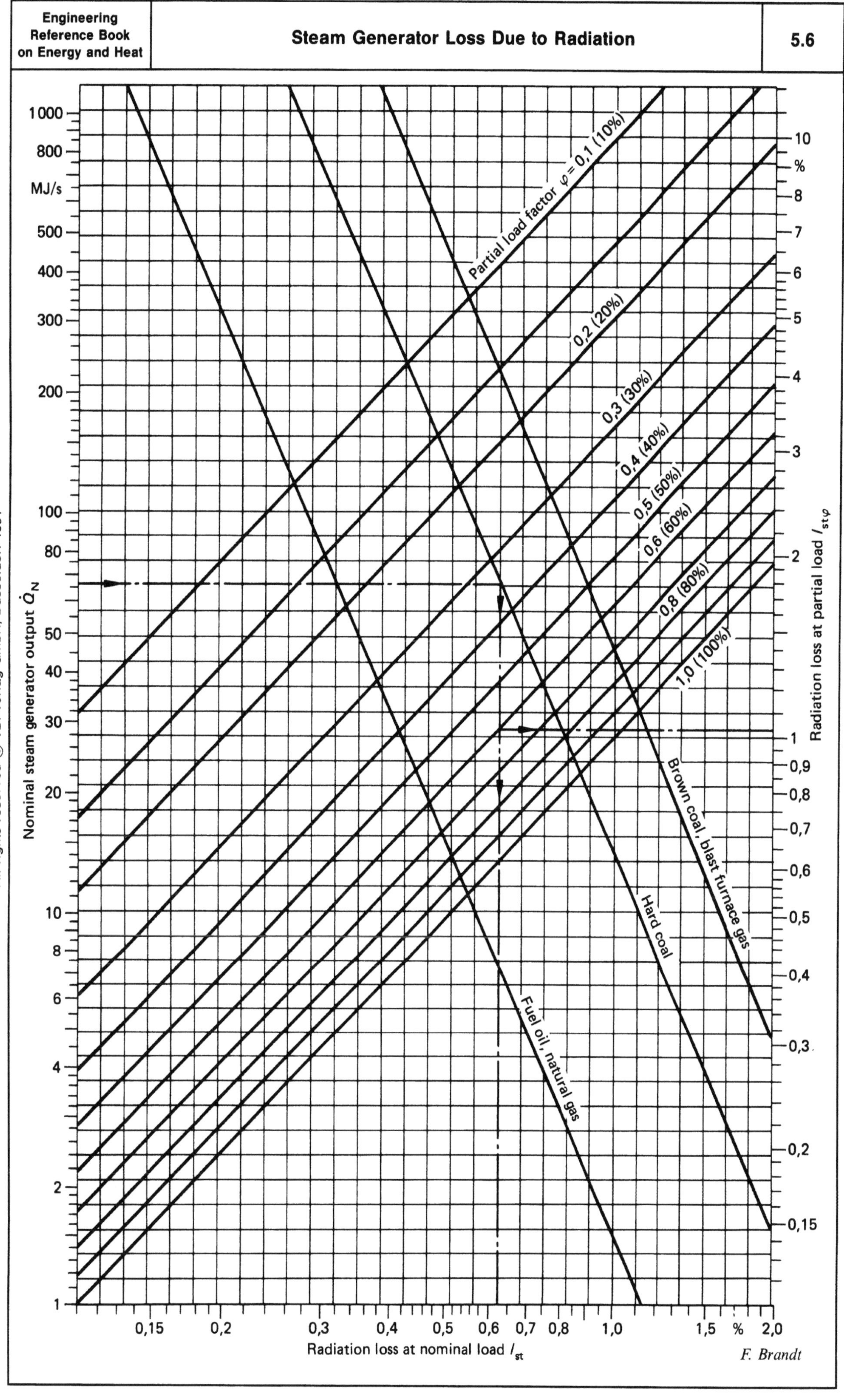

F. Brandt

Explanation

With worksheet No 5.6, one can determine the loss due to radiation of steam generators at nominal load l_{St} and partial load $l_{St\varphi}$ dependent on the nominal steam generator output $\dot{Q}_N$, the type of fuel and the partial load factor φ.

Definitions:

l_{St} in %	Radiation loss at nominal load
$l_{St\varphi}$ in %	Radiation loss at partial load
C in m²/(MW)$^{0.7}$	Surface radiation factor
q in MW/m²	Average heat flux density
$\dot{Q}_N$ in MW	Nominal steam generator output
$\dot{Q}_{N\varphi}$ in MW	Partial steam generator output
φ in %	Partial load factor

Table:

Type of fuel	Surface radiation factor C in m²/(MW)$^{0.7}$	Average heat flux density q in MW/m²
Brown coal, Blast furnace gas	77	405
Hard coal	56	392
Fuel oil, natural gas	33	347

Formulae for calculation:

$$l_{St} = C\,q\,\dot{Q}_N^{-0.3}\,10^{-6} \tag{1}$$

$$l_{St\varphi} = l_{St}/\varphi \tag{2}$$

$$\varphi = \dot{Q}_{N\varphi}/\dot{Q}_N \tag{3}$$

Example at nominal load:

Known values: $\dot{Q}_N$ = 66.5 MW, hard coal

Result: l_{St} = 0.0062 = 0.62%

Example at partial load:

Known values: $\dot{Q}_{N\varphi}$ = 40.0 MW, hard coal, φ = 0.60

Result: $l_{St\varphi}$ = 0.0103 = 1.03%

Bibliography

DIN 1942 „Abnahmeversuche an Dampferzeugern". Edition 1975

°C
0
200
400
600
800
1000
Flue gas temperature ϑ
$x_{H_2O} = 0.3$
0.25
0,20
0.15
0.10
0,05
0
kJ/(kgK)
1,7
1,6
1,5
1,4
1,3
1,2
1,1
1,0
0,9
$x_{CO_2} = 0.2$
0,1
0
Integral specific heat of flue gas $\overline{c}_{pG}$

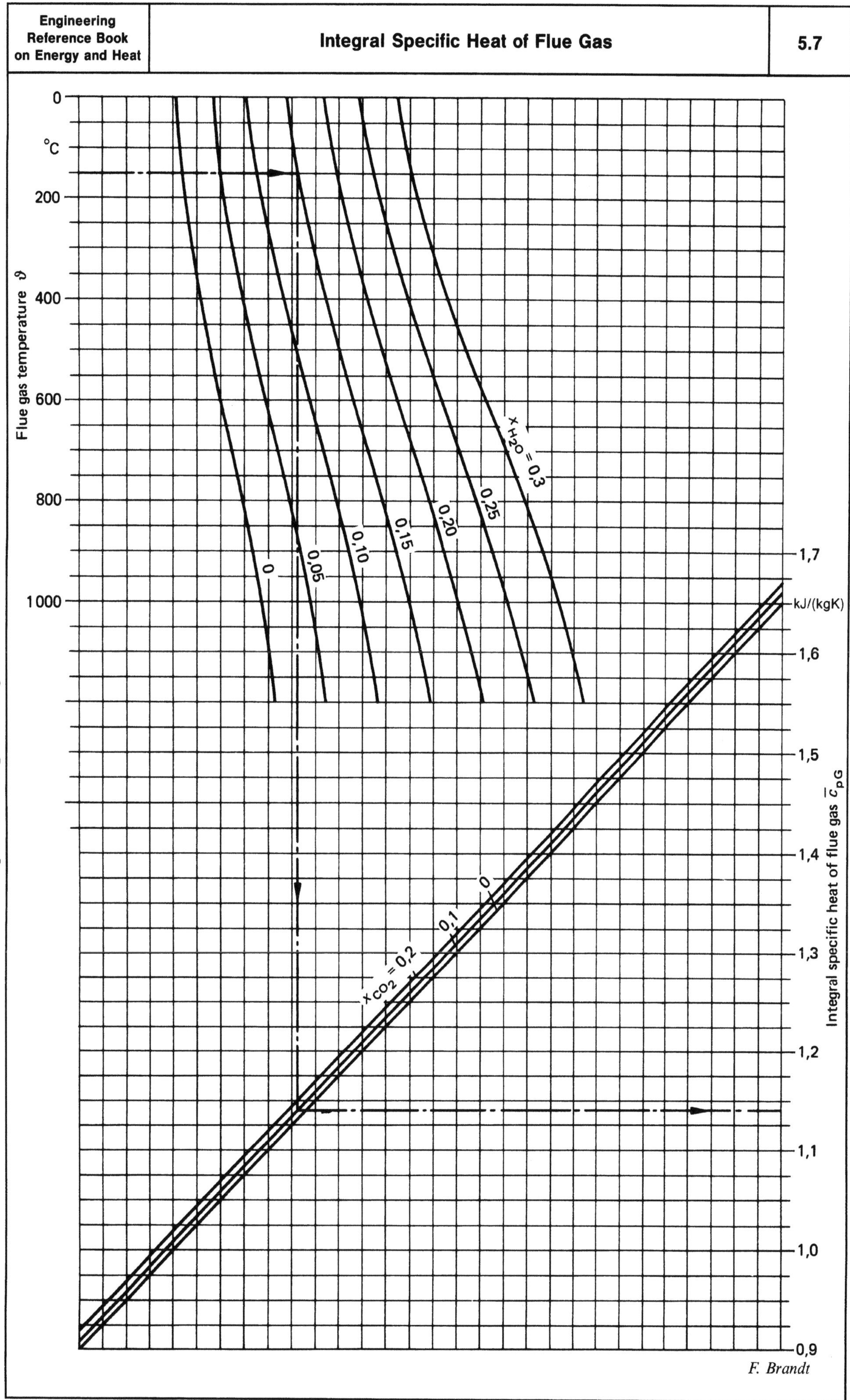

F. Brandt

Explanation

With worksheet No 5.7, one can determine the integral specific heat of the flue gas $\bar{c}_{pG}$ dependent on the flue gas temperature ϑ, the H_2O-content x_{H_2O} and the CO_2-content x_{CO_2} of the humid flue gas.

Definitions:

$\bar{c}_{pG}$ in kJ/(kgK)	Integral specific heat of the flue gas
$\bar{c}_{pLT}$ in kJ/(kgK)	Integral specific heat of dry air
$P_{1,2}$ in kJ/kg	Polynomials (see table)
x_{H_2O} in kg/kg	H_2O-content of humid flue gas
x_{CO_2} in kg/kg	CO_2-content of humid flue gas
μ_{H_2O} in kg/kg	Specific steam content of the flue gas
μ_G in kg/kg	Specific flue gas mass from sheet No 4.3.1 to 4.3.3
μ_{CO_2} in kg/kg	Specific CO_2-content of flue gas
h in kJ/kg	Specific enthalpy of flue gas
ϑ in °C	Temperature of flue gas
J in kJ/kg	Related enthalpy of flue gas

Formulae for calculation:

$$\bar{c}_{pG} = \bar{c}_{pLT} + P_1\, x_{H_2O} + P_2\, x_{CO_2} \tag{1}$$

$$x_{H_2O} = \frac{\mu_{H_2O}}{\mu_G} \tag{2}$$

$$x_{CO_2} = \frac{\mu_{CO_2}}{\mu_G} \tag{3}$$

$$h = c_{pG}\,\vartheta \tag{4}$$

$$J = \mu_G\, \bar{c}_{pG}\, \vartheta$$
$$= (\bar{c}_{pLT}\, \mu_G + P_1\, \mu_{H_2O} + P_2\, \mu_{CO_2})\, \vartheta \tag{5}$$

$$\bar{c}_{pLT} = a + \frac{b}{2}\vartheta + \frac{c}{3}\vartheta^2 + \frac{d}{4}\vartheta^3 + \frac{e}{5}\vartheta^4 + \frac{f}{6}\vartheta^5 \tag{6}$$

$$P_1 = a_1 + \frac{b_1}{2}\vartheta + \frac{c_1}{3}\vartheta^2 + \frac{d_1}{4} + \frac{e_1}{5}\vartheta^4 \tag{7}$$

$$P_2 = a_2 + \frac{b_2}{2}\vartheta + \frac{c_2}{3}\vartheta^2 + \frac{d_2}{4}\vartheta^3 + \frac{e_2}{5}\vartheta^4 \tag{8}$$

Table:

	$\bar{c}_{pLT}$		P_1		P_2
a	0.1004173 E + 01	a_1	0.8554535 E 00	a_2	− 0.1002311 E 00
b	0.1919210 E − 04	b_1	0.2036005 E − 03	b_2	0.7661864 E − 03
c	0.5883438 E − 06	c_1	0.4583082 E − 06	c_2	− 0.9259622 E − 06
d	− 0.7011184 E − 09	d_1	− 0.2798080 E − 09	d_2	0.5293496 E − 09
e	0.3309525 E − 12	e_1	0.5634413 E − 13	e_2	− 0.1093573 E − 12
f	− 0.5673876 E − 16				

Example:

Known values: $\vartheta = 150\,°C$; $x_{H_2O} = 0.15$; $x_{CO_2} = 0.10$

$\bar{c}_{pLT} = 1.009$ kJ/(kgK)

$P_1 = 0.874$

$P_2 = -0.0493$

Result: $\bar{c}_{pG} = 1.135$ kgJ/(kgK)

An adequate operation of steam generators and turbines is possible only, if minimum qualitative requirements concerning feed water, boiler water and steam are fulfilled. These minimum qualitative requirements have been fixed in guidelines, which have been worked out by the responsible technical organisations and from time to time are revised according to the state of technical knowledge. For steam generators up to a permissible operating pressure of 68 bars, these guidelines are published by the "Vereinigung der Technischen Überwachungsvereine" (VdTÜV) in Western Germany. For steam generators with higher operating pressure these guidelines are published by the "Vereinigung der Großkraftwerksbetreiber" (VGB). Parts of these guidelines are published in this chapter according to the VdTÜV-guideline April, 1983 edition and the VGB-guideline October 1980 edition.

1. VdTÜV-Guidelines

Table 1: Feed water with dissolved salts for boilers with internal circulation (water tube- and shell boilers)

Permissible operating pressure	bar	≤ 1	$> 1 \leq 68$
General appearance	–	clear, colourless, free of undissolved materials	
pH value at 25 °C	–	> 9	> 9
Conductivity at 25 °C	μS/cm	values given for boiler water are decisive	
Sum of alkalines ($Ca^{2+} + Mg^{2+}$)	mmol/l	< 0.015	< 0.010
Oxygen (O_2)	mg/l	< 0.1	< 0.02
Carbon dioxide (CO_2) bound	mg/l	< 25	< 25
Total iron (Fe)	mg/l	–	< 0.03 [1]
Total copper (Cu)	mg/l	–	< 0.005 [1]
Silica (SiO_2)	mg/l	values given for boiler water are decisive	
$KMnO_4$ consumption	mg/l	< 10	< 10
Oil, grease	mg/l	< 3	< 1

[1] For shell boiler ≤ 22 bar: Fe < 0.05 mg/l, Cu < 0.01 mg/l.

Table 2: Boiler water made of feed water with dissolved salts

Permissible operating pressure	bar	≤ 1	$> 1 \leq 22$ [1]	$> 22 \leq 44$	$> 44 \leq 68$
General appearance	–	clear, colourness, free of undissolved materials			
pH value at 25 °C	–	10.5 to 12	10.5 to 12	10 to 11.8	10 to 11
Acid capacity to pH 8.2 ($K_{S8.2}$)	mmol	1 to 12	1 to 12	0.5 to 6	0.1 to 1
Conductivity at 25 °C	μS/cm	< 5 000	< 10 000	< 5 000	< 2 500
Silica (SiO_2)	mg/l	–	values given for boiler water are decisive depending on the pressure according to Fig. 1		< 10
Phosphate (PO_4) [2]	mg/l	10 to 20	10 to 20	5 to 15	5 to 15

[1] For steam generators with superheater in the pressure range $> 1 \leq 22$ bar the boiler water guidelines of the pressure range $> 22 \leq 44$ bar are valid.

[2] The addition of phosphate is recommended but not always obligatory.

H. E. Hömig

Explanation

Acid- and base capacity

The former used terms "p-alkalinity" and "m-alkalinity" resp. "minus p-alkalinity" and "minus m-alkalinity" have been replaced by the terms "acid capacity" ($K_{S8.2}$ and $K_{S4.3}$) resp. "base capacity" ($K_{B8.2}$ and $K_{B4.3}$). The units are mmol/l and mol/m³. For tests during boiler operation, instead of using the pH value, the change point of the dye-indicators phenolphthalein and methyl orange may be used. The actual and former units are connected in the following way: mmol/l $\triangleq$ mval/l.

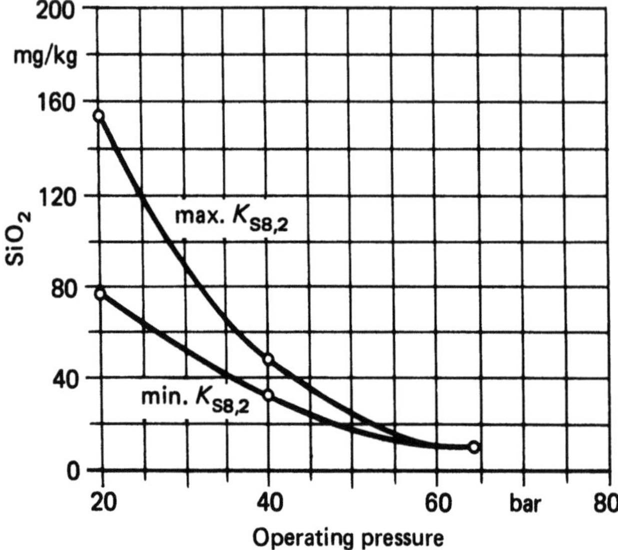

Figure 1: Permissible SiO$_2$ content vs. operating pressure and $K_{S8.2}$ value

Example:

Acid capacity up to a pH value 8.2 ($K_{S8.2}$):

$$K_{S8.2} = 3.5 \text{ mmol/l} \triangleq p\text{-alkalinity} = 3.5 \text{ mval/l}$$

With the acid- and base capacity, one can determine the concentrations of hydroxide, carbonate, hydrogen carbonate and carbon dioxide in the water.

Example:

Calculation of the hydrogen carbonate concentration in water with a pH value < 8.2:

$$K_{S4.3} (\text{mmol/l}) \cdot 61 = \text{mg HCO}_3/\text{l}$$

Raw water with $K_{S4.3} = 2.7$ mmol/l

$$\text{mg HCO}_3/\text{l} = K_{S4.3} \cdot 61 = 2.7 \cdot 61 \cong 165$$

Besides the pH value, with the "acid capacity up to a pH value of 8.2" ($K_{S8.2}$), there exists another way to express the alkalinity of water. A direct relation between pH and $K_{S8.2}$ value exists only for water with a low salt content.

Bibliography

VdTÜV-Richtlinien für Speisewasser, Kesselwasser und Dampf von Dampferzeugern bis 68 bar zulässigem Betriebsüberdruck. Edition April 1983. TÜV Rheinland, Köln

Table 3: Demineralized feed water at alkaline operation of boilers and water for spray attemperation

		Type of boiler		
		Unit shell and circulating boilers		Once-through boilers and spray water [1]
Permissible operating pressure	bar	≤ 1	$>1 \leq 68$	≤ 68
General appearance		clear, colourness, free of undissolved materials		
pH value at 25 °C	–	>9	>9	>9
Conductivity at 25 °C measured at local continuous-flow sampling point downstream of strong-acid cation exchanger	μS/cm	<0.2	<0.2	<0.2
Oxygen (O_2)	mg/l	<0.1	<0.1	no value
Total iron (Fe)	mg/l	–	<0.03	<0.02
Total copper (Cu)	mg/l	–	<0.005	<0.003
$KMnO_4$ consumption	mg/l	<10	<3	<3
Oil, grease	mg/l	<3	<1	n. n.
Silica (SiO_2)	mg/l	–	<0.02	<0.02

1) For once-through boilers, whose feed water has been treated with oxidising agents, the "VGB-Richtlinien für Kesselspeisewasser, Kesselwasser und Dampf von Wasserrohrkesseln ab 64 bar Betriebsüberdruck" are valid.

Table 4: Boiler water made of demineralized feed water

		With addition of solid and gaseous oxidising agents	With addition of gaseous alkalising agents *only*
Permissible operating pressure	bar	≤ 68	
General appearance		clear, colourless, free of undissolved materials	
pH value at 25 °C	–	9.5 to 10.5 [1]	>7
Conductivity at 25 °C measured at local continuous-flow sampling point downstream of strong-acid cation exchanger	μS/cm	<150	<3
Without strong-acid cation exchanger	μS/cm	< 50	–
Phosphate (PO_4)	mg/l	< 6	–
Silica (SiO_2)	mg/l	< 4	<4

1) According to TRD 611, for shell boilers instead of the solid alkalisation agents sodium or potassium hydroxide tri sodium phosphate is recommended.

H. E. Hömig

Table 1: Requirements on demineralized feed water for once-through and circulation boilers, measured at boiler entrance, as well as requirements for water spray attemperators in continuous operation

	Unit	Alkaline operation	Neutral operation
General appearance		clear and colourless	
Conductivity at 25 °C, measured at local continuous-flow sampling point	μS/cm	not specified	<0.25
Conductivity at 25 °C, measured at local continuous-flow sampling point down-stream of strong-acid cation exchanger	μS/cm	<0.20	<0.20
pH value at 25 °C (for once-through boilers and spray attemperators, only volatile alkalisation agents are admissible)		>9	>6.5 simultaneously keeping the conductivity limits
Oxygen (O_2)	mg/l	not specified (see explanation)	>0.050 (see explanation)
Total iron (Fe)	mg/l	<0.020	
Total copper (Cu)	mg/l	<0.003	
Silica (SiO_2)	mg/l	<0.020	
Organic substances		see explanation	

Table 2: Requirements on boiler water in circulation boilers in continuous operation, fed with feed water according to table 1

	Unit	Combined treatment with solid and volatile alkalising agents [1]	
Pressure stage	bar	≤125	>125
Permissible operating pressure	bar	≤136	>136
pH value at 25 °C [2]		9.5 to 10.5	in the range 9 to 10 preferably >9.5
Conductivity at 25 °C, measured at local continuous-flow sampling point down-stream of strong-acid cation exchanger	μS/cm	<150	< 50
With phosphate treatment: phosphate as PO_4^{3-}	mg/l	<6	
Silica (SiO_2)		depends on the pressure according to Fig. 1	

1) Two alternative modes of operation are possible, if the feed water is within the limits of table 1 and the conductivity of the boiler water is in the range of
   <5 μS/cm with local heat flux densities <250 kW/m²
   <3 μS/cm with local heat flux densities >250 kW/m²:
   a) *alkaline mode of operation* with only volatile alkalising agents
   b) *neutral mode of operation* with correction of the boiler water pH value at 25 °C by caustic soda to pH = 7 to 8.
2) The suggested pH values have to be achieved by caustic soda. On phosphate treatment, the additional use of caustic soda is only necessary, if the suggested pH value can not be achieved by the Na_3PO_4-agent.

H. E. Hömig

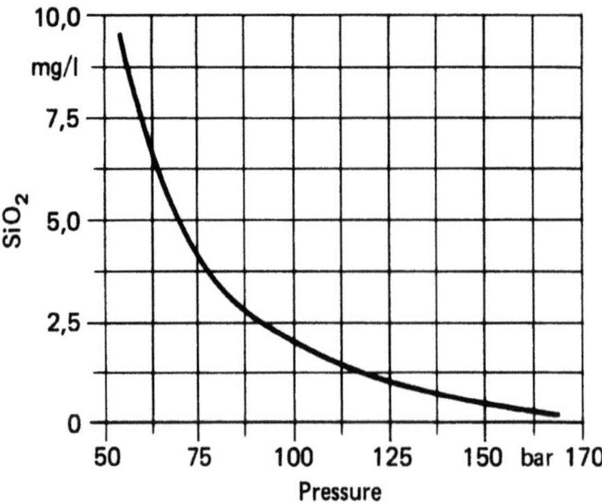

Figure 1: Maximum permissible silica concentration in boiler water (pH value ≤ 10) vs. pressure with consideration of the limits for the steam (0.020 mg/l SiO$_2$; table 3). At a pH value $= 11$ the data may be doubled.

Oxygen

Below the temperature, at which spontaneous magnetite occurs, oxygen favours the corrosion of ferrous material in electrolytical water. In this case, the velocity of corrosion is mainly influenced by the concentration of oxygen. This is the reason, why the concentration of oxygen in feed water with dissolved salts has to be closely limited (table 4). With demineralized feed water, on ferrous material, oxygen acts more or less as an inhibitor. By a neutral mode of operation, the feed water is treated by oxidising agents to achieve a minimum concentration of oxygen (table 1). From the point of view of corrosion control, even in alkaline operation, the former limits of an oxygen concentration > 0.02 mg/l have to be abolished.

Here, the definition of feed water with dissolved salts and demineralized feed water has to be set very closely. If the electrolyte content of the feed water has been risen over a period of several days because of leaks in the condenser, impurities of the condensate backflow or ionic leakage in the water treatment plant, the oxygen concentration according to table 4 is valid.

Organic substances

Organic substances, which enter the steam generator with the feed water, may – after accumulation – increase the inclination to foaming of the boiler water of circulation boilers, cause the incomplete separation of feed water droplets, and thus spoil the quality of steam indirectly. Decomposition products of organic substances, which occur during boiler operation, may influence the pH value of the boiler water and, if they are volatile, influence the quality of steam. In the case of demineralized feed water operation, this can be indicated by differences in the conductivity of feed water and steam, if the steam does not contain minerals.

In the last edition of this guideline (April 1972), limits for the use of potassium permanganate (< 5 mg/l) have been fixed. Limits for the oil content of the feed water (< 0.5 and < 0.3 mg/l), caused by impurities of the raw water or by small leaks, for example in the oil preheater, have been established. Due to the increasing pollution by organic substances of the surface water and the shore filtration water resp. the back flow of polluted condensate, the imposition of general limits is impossible. Especially in the case of back flow of polluted condensate, continuous supervision is necessary, because discontinuous unspecific analytical methods do not give sufficient information. Therefore, in each individual case, criteria for judgement on the possible influence of oil, organic substances resp. their decomposition products have to be worked out according to the specific operating experiences and their limits have to be fixed.

Bibliography

VGB-Richtlinien für Kesselspeisewasser, Kesselwasser und Dampf von Wasserrohrkesseln der Druckstufen ab 64 bar. Edition October 1980, VGB-Kraftwerkstechnik, No 10, 1980, p. 793/800

Table 3: Requirements on steam for turbines at continuous operation

	Unit	Alkaline operation	Neutral operation
Conductivity at 25 °C, measured at local continuous-flow sampling point	μS/cm	not specified	<0.025
Conductivity at 25 °C, measured at local continuous-flow sampling point downstream of strong-acid cation exchanger [1]	μS/cm	<0.20	<0.20
Silica (SiO_2)	mg/kg	<0.020	
Total iron (Fe)	mg/kg	<0.020	
Total copper (Cu)	mg/kg	<0.003	
Sodium (Na)	mg/kg	<0.010	

1) In the $\leq$ 80 bar operating pressure range, normally, it is not possible to achieve a steam conductivity <0.2 μS/cm, if the maximum allowable conductivity of boiler water according to table 5 has been adopted. In such cases, check, whether turbine operation will permit a higher conductivity of the steam. Failing this, it will be necessary, to reduce the conductivity of the boiler water compared to the data of table 5 (special arrangements with turbine- and boiler suppliers).

Table 4: Requirements on feed water with dissolved salts for circulation boilers operating at 64 and 80 bars, measured at feed water [1] inlet during continuous opeation

General appearance		Clear and Colourless
Conductivity	μS/cm	not specified; the values for boiler water according to table 5 have to be considered
pH value at 25 °C [2])		>9
Oxgen (O_2)	mg/l	<0.020
Total iron (Fe)	mg/l	<0.030
Total copper (Cu)	mg/l	<0.005
Silica (SiO_2)	mg/l	the values for boiler water according to table 5 have to be considered
Sum o alkalines ($Ca^{2+} + Mg^{2+}$)	mmol/l	<0.005
Organic substances	mg/l	see explanations

1) With local heat flux densities >250 kW/m^2, the use of demineralized feed water is recommended.
2) With regard to pH-value-adjusted see section "alcaline operation" and explanation in section "pH-value".

Table 5: Requirements on boiler water in circulation boilers operating at 64 and 80 bars during continuous operation, fed with feed water [1] containing dissolved salts according to table 4

Pressure stage	bar	64	80
Permissible operating pressure	bar	68	87
Conductivity at 25 °C, measured at local continuous-flow sampling point	μS/cm	<2 500	<300
Acid capacity $K_{s\,8.2}$ (former p-alkalinity in mval/l) [2]	mmol/l	0.01 to 1.0	<0.3
pH value at 25 °C		10 to 11	9.5 to 10.5
At phosphate treatment: Phosphate as (PO_4^{3-})	mg/l	5 to 15	2 to 6
Silica (SiO_2)	mg/l	depends on the pressure according to Fig. 1	

1) With local heat flux densities >250 kW/m^2, the use of demineralized feed water is recommended. The feed water and boiler water values according to table 1 and 2 have to be considered.
2) See DIN 38 409, part 7, May 1979 edition.

Explanation: see worksheet No 6.3

H. E. Hömig

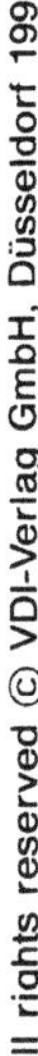

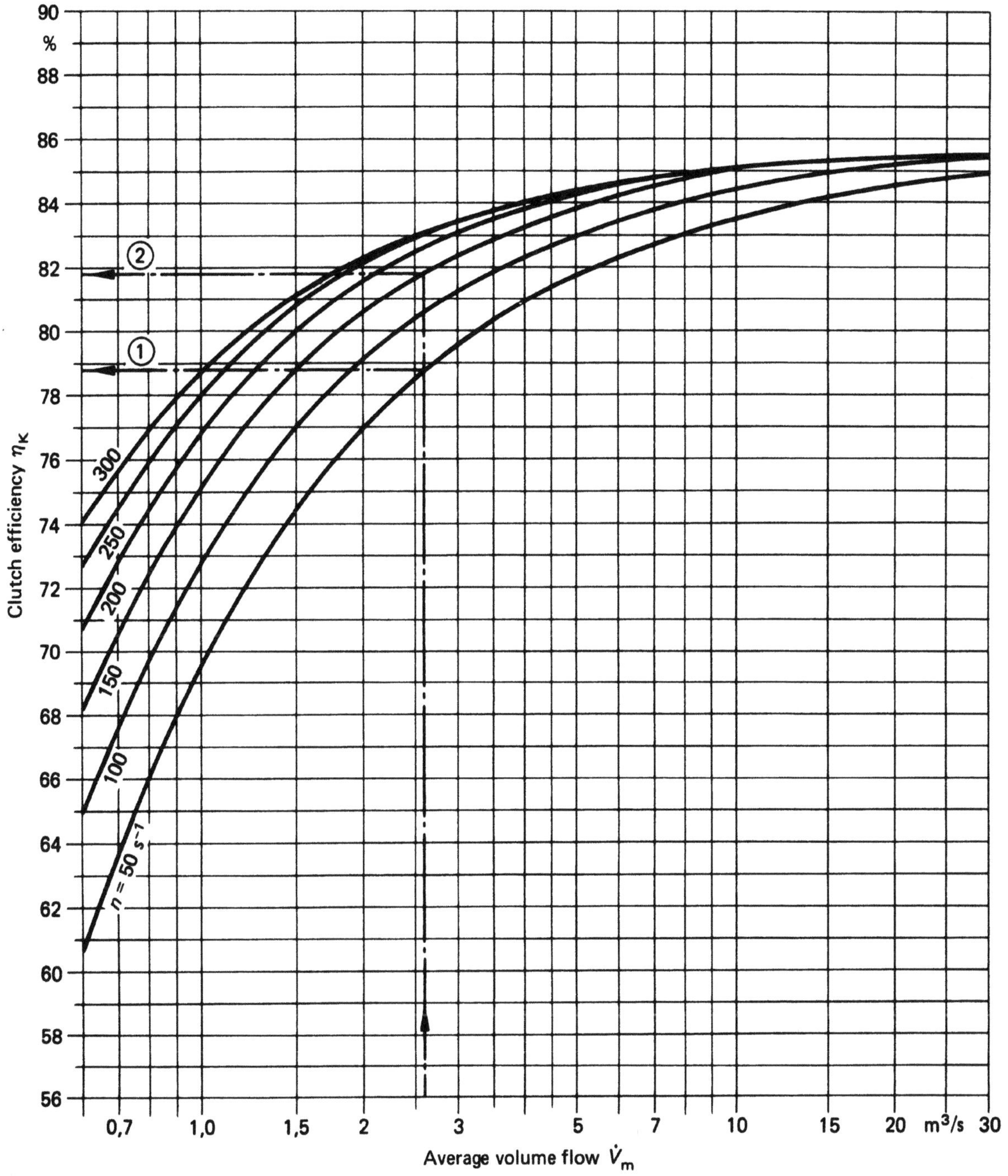
90
%
88
86
84
82
80
78
76
74
72
70
68
66
64
62
60
58
56
Clutch efficiency η_K
②
①
300
250
200
150
100
$n = 50\ s^{-1}$
0,7
1,0
1,5
2
3
5
7
10
15
20
m³/s
30
Average volume flow $\dot{V}_m$

A. Mellgren

Explanation

With worksheet No 7.1, one can determine the clutch efficiency of back pressure turbines dependent on the average volume flow $\dot{V}_m = \sqrt{\dot{V}_1 \cdot \dot{V}_2}$ and the number of revolutions.

Definitions:

$\dot{V}_m$ in m³/s	average volume flow
$\dot{V}_1$ in m³/s	volume flow at turbine entrance
$\dot{V}_2$ in m³/s	volume flow at turbine exit
n in 1/s	number of revolutions

Example:

Known values:

Volume flow at the entrance	$\dot{V}_1$	$= 0.90 \ \text{m}^3/\text{s}$
Volume flow at the exit	$\dot{V}_2$	$= 7.50 \ \text{m}^3/\text{s}$
Average volume flow	$\dot{V}_m$	$= 2.60 \ \text{m}^3/\text{s}$

Result:

		Line ①	Line ②	
Number of revolutions	n	50	150	s^{-1}
Clutch efficiency	η_K	78.1	81.8	%

A gear efficiency of about 98% has to be considered if necessary.

Explanation

With worksheet No 7.2.1, one can determine the specific low pressure exit energy or the exhaust steam data of condensing steam turbines without reheating dependent on the design data. Basic assumptions for the worksheet is a turbine efficiency of 82%, and, it is only valid for high speed turbines (rated speed 3600 1/min) and an output at the clutch up to 20 MW.

Definitions:

h in kJ/kg	specific enthalpy
k in kJ/kg	specific (exit) energy
$\dot{m}$ in kg/s	mass flow
p in bar	pressure
ϑ in °C	temperature
A in m²	area at the low pressure exit
P in MW	output
$\dot{V}$ in m³/s	volume flow
VW (−)	number of preheaters

Indexes:

mV	with preheater
oV	without preheater
s	isentropic
F	live steam
K	clutch
V	pre-heating
ω	exit of the low pressure turbine

Procedure of calculation:

1) Determination of the isentropic enthalpy difference Δh_s. At first, the drop between live steam conditions and a basic exhaust steam pressure has to be determined by the lines 1 and 2.

   At different exhaust steam pressure the additional drop Δh_2 has to be determined by the lines 3 and 4.

   Total enthalpy difference: $\Delta h_s = \Delta h_{s1} - \Delta h_{s2}$

   It is possible to determine Δh_s directly by a h, s-diagram.

2) The mass flow of live steam without preheater $\dot{m}_{FoV}$ has to be determined by the output at the clutch P_K and the isentropic enthalpy difference Δh_s (lines 5 and 6).

3) With the number of preheaters VW and the difference of temperature between the temperature at preheater exit and the temperature of the condensate $\Delta\vartheta_V$ the mass flow of live steam with pre-heating can be determined (line 7 and 8).

4) With line 9, one can determine the exhaust steam mass flow $\dot{m}_\omega$.

5) From the exhaust steam pressure p_ω the exhaust steam volume flow $\dot{V}_\omega$ may be determined (line 10).

6) The specific low pressure exit energy k_ω in dependence of the exit area A has to be determined by $\dot{V}_\omega$ (line 11).

Example:

Known values:

Output at the clutch	P_K	$= 12\,\text{MW}$
Live steam pressure	p_F	$= 40\,\text{bar}$
Live steam temperature	ϑ_F	$= 420\,°C$
Steam pressure at the exit	p	$= 0.18\,\text{bar}$
Number of preheaters	VW	$= 2$
Temperature at preheaters exit	ϑ_V	$= 152\,°C$
Temperature of condensate	ϑ_ω	$= 58\,°C$

Results:

1. $\Delta h_{s1}\,(p_F = 40,\ \vartheta_F = 420,\ p_\omega = 0.12) \quad = \quad 1\,080\,\text{kJ/kg}$
   $\Delta h_{s2}\,(p_\omega = 0.12 \rightarrow p_\omega = 0.18) \quad = \quad 50\,\text{kJ/kg}$
   $\Delta h_s = \Delta h_{s1} - \Delta h_{s2} \quad = \quad 1\,030\,\text{kJ/kg}$

2. $\dot{m}_{FoV}\,(P_K = 12,\ \Delta h_s = 1\,030) \quad = \quad 14.2\,\text{kg/s}$

3. $\dot{m}_{FmV}\,(\text{VW} = 2,\ \Delta\vartheta_V = 94) \quad = \quad 14.9\,\text{kg/s}$

4. $\dot{m}_w\,(\dot{m}_{FoV} = 14.2,\ \text{VW} = 2) \quad = \quad 12.7\,\text{kg/s}$

5. $\dot{V}_\omega\,(\dot{m}_\omega = 12.7,\ p_\omega = 0.18) \quad = \quad 95\,\text{m}^3/\text{s}$

6. $k_\omega\,(\dot{V}_\omega = 95,\ A = 0.40\,\text{m}^2) \quad = \quad 32\,\text{kJ/kg}$

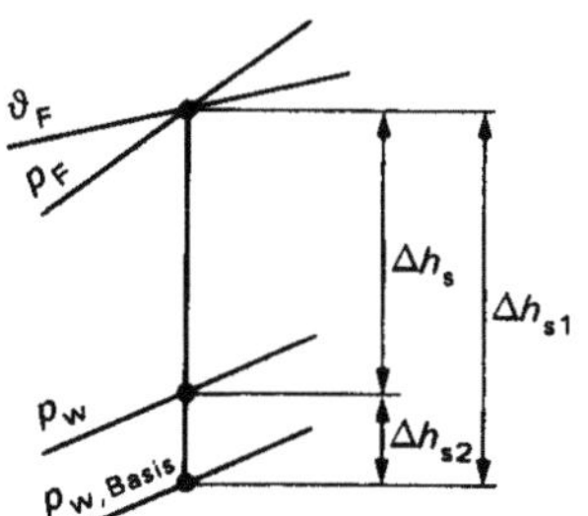

Explanation

With worksheet No 7.2.2, one can determine the specific low pressure exit energy or the exhaust steam data of condensing steam turbines without reheating dependent on the design data. Basic assumptions for the worksheet is a turbine efficiency of 85%, and, it is valid for normal speed turbines (rated speed 3 000 1/min up to 3 600 1/min) and an output at the clutch between 10 MW and 110 MW.

Further explanation see worksheet No 7.2.1.

Example:

Known values:

Output at the clutch	P_K	$= 40$ MW
Live steam pressure	p_F	$= 63$ bar
Live steam temperature	ϑ_F	$= 485\,°C$
Exhaust steam pressure	p	$= 0.10$ bar
Number of preheaters	VW	$= 4$
Temperature at preheaters exit	ϑ_V	$= 200\,°C$
Condensate temperature	ϑ_ω	$= 46\,°C$

Results:

1. $\Delta h_{s1}\,(p_F = 63,\ \vartheta_F = 485,\ p_\omega = 0.08)$ $=$ 1 240 kJ/kg
   $\Delta h_{s2}\,(p_\omega = 0.08 \rightarrow p_\omega = 0.10)$ $=$ 28 kJ/kg
   $\Delta h_s = \Delta h_{s1} - \Delta h_{s2}$ $=$ 1 212 kJ/kg

2. $\dot{m}_{FoV}\,(P_K = 40,\ \Delta h_s = 1\,212)$ $=$ 39 kg/s

3. $\dot{m}_{FmV}\,(VW = 4,\ \Delta\vartheta_V = 154)$ $=$ 44 kg/s

4. $\dot{m}_w\,(\dot{m}_{FoV} = 33,\ VW = 4)$ $=$ 33 kg/s

5. $\dot{V}_\omega\,(\dot{m}_\omega = 33,\ p_\omega = 0.10)$ $=$ 430 m³/s

6. $k_\omega\,(\dot{V}_\omega = 430,\ A = 2.00\ \mathrm{m}^2)$ $=$ 27 kJ/kg

Additional information of this book

(Engineering Reference Book on Energy and Heat; 978-3-642-51125-7;
978-3-642-51125-7_OSFO3) is provided:

http://Extras.Springer.com

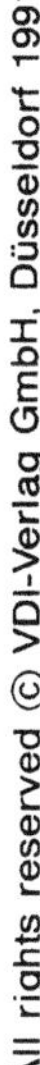

A. Mellgren, E. Koch

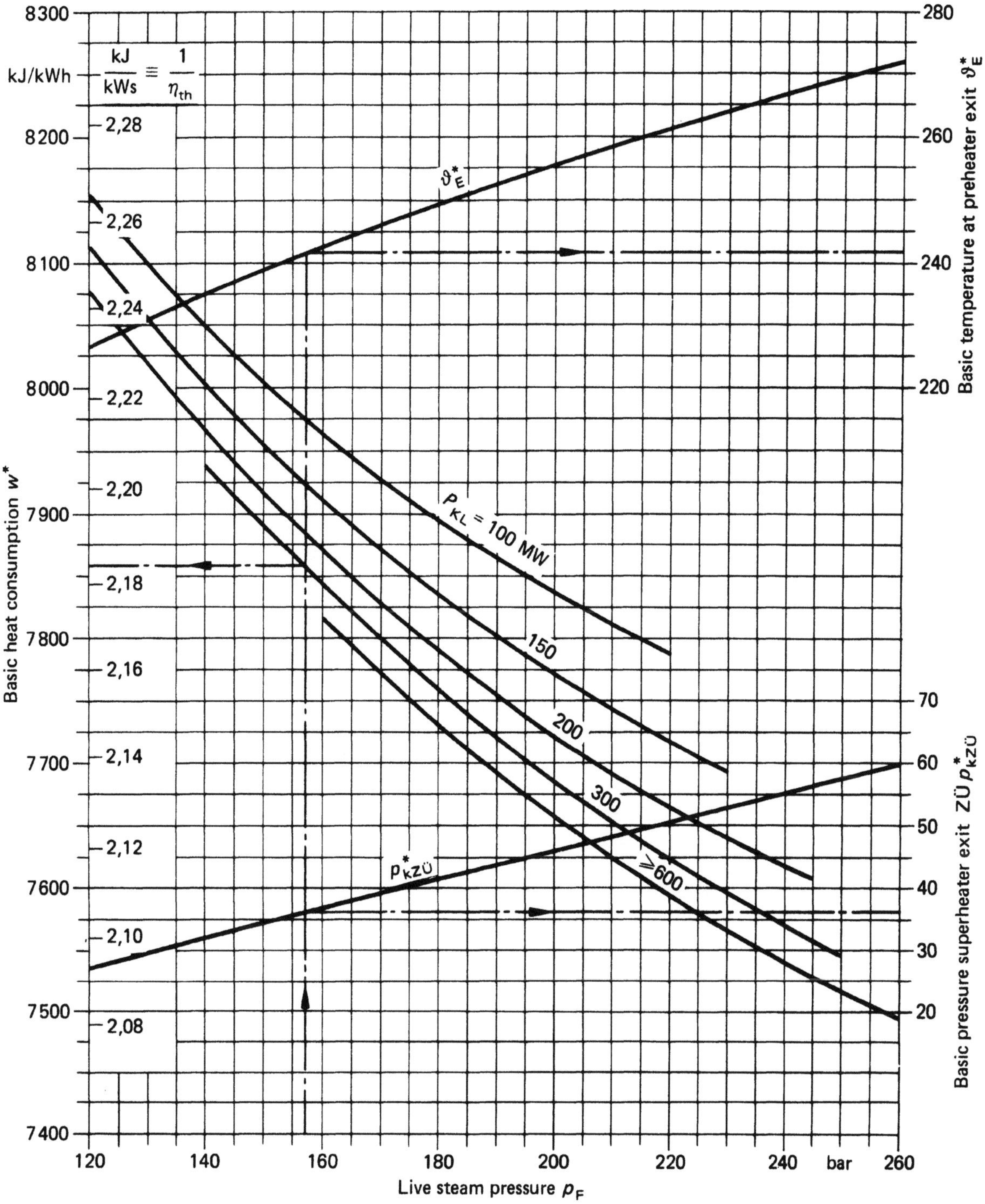

Explanation

With the worksheets No 7.3 to 7.5 it is possible, to estimate the heat consumption of condensing turbines with various design data and single reheating. Because of the great number of parameters, the diagrams have been calculated so, that in the first sheet the basic heat consumption can be determined for special design data. By means of correction curves in the following worksheets, this basic heat consumption can be converted to the desired design data.

Worksheet No 7.3 shows the basic heat consumption dependent on the live steam pressure and the design output.

A certain reheating pressure, and, with this a certain reheating temperature, are assigned to the live steam (last pre-heating stage).

The other parameters have been fixed for the following values:

Live steam pressure	ϑ_F	$= 540\,°C$
Superheater temperature	$\vartheta_{ZÜ}$	$= 540\,°C$
Condensing pressure	p_K	$= 0.035$ bar
Specific energy at the exit	k_ω	$= 21$ kJ/kg
Pressure loss in per cent	$\left(\dfrac{\Delta p}{p}\right)_{ZÜ}$	$= 10\%$
Generator efficiency	η_{Gen}	$= 98.8\%$
Number of preheater stages	z	$= 7$
Rated speed	n	$= 3\,000\ \text{min}^{-1}$

Worksheet No 7.4 shows in the lower diagram the influence of live steam temperature, superheater temperature and pressure loss in the superheater. The upper diagram shows the influence of the temperature at preheater exit and the number of preheater stages.

Worksheet No 7.5 shows the influence of condenser pressure and the energy at the exit. This energy depends on company specifications and is normally unknown. In case of a known curve for exit energy, depending on the volume flow of exhaust steam, mass flow of live steam can be determined by the curves "(specific) steam rate" in the lower diagram. Thus, with the curve "percentage of steam flow" in the upper diagram, the mass flow of condenser steam and therefore the volume flow of exhaust steam can be estimated.

Definitions:

η in %	efficiency
h in kJ/kg	specific enthalpy
k in kJ/kg	specific (exit) energy
$\dot{m}$ in kg/s	mass flow
n in l/min	rated speed
p in bar	pressure
ϑ in °C	temperature
w in kJ/kWh	heat consumption
$z\ (-)$	number of preheaters
K in %	factor
$\dot{M}$ in kg/kWs	mass flow (related to the output)
P in MW	output
$\dot{Q}$ in MW	heat flow

Indexes:

ab	discharge
g	total
kZÜ	at superheater entrance
n	mechanical
s	at the state of saturation
th	thermal
ZÜ	delivered
ω	at the exit
E	at preheater exit
F	live steam
Gen	generator
HD	high pressure turbine
KL	generator terminal
K	condenser
MD	medium pressure turbine
ND	low pressure turbine
P	pump
Spt	turbine for driving feed pumps
V	preheater
ZÜ	at superheater exit

Example: Determination of the heat consumption of a turbo set with superheater and the following design data:

Generator terminal output	P_{KL}	$= 300$ MW
Live steam pressure	p_F	$= 157$ bar
Live steam temperature	ϑ_F	$= 525\,°C$
Temperature at superheater exit	$\vartheta_{ZÜ}$	$= 535\,°C$
Number of preheaters	z	$= 6$
Temperature at preheater exit	ϑ_E	$= 235\,°C$
Pressure in the condenser	p_K	$= 0.059$ bar
Specific exit energy	k_ω	$= 37.5$ kJ/kg
Pressure loss in per cent	$\left(\dfrac{\Delta p}{p}\right)_{ZÜ}$	$= 8\%$
Efficiency of the generator	η_{Gen}	$= 98.4\%$

With worksheet No 7.3 one can calculate:

Basic heat consumption at	w^*	$= 7\,860$ kJ/kWh
Basic temperature at preheater exit	ϑ_E^*	$= 242\,°C$
Basic pressure superheater exit	$p_{kZÜ}^*$	$= 36.3$ bar

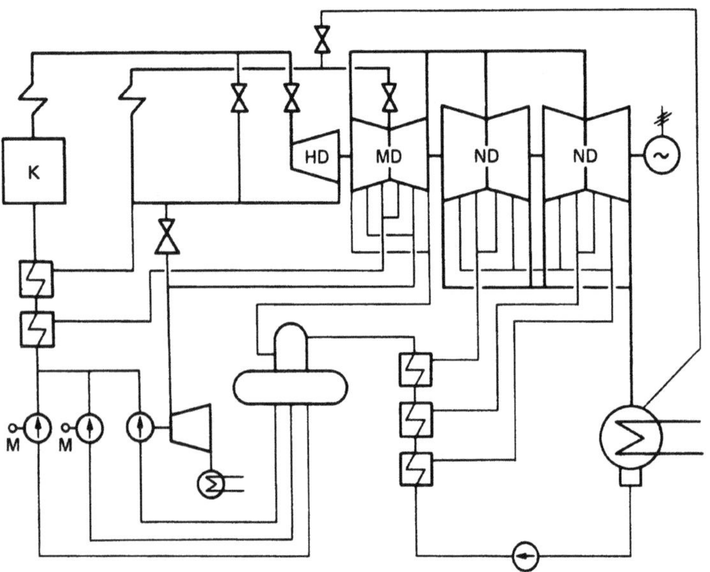

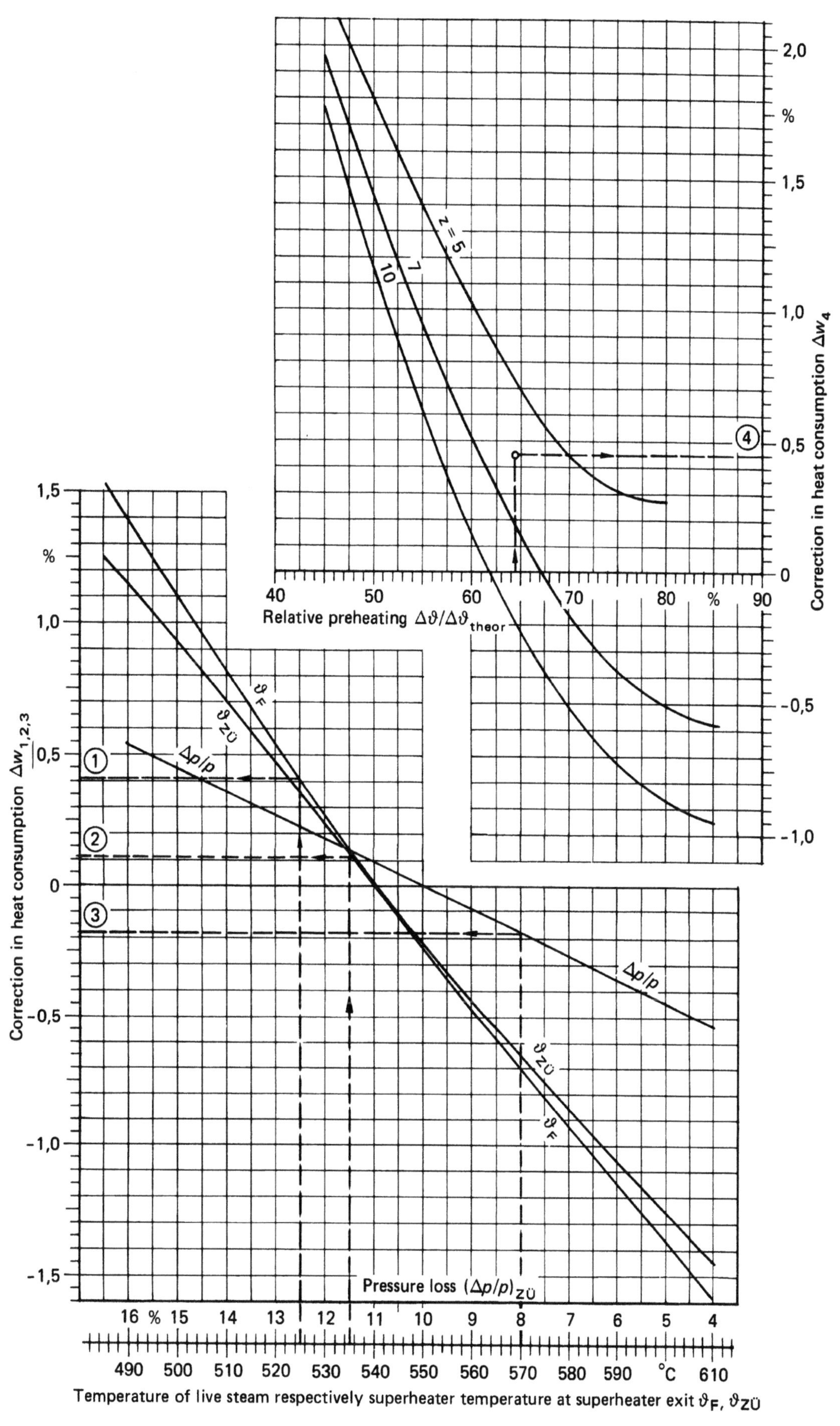

A. Mellgren, E. Koch

Explanation

Correction at changes of design data:

From sheet No 7.4, one can get the following corrected data:

1. Live steam pressure $\qquad \vartheta_F \quad = 525\,°C: \quad \Delta w_1 = +0.41\%$
2. Superheater temperature $\qquad \vartheta_{z\ddot{U}} \quad = 535\,°C: \quad \Delta w_2 = +0.11\%$
3. Pressure loss per cent content $\qquad \left(\dfrac{\Delta p}{p}\right)_{z\ddot{U}} = 8\%: \qquad \Delta w_3 = -0.18\%$

4. Temperature at preheater exit ϑ_E:

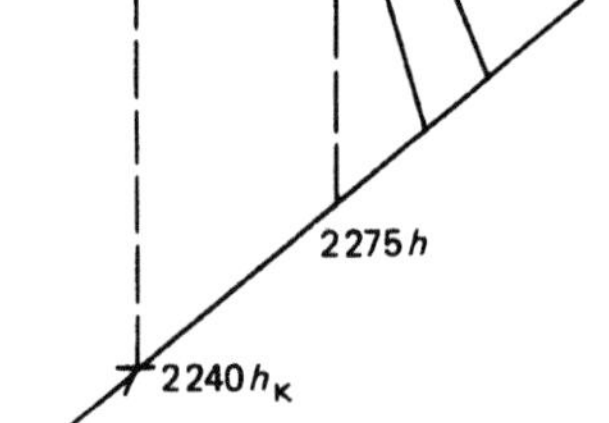

   Saturation temperature of live steam $\quad \vartheta_{sF}$ at 157 bar $= 345.8\,°C$

   Condensing temperature $\qquad\qquad \vartheta_K$ at 0.059 bar $= \underline{\ \ 35.9\,°C}$

$\qquad\qquad\qquad\qquad\qquad\qquad\qquad\quad \Delta\vartheta_{\text{theor.}} \qquad = 309.9\,°C$

   Temperature at preheater exit $\qquad\quad \vartheta_E \qquad\qquad = 235.0\,°C$

$\qquad\qquad\qquad\qquad\qquad\qquad\qquad\quad \vartheta_K$ at 0.059 bar $= \underline{\ \ 35.9\,°C}$

$\qquad\qquad\qquad\qquad\qquad\qquad\qquad\quad \Delta\vartheta_V \qquad\qquad = 199.1\,°C$

   Relative pre-heating $\qquad\qquad\qquad \Delta\vartheta_V/\Delta\vartheta_{\text{theor.}} \ = 0.642\ (64.2\%)$

   Number of preheater stages $\qquad\quad n \qquad\qquad\quad = 6$

$$\Delta w_4 = +0.45\%$$

From worksheet No 7.5:

5. Pressure in the condenser p_K:

   Determination of enthalpy rise during pre-heating

   Pressure of feed water $p_P = 1.25 \cdot p_F = 196$ bar
   (The factor 1.25 considers the pressure loss in the steam generator and the preheaters)

   at $p_P = 196$ bar and $\vartheta_E = 235\,°C \qquad h_E \qquad = 1\,017\ \text{kJ/kg}$

   at $p_K = 0.059$ bar $\qquad\qquad\qquad\quad h' \qquad = \underline{\ \ 150\ \text{kJ/kg}}$

   Pre-heating $\qquad\qquad\qquad\qquad\quad \Delta h_V \qquad = \ \ \ 867\ \text{kJ/kg}$

   Factor $\qquad\qquad\qquad\qquad\qquad\quad K \qquad\ = 77.7\%$

The isentropic drop of enthalpy can be determined by the h, s-diagram (sheet No 2.1.1)

   at $p_F = 157$ bar and $\vartheta_F = 525\,°C \qquad h_F \qquad = 3\,372\ \text{kJ/kg}$

   at $p_{kz\ddot{U}} = 36.3$ bar $\qquad\qquad\qquad\quad h_{kz\ddot{U}} \quad = \underline{2\,965\ \text{kJ/kg}}$

$\qquad\qquad\qquad\qquad\qquad\qquad\qquad\ \Delta h_{HD} \quad = \ \ \ 407\ \text{kJ/kg}$

   at $p_{z\ddot{U}} = 33.4$ bar and $\vartheta_{z\ddot{U}} = 535\,°C \quad h_{z\ddot{U}} \quad = 3\,531\ \text{kJ/kg}$

   $p_K = 0.059$ bar $\qquad\qquad\qquad\qquad\ h_K \qquad = \underline{2\,240\ \text{kJ/kg}}$

$\qquad\qquad\qquad\qquad\qquad\qquad\qquad\ \Delta h_{MD+ND} = 1\,291\ \text{kJ/kg}$

   Total drop of enthalpy $\Delta h_g \ = (407 + 1\,291)\,\text{kJ/kg} = 1\,698\ \text{kJ/kg}$

   at $p_K = 0.059$ bar versus $p_K^* = 0.035$ bar $\quad \Delta h_K \qquad = 64.5\ \text{kJ/kg}$

$$\Delta w_5 = \frac{K \cdot \Delta h_K}{\Delta h_g} = \frac{77.7 \cdot 64.5}{1\,698} = +2.95\%$$

6. Specific energy at the exit $k_\omega = 37.5\ \text{kJ/kg}$:

$$\Delta w_6 = \frac{K \cdot \Delta h_a}{0.8 \cdot \Delta h_g} = \frac{77.7\,(37.5 - 21.0)}{0.8 \cdot 1\,698} = +0.94\%$$

7. Generator efficiency $\eta_{Gen} = 98.4\%$

$$\Delta w_7 = \eta_{Gen}^* - \eta_{Gen} = +0.40\%.$$

   Total correction factor $\qquad\qquad\qquad \displaystyle\sum_{i=1}^{7} \Delta w_i = +5.08\%$

   Corrected heat consumption

$$w = 7\,860\ \text{kJ/kWh} \cdot 1.051 = 8\,261\ \text{kJ/kWh} = 2.295\ \text{kJ/kWs}.$$

Explanation

With worksheet No 7.6, one can determine the specific absorption capacity of nozzles charged by steam at supercritical pressures in dependence of the temperature and the pressure in front of the nozzle. Here $\varepsilon \leq \varepsilon_s$ with

$$\varepsilon = \frac{p_2}{p_1} \quad \text{and} \quad \varepsilon_s = \frac{p_{2s}}{p_1}.$$

is valid.

For superheated steam holds $\varepsilon_s = 0.546$ and for saturated steam holds $\varepsilon_s = 0.577$. The last value holds approximately for wet steam.

For the wet steam region, in the diagram, the specific steam content x has to be used as parameter.

Definitions:

$\dot{m}$ in kg/s	mass flow
p in bar	pressure
ϑ in °C	temperature
x in $(-)$	steam content
ε in $(-)$	pressure relationship
α in $kg \cdot 10^{-4}/(s\,mm^2\,bar)$	absorption capacity

Indexes:

1   in front of the nozzle
2   after the nozzle
s   critical condition

Example:

		Line ①	Line ②	
Pressure in front of the nozzle	p_1	2.35	73.5	bar
Temperature	ϑ_1	saturated steam	450	°C
Results in: specific absorption capacity	α/p_1	1.5	1.2	$\dfrac{kg\,10^{-4}}{s\,mm^2\,bar}$
Total absorption capacity	α	3.53	88.2	$\dfrac{kg\,10^{-4}}{s\,mm^2\,bar}$

Additional information of this book

(Engineering Reference Book on Energy and Heat; 978-3-642-51125-7;
978-3-642-51125-7_OSFO4) is provided:

http://Extras.Springer.com

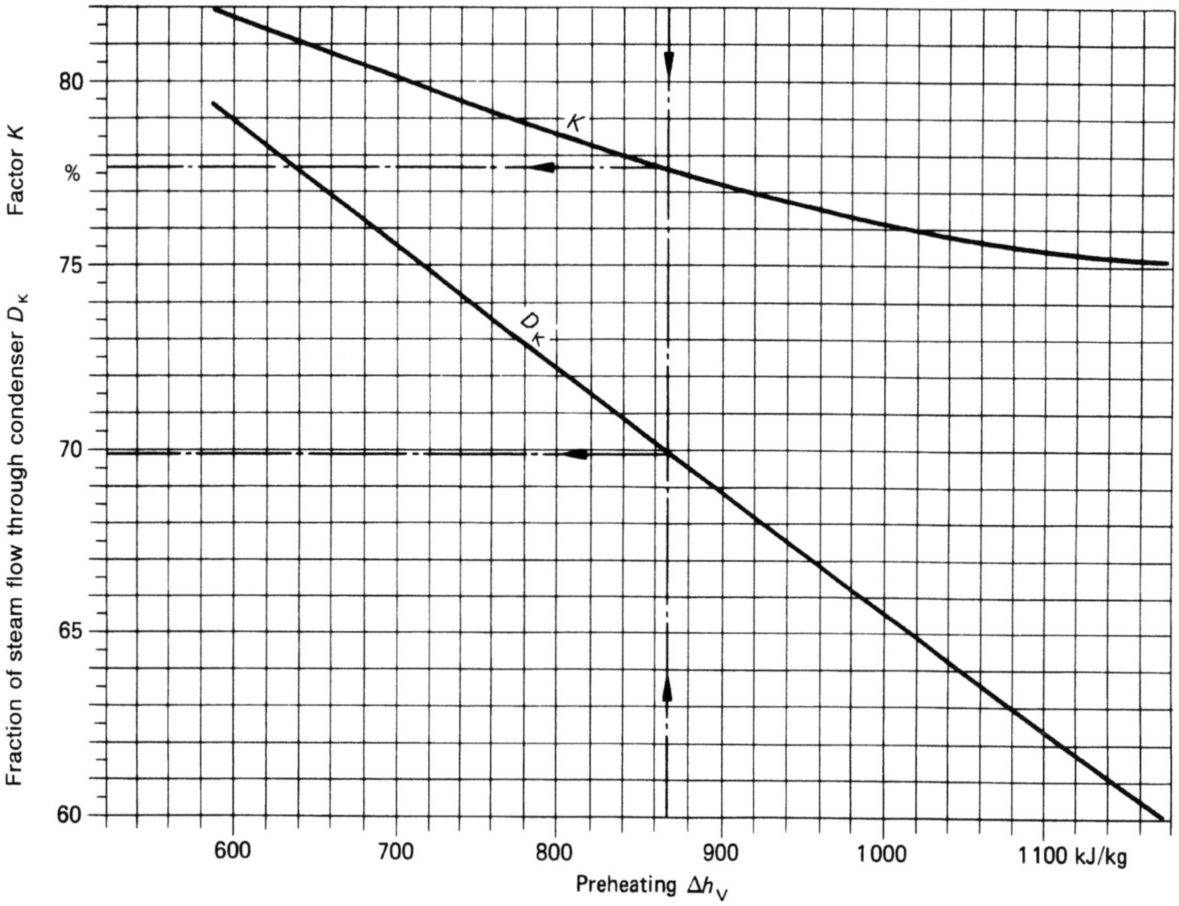

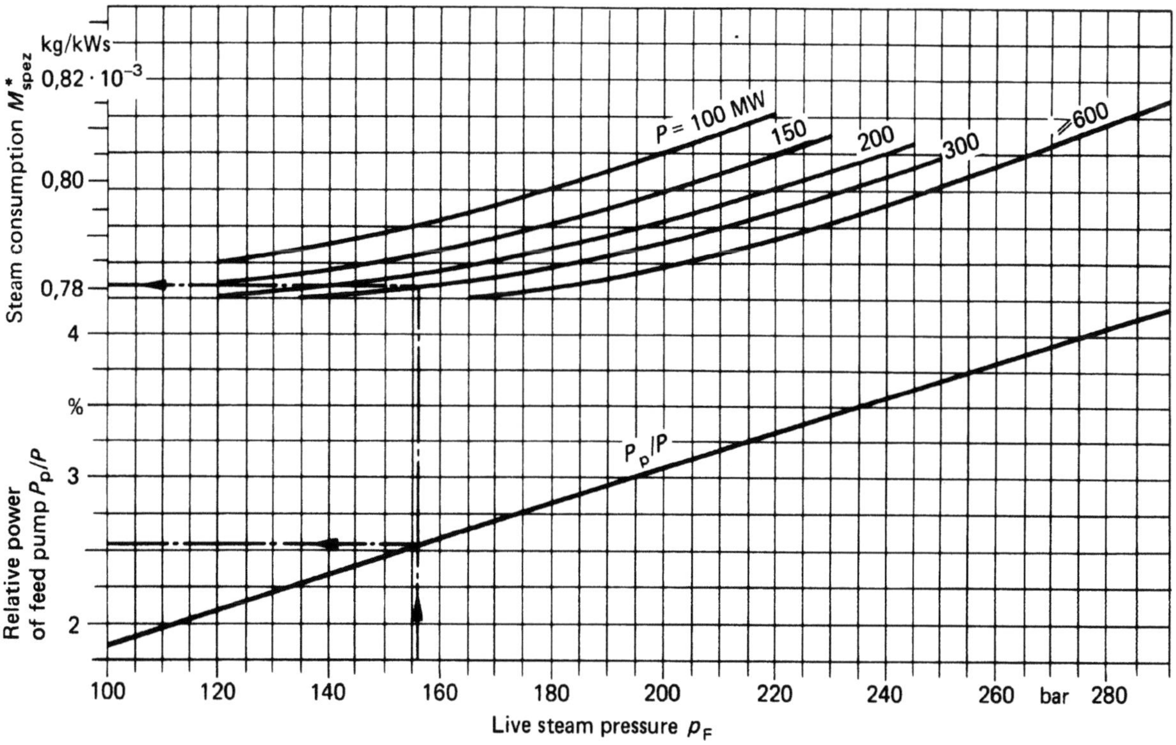

A. Mellgren, E. Koch

Explanation

Determination of the live steam and condenser steam mass flow

The specific steam rate, and thus, the mass flow of live steam can be determined from the equation:

$$w = \dot{M}_{\text{spec.}} \left[(h_F - h_E) + 0.9 \, (h_{Z\ddot{U}} - h_{\omega HD}) \right]$$

The factor 0.9 considers the bleeding at superheater exit. With the isentropic enthalpy drop in the high pressure turbine $\Delta h_{HD} = 407$ kJ/kg and an assumed internal efficiency of the turbine of 85%, one gets:

$$h_{\omega HD} = h_F - 0.85 \cdot \Delta h_{HD} =$$
$$= (3\,372 - 0.5 \cdot 407)\,\text{kJ/kg} = 3\,026\,\text{kJ/kg}$$

and

$$\dot{M}_{\text{spec.}} = \frac{2.295}{(3\,372 - 1\,017) + 0.9\,(3\,531 - 3\,026)}\,\text{kg/kWs} =$$
$$= 0.817 \cdot 10^{-3}\,\text{kg/kWs}.$$

So, the live steam mass flow may be calculated to:

$$\dot{m}_F = 300\,000 \cdot 0.817 \cdot 10^{-3}\,\text{kg/s} = 245\,\text{kg/s}.$$

From the upper diagram in worksheet No 7.5, one can get the part of the steam flow through the condenser $D_k = 69.9\%$, and thus, the mass flow through the condenser

$$\dot{m}_K = 245 \cdot 0.699\,\text{kg/s} = 171.3\,\text{kg/s}.$$

As shown below, the mass flow of steam through the condenser may be calculated otherwise by the specific heat consumption. From the equation

$$\eta_{th} = \frac{1}{w} = \frac{\dot{Q}_{zu} - \dot{Q}_{ab}}{\dot{Q}_{zu}} \cdot \eta_m \cdot \eta_{Gen} = \frac{P_{Kl}}{\dot{Q}_{zu}}$$

one gets:

$$\dot{Q}_{ab} = \dot{m}_K \cdot \Delta h_{ab} = P_{Kl} \left(w - \frac{1}{\eta_m \, \eta_{Gen}} \right)$$

η_{th} is the thermal and η_m the mechanical efficiency. $\dot{Q}_{zu}$ is the heat delivered to the turbine, while $\dot{Q}_{ab}$ is the total heat and Δh_{ab} the specific heat, withdrawn by the condenser. With an assumed combined internal efficiency of 90% of the medium- and low pressure turbines, one

can get:

$$\Delta h_{ab} = h_{Z\ddot{U}} - 0.9 \cdot \Delta h_{MD+ND} - h' =$$
$$= (3\,531 - 0.9 \cdot 1\,291 - 150)\,\text{kJ/kg} = 2\,219\,\text{kJ/kg}.$$

With the product $\eta_m \cdot \eta_{Gen} = 0.977$, one can get:

$$\dot{m}_K = \frac{300\,000 \left(2.295 - \dfrac{1}{0.977} \right)}{2\,219}\,\text{kg/s} = 171.9\,\text{kg/s}.$$

Both results agree sufficiently.

Power station with branch turbine

The blading of the branch turbine has a lower efficiency than the equivalent blading of the main turbine. On the other hand, the steam flow through the condenser, and thus, the energy at the exit of the main turbine, is lower than before. In addition, the branch turbine output will not be reduced by generator losses. Assuming the gains and losses are equal, the heat consumption of the plant with branch turbine is the same as that one without branch turbine, if the heat consumption is defined as follows:

$$w = \frac{\dot{Q}_{zu}}{P_{Kl} + P_{Spt}}.$$

If the live steam mass flow is unknown, the specific steam rate has to be multiplied by the total output. From the lower diagram in worksheet No 7.5, one can get the relative energy consumption of the feed water pump of 2.55% equal to 7.65 MW. Thus

$$\dot{m}_F = 307\,650 \cdot 0.817 \cdot 10^{-3}\,\text{kg/s} = 251.4\,\text{kg/s},$$
$$\dot{m}_K = 251.4 \cdot 0.699\,\text{kg/s} = 175.7\,\text{kg/s}.$$

This is the sum of exhaust steam mass flow. At an isentropic drop of enthalpy of 925 kJ/kg (branch steam pressure 10 bar, $h = 3\,200$ kJ/kg, expansion to 0.059 bar) and an efficiency at the turbine clutch of 80%, the steam rate of the feed water turbine is:

$$\dot{m}_{Spt} = \frac{7\,650}{925 \cdot 0.8}\,\text{kg/s} = 10.3\,\text{kg/s}.$$

Thus, the mass flow of steam through the main condenser is 165.4 kg/s.

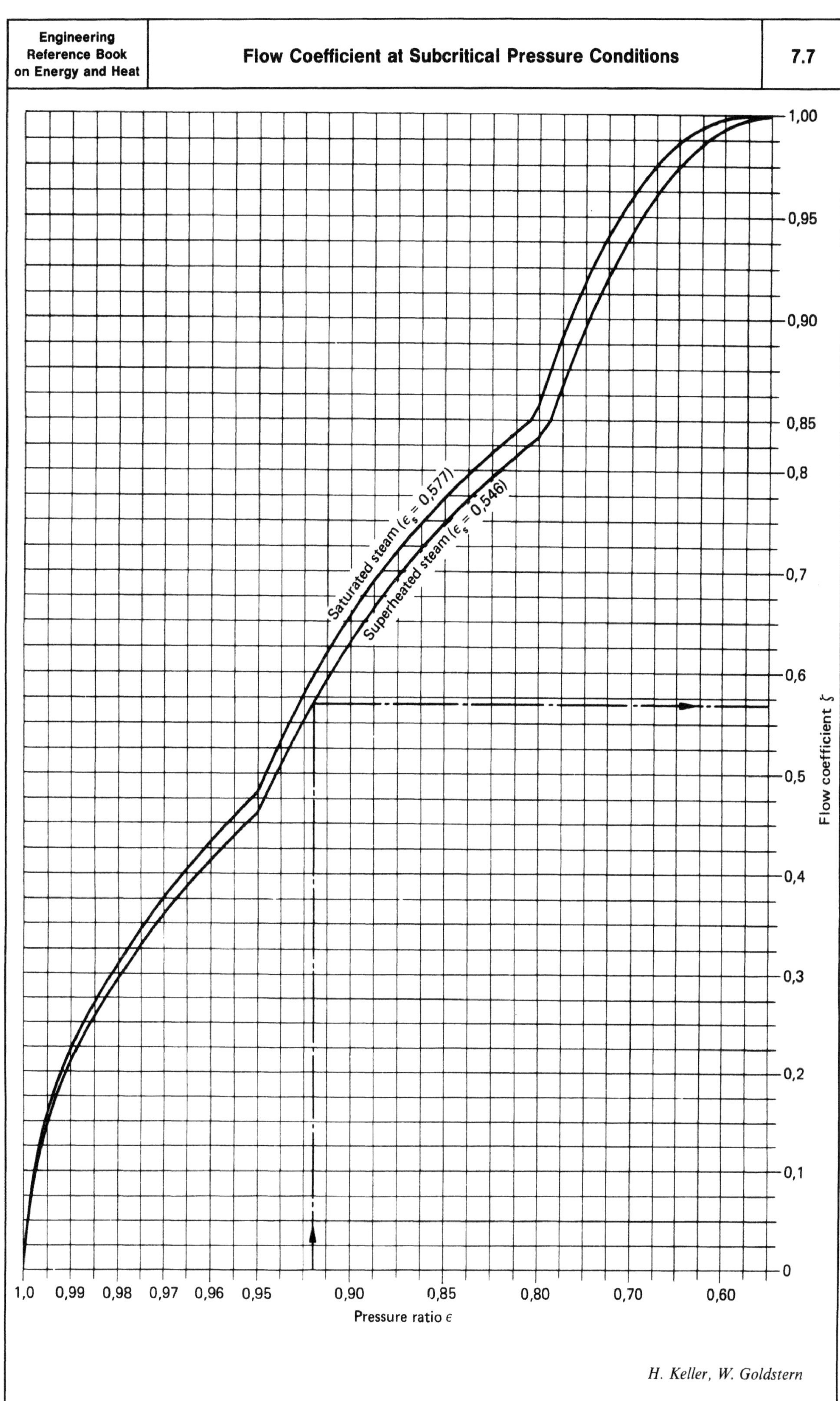

H. Keller, W. Goldstern

Explanation

The specific absorption capacity of nozzles from worksheet No 7.6 has to be multiplied by the flow coefficient ξ to get the reduced flow rate at subcritical pressure.

Definitions:

p in bar   pressure
ε in $(-)$ pressure ratio
$\varkappa$ in $(-)$ polytropic exponent
ξ in $(-)$ flow coefficient

Indexes:

1   in front of the nozzle
2   after the nozzle
s   critical condition

Basic equations:

$$\xi = \sqrt{\frac{\varepsilon^{\frac{2}{\varkappa}} - \varepsilon^{\frac{\varkappa+1}{\varkappa}}}{\varepsilon_s^{\frac{2}{\varkappa}} - \varepsilon_s^{\frac{\varkappa+1}{\varkappa}}}}$$

$$\varepsilon = \frac{p_2}{p_1} \quad \text{and} \quad \varepsilon_s = \frac{p_{2s}}{p_1}$$

		Saturated steam	Superheated steam
Polytropic exponent	$\varkappa$	1.14	1.3
Critical pressure ratio	s	0.577	0.546

Example:

Pressure in front of the nozzle	$p_1 = 73.5$ bar
Pressure after the nozzle	$p_2 = 67.6$ bar
Pressure ratio	$\varepsilon = 0.92$
Critical pressure ratio (superheated steam)	$\varepsilon_s = 0.546$

Results in:

Flow coefficient	$\xi = 0.570$

From sheet No 7.6, example No ②:

Total absorption capacity $\alpha = 88.2 \dfrac{\text{kg } 10^{-4}}{\text{s mm}^2 \text{ bar}}$

Reduced flow rate $\xi \cdot \alpha = 50.3 \dfrac{\text{kg } 10^{-4}}{\text{s mm}^2 \text{ bar}}$

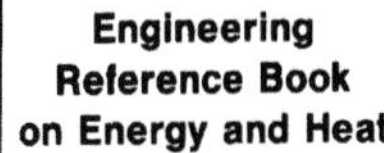

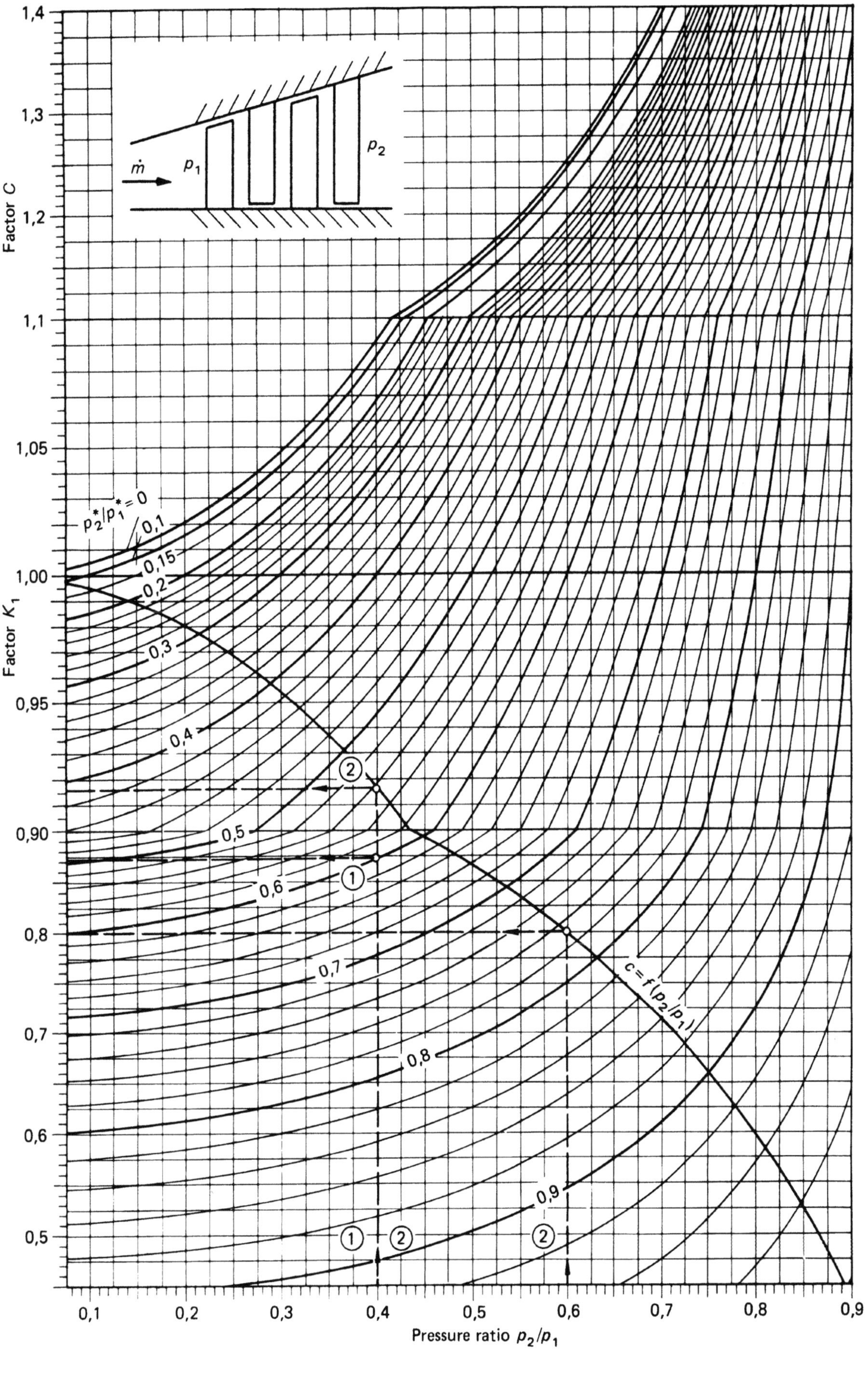

W. Badenhausen

Explanation

Definitions:

$\dot{m}$ in kg/s	mass flow
p in bar	pressure
v in m³/kg	specific volume
T in K	temperature
x in $(-)$	specific steam content
K_1 in $(-)$	factor from sheet 7.8.1
C in $(-)$	factor from sheet 7.8.1
K_2 in $(-)$	factor from sheet 7.8.2

Indexes:

1 in front of the nozzle
2 after the nozzle
* value at operating condition

Basic equations:

The ellipse law is valid for a group of steam turbine blades (single or several rows), in which the steam does not reach critical flow conditions:

$$\frac{\dot{m}^*}{\dot{m}} = \sqrt{\frac{p_1^{*2} - p_2^{*2}}{p_1^2 - p_2^2} \cdot \frac{p_1 \cdot v_1}{p_1^* \cdot v_1^*}} \,.$$

Also the temperature correction $\sqrt{(T_1 \cdot x_1)/(T_1^* \cdot x_1^*)}$, instead of the $(p\,v)$ correction, may be used with sufficient accuracy. (In the superheated steam region $x = 1$ is valid)

$$\frac{\dot{m}^*}{\dot{m}} = \sqrt{\frac{p_1^{*2} - p_2^{*2}}{p_1^2 - p_2^2} \cdot \frac{T_1 \cdot x_1}{T_1^* \cdot x_1^*}} \,.$$

Determination of the mass flow (sheet No 7.8.1).

Neglecting the $(p\,v)$ correction, the ellipse law leads to:

$$\dot{m}^* = \dot{m} \,\frac{p_1^*}{p_1} \cdot K_1$$

with

$$K_1 = \sqrt{\frac{1 - \left(\dfrac{p_2^*}{p_1^*}\right)^2}{1 - \left(\dfrac{p_2}{p_1}\right)^2}} = \frac{C^*}{C}$$

The factor K_1 can either be determined directly from the worksheet dependent on the pressure ratio p_2/p_1 with p_2^*/p_1^* as a parameter, or calculated as the quotient of the factors C^* and C. C^* and C can be determined from the curve $C = f(p_2/p_1)$ in the worksheet.

Example:

	Design	Operation	
Pressure in front of the group of blades	$p_1 = 100$	$p_1^* = 80$	bar
Pressure after the group of blades	$p_2 = 40$	$p_2^* = 48$	bar
Mass flow	$\dot{m} = 70$	$\dot{m}^* =$ unknown	kg/s

The factor K_1 can either be determined by the pressure ratios $p_2/p_1 = 0.4$ and $p_2^*/p_1^* = 0.6$ from the diagram according to ① to $K_1 = 0.873$, or can be calculated according to ② from $C(0.4) = 0.914$ and $C^*(0.6) = 0.8$ to

$$K_1 = \frac{0.8}{0.916} = 0.873\,.$$

The mass flow is then $\dot{m}^* = 70 \cdot \dfrac{80}{100} \cdot 0.873 = 48.9$ kg/s.

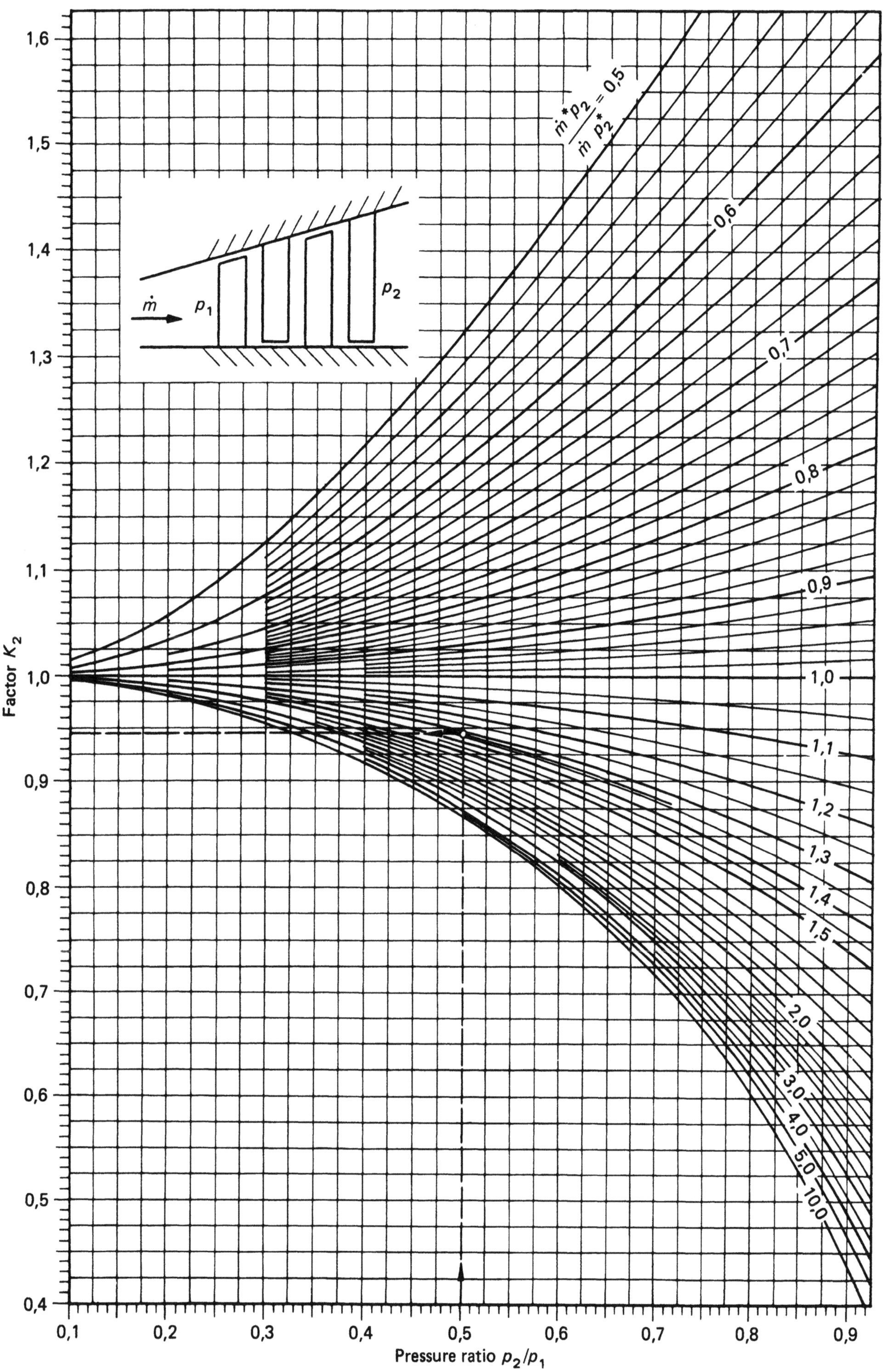

W. Badenhausen

Explanation

The worksheet 7.8.2 relies as well on the ellipse law. For further explanations see worksheet No 7.8.1.

Determination of pressure in front of a group of blades

Neglecting the $(p\,v)$ correction, the ellipse law leads to:

$$p_1^* = p_1 \frac{\dot{m}^*}{\dot{m}} \cdot K_2$$

with

$$K_2 = \sqrt{1 - \left(\frac{p_2}{p_1}\right)^2 \left[1 - \left(\frac{\dot{m}}{\dot{m}^*} \frac{p_2^*}{p_2}\right)^2\right]}$$

Example:

	Design	Operation	
Pressure in front of the group of blades	$p_1 = 20$	$p_1^* = \text{unknown}$	bar
Pressure after the group of blades	$p_2 = 10$	$p_2^* = 9$	bar
Mass flow	$\dot{m} = 50$	$\dot{m}^* = 60$	kg/s

With

$$\frac{\dot{m}^*}{\dot{m}} \cdot \frac{p_2}{p_1} = \frac{60 \cdot 10}{50 \cdot 9} = 1.333 \quad \text{and}$$

$$\frac{p_2}{p_1} = \frac{10}{20} = 0.5,$$

from the diagram one gets

factor $K_2 = 0.944$.

The pressure in front of the group of blades is

$$p_1^* = 20 \cdot \frac{60}{50} \cdot 0.944 = 22.65 \text{ bar}.$$

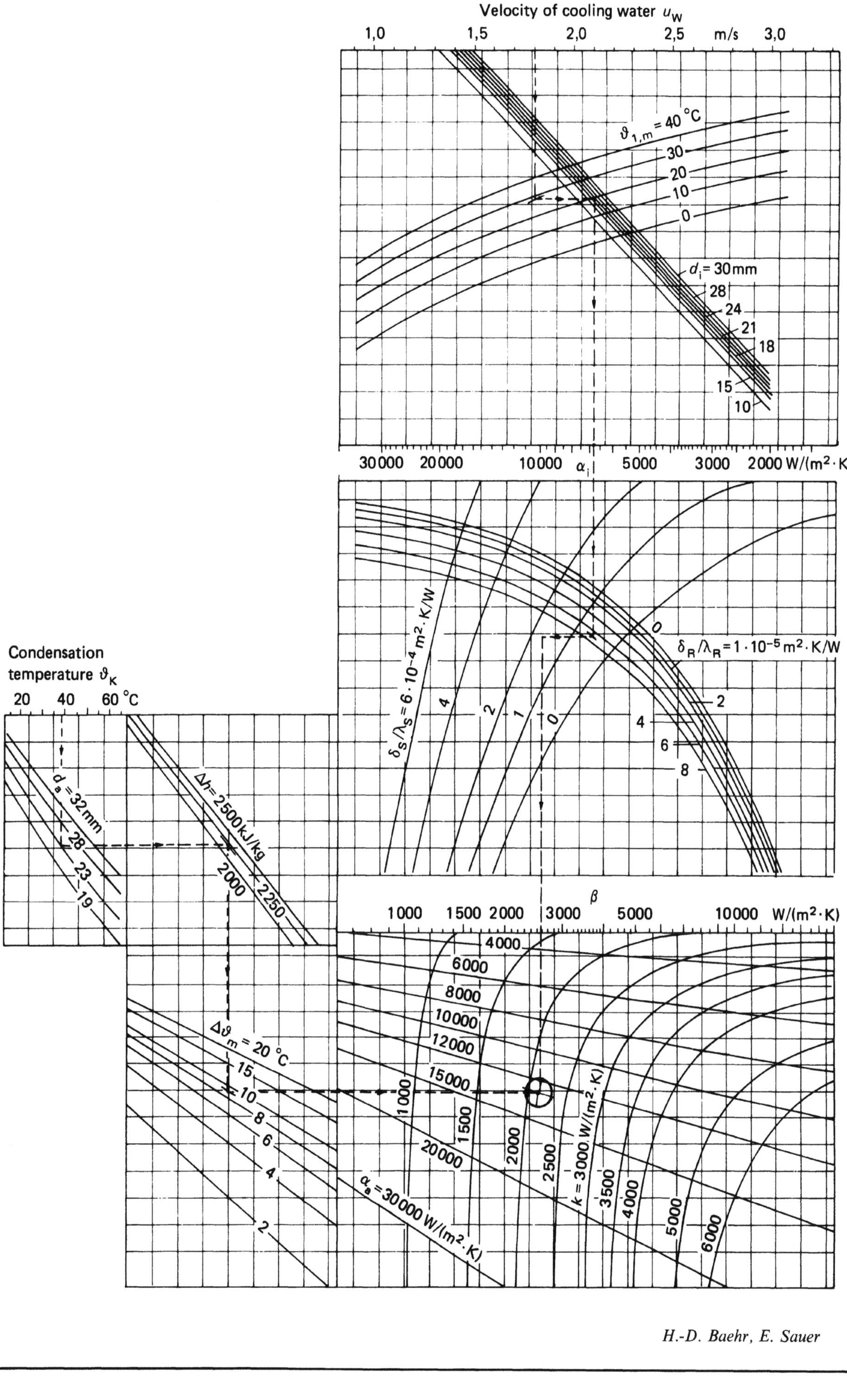

Heat Transfer Coefficient of Horizontal Condenser Tubes
According to *Kraußold/Nußelt/Neumann*

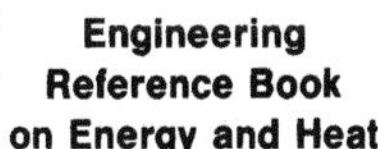

H.-D. Baehr, E. Sauer

Explanation

From the worksheet No 8.1, one can determine the heat transfer coefficient k of horizontal tube banks in steam condensers. The cooling water flows through the tubes and the steam condenses at the outer surface of the tubes.

Definitions:

$c_{p,w}$ in kJ/(kg·K)	Specific heat capacity of the cooling water
d_i in m	Inner diameter of the tubes
Pr in (−)	Prandtl number
Re in (−)	Reynolds number
u_w in m/s	Velocity of the coolant inside of the tubes
α_a in W/(m²·K)	Heat transfer coefficient (at the out side of the tubes)
α_i in W/(m²·K)	Heat transfer coefficient (at the inside of the tubes)
η in Pa s	Dynamic viscosity
v in m²/s	Kinematic viscosity: $v = \eta/\varrho$
λ in W/(m·K)	Thermal conductivity

Basic equations:

The heat transfer coefficient inside the tubes α_i has to be calculated from:

$$\frac{\alpha_i \cdot d_i}{\lambda} = \mathrm{Nu}_i = 0.024 \cdot \mathrm{Re}^{0.8} \cdot \mathrm{Pr}^{1/3} \qquad (1)$$

$$\mathrm{Re} \quad \text{Reynolds number} = \frac{u_w \cdot d_i}{v} \qquad (2)$$

$$\mathrm{Pr} \quad \text{Prandtl number} = \frac{\eta \cdot c_p}{\lambda} \qquad (3)$$

The heat transfer coefficient outside the tubes α_a has to be calculated according to the equation for film condensation on horizontal tubes from *W. Nusselt*, modified by *F. Neumann* [1].

$$\alpha_a = 0.726 \left[\frac{\Delta h \cdot \lambda^3 \cdot \varrho^2 \cdot g \cdot (a+1)}{\eta \cdot d_a \cdot \Delta \vartheta_m} \right]^{1/4} \qquad (4)$$

$$a = \frac{\alpha_a}{\beta} \qquad (5)$$

$$\frac{1}{\beta} = \frac{1}{\alpha_i} + \frac{\delta_s}{\lambda_s} + \frac{\delta_R}{\lambda_R}. \qquad (6)$$

At the condensation of steam mixed with air, α_a is smaller than at the condensation of pure steam. In order to take into account the influence of the amount of air, (which value is seldom known) the heat transfer coefficient α_a in the diagram has to be used with only 80% of its theoretical value.

The values for the thermal resistance δ_R/λ_R have been listed in table A:

Pipe material according to DIN 1785	δ_R/λ_R in 10^{-5} m²·K/W for wall thickness δ_R			
	0.7 mm	1.0 mm	1.5 mm	2.0 mm
SD–Cu	0.323	0.461	0.691	0.922
SB–Cu	0.42	0.599	0.898	1.197
Cu Zn 28 Sn	0.642	0.917	1.376	1.835
Cu Zn 20 Al	0.7	1.0	1.5	2.0
Cu Al 15 As	0.833	1.191	1.786	2.381
St 35.8	1.346	1.923	2.885	3.846
Cu Ni 10 Fe	1.522	2.174	3.261	4.348
Cu Ni 20 Fe	2.09	2.99	4.478	5.97
Cu Ni 30 Fe	2.399	3.413	5.12	6.826
X5 Cr Ni 189 X5 Cr Ni Mo 1812	4.795	6.849	10.274	13.659
Titanium	4.516	6.452	9.677	12.903

Standard value for the thermal resistance δ_s/λ_s of fouling layers in 10^{-4} m²·K/W according to *McAdams* [2] have been listed in table B:

See water	0.86	Pure river water	1.7
Drinking water	1.7	Muddy river water	3.5
Water from large lakes	1.7	High hardness water (hardness 10° d.H)	5.7

According to *Short* and *Brown* [3], a negative influence of the tube bundles on the heat transfer coefficient α_a is not known. So, α_a for the whole tube bank may be calculated in the same way as for the single horizontal tube.

Bibliography

[1] *Neumann, F.:* Vereinfachte Berechnung der Wärmedurchgangszahl von Kondensatoren. Z. VDI 91 (1949), p. 331/335

[2] *McAdams, W. H.:* Heat Transmission, 3rd edition, p. 189. New York, Toronto, London: McGraw-Hill Book Comp. 1954

[3] *Short, B. E., Brown, H. E.:* Proceedings Gen. Disc. Heat Transfer, London 1951, p. 27

[4] *Baehr, H. D.:* Handbuch der Kältetechnik. Abschnitte Wärmeleitung und Wärmestrahlung, Vol. III, Springer-Verlag, 1959

Example:

Known values:

Temperature of cooling water	ϑ_1'	$= 21.5\,°C$
Mass flow of cooling water	$\dot{m}_W$	$= 16\,750$ kg/s
Condensation temperature	ϑ_K	$= 38\,°C$
Mass flow of exhaust steam	$\dot{m}_D$	$= 382.25$ kg/s
Enthalpy of exhaust steam	h''	$= 2\,330$ kJ/kg
Enthalpy of condensate	h'	$= 152$ kJ/kg
Difference of enthalpy in the condenser	Δh	$= h'' - h'$
Water quality		pure river water

Design values:

Material		X 5 Cr Ni 189
Velocity of coolant in the condenser tubes	u_W	$= 1.8$ m/s
Inner diameter of the tubes	d_i	$= 21$ mm
Outer diameter of the tubes	d_a	$= 23$ mm

Result:

Exit temperature of cooling water

$$\vartheta_1'' = \frac{\dot{m}_D \cdot (h'' - h')}{\dot{m}_W \cdot c_{p,W}} + \vartheta_1' = 33.37\,°C$$

Average temperature of the cooling water (for the calculation of the properties of media)

$$\vartheta_{1,m} = \frac{\vartheta_1' + \vartheta_1''}{2} = 27.4\,°C$$

Heat transfer coefficient to the cooling water

$$\alpha_i = 6\,800 \ W/(m^2 \cdot K)$$

Wall thickness of the pipe $\quad \delta_R = \dfrac{d_a - d_i}{2} = 1$ mm

Thermal resistance coefficient of the pipe wall (according to table A)

$$\delta_R/\lambda_R = 6.849 \cdot 10^{-5}\,m^2 \cdot K/W$$

Thermal resistance coefficient of the fouling layer (according to table B)

$$\delta_s/\lambda_s = 1.7 \cdot 10^{-4}\,m^2 \cdot K/W$$
$$\beta = 2\,595 \ W/(m^2 \cdot K)$$

Logarithmic mean temperature difference

$$\Delta\vartheta_m = \frac{\vartheta_1'' - \vartheta_1'}{\ln \dfrac{\vartheta_K - \vartheta_1'}{\vartheta_K - \vartheta_1''}} = 9.34\,°C$$

Heat transfer coefficient to steam

$$\alpha_a = 13\,000 \ W/(m^2 \cdot K)$$

Heat transfer coefficient $\quad k = 2\,164 \ W/(m^2 \cdot K).$

Example:

Known values:

Density of the cooling water	ϱ_W	$= 1\,000\ \text{kg/m}^3$
Specific heat capacity of the cooling water	$c_{p,W}$	$= 4.186\ \text{kJ/(kg} \cdot \text{K)}$
Temperature of condensation	ϑ_K	$= 38\,°\text{C}$
Temperature of cooling water at the entrance	ϑ'_1	$= 21.5\,°\text{C}$
Mass flow of cooling water	$\dot{m}_W$	$= 16\,750\ \text{kg/s}$
Mass flow of exhaust steam	$\dot{m}_D$	$= 382.25\ \text{kg/s}$
Difference of enthalpy in the condenser	Δh	$= 2\,178\ \text{kJ/kg}$

Selected:

Material	X5 Cr Ni 189	
Inner diameter of the tubes	d_i	$= 21\ \text{mm}$
Outer diameter of the tubes	d_a	$= 23\ \text{mm}$
Wall thickness of the tubes	δ_R	$= 1\ \text{mm}$
Velocity of coolant in the condenser tubes	u_W	$= 1.8\ \text{m/s}$
Number of water ways	Z_W	$= 1$

Results:

Correction factor for material and wall thickness of the condenser tubes	C_M	$= 0.73$
Fouling factor	C_{CL}	$= 0.9$
Correction factor for the temperature of the cooling water at the entrance	C_T	$= 1.006$
Heat transfer coefficient	k	$= 2\,403\ \text{W/(m}^2 \cdot \text{K)}$
Maximum temperature difference in the condenser	$(\vartheta_K - \vartheta'_1)$	$= 16.5\,°\text{C}$
Rise in temperature $(\vartheta''_1 - \vartheta'_1)$ of the coolant	$\dfrac{\dot{m}_D \cdot \Delta h}{\dot{m}_W \cdot c_{p,W}}$	$= 11.87\,°\text{C}$
Exit temperature of cooling water	ϑ''_1	$= 33.37\,°\text{C}$
Effective length of the cooling tubes	L	$= 19.1\ \text{m}\ .$

Bibliography

[1] Standards for Steam Surface Condensers, seventh Edition. Heat Exchange Institute 1230 Keith Building Cleveland, Ohio 44115

Explanation

With worksheet No 8.2, one can determine the effective length L of cooling tubes in water cooled surface condensers for steam turbines.

Definitions:

k in $W/(m^2 \cdot K)$	Heat transfer coefficient
C_T in $(-)$	Correction factor for cooling water temperatures different from $21\,°C$
C_M in $(-)$	Correction factor for pipe wall thickness different from 1.24 mm and pipe materials different from Cu Zn 28 Sn
C_{CL} in $(-)$	Fouling factor
$\dot{Q}$ in MW	Transmitted heat flow
L in m	Effective length of cooling tubes

Procedure of calculation:

At first, the heat transfer coefficient k has to be determined according to the guidelines of the Heat Exchange institute [1].

$$k = 6.47878 \, \frac{W}{m \cdot K \cdot mm} \cdot (441.325\,mm - d_a) \, \sqrt{u_W \cdot s/m} \cdot C_T \cdot C_M \cdot C_{CL} \qquad (1)$$

$$C_T = 1.395 - \exp\left(\frac{-\vartheta_1'}{22.61\,°C}\right) - \frac{(\vartheta_1' - 21\,°C)}{166\,°C} \quad \text{für} \;\; 5\,°C < \vartheta_1' < 40\,°C \qquad (2)$$

Correction factor C_M							
Wall thickness δ_R in mm	0.56	0.71	0.89	1.24	1.65	2.11	2.77
CuZn28Sn SB-Cu Aluminium	1.06	1.04	1.02	1.00	0.96	0.92	0.87
CuZn20Al CuAl5As Muntzmetal	1.03	1.02	1.00	0.97	0.94	0.90	0.84
CuNi10Fe	0.99	0.97	0.94	0.90	0.85	0.80	0.74
CuNi30Fe	0.93	0.90	0.87	0.82	0.77	0.71	0.64
Carbon steel	1.00	0.98	0.95	0.91	0.86	0.80	0.74
X10Cr13	0.88	0.85	0.82	0.76	0.70	0.65	0.59
X5CrNi189 u.X5CrNiMo1812	0.83	0.79	0.75	0.69	0.63	0.56	0.49
X8Cr17	0.78	0.76	0.74	0.69	0.65	0.60	0.54
Titanium	0.85	0.81	0.77	0.71	–	–	–

C_{CL} Fouling factor

The fouling factor considers the fouling of condenser tubes during operation.

$\quad C_{CL} = 0.85$ for tubes made of brass, aluminium, bronze or copper nickel alloys

$\quad C_{CL} = 0.9$ to 0.95 for tubes made of stainless steel and titanium.

Next, from the equation for the transmitted heat flow

$$\dot{Q} = k \cdot A \cdot \Delta\vartheta_m \qquad (3)$$

$$\dot{Q} = \dot{m}_W \cdot c_{p,W} \cdot (\vartheta_1'' - \vartheta_1') \qquad (4)$$

$$\dot{Q} = \dot{m}_D \cdot \Delta h \qquad (5)$$

the effective length of the cooling tubes can be calculated

$$L = \frac{\varrho_W \cdot c_{p,W} \cdot u_W \cdot d_i^2}{4 \cdot Z_W \cdot k \cdot d_a} \cdot \ln \frac{\vartheta_K - \vartheta_1'}{\vartheta_K - \vartheta_1''} \qquad (6)$$

Additional information of this book

(Engineering Reference Book on Energy and Heat; 978-3-642-51125-7;
978-3-642-51125-7_OSFO5) is provided:

http://Extras.Springer.com

Engineering Reference Book on Energy and Heat	Number of Cooling Tubes and Size of Cooling Area in the Condenser	8.3

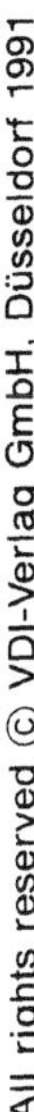

E. Sauer

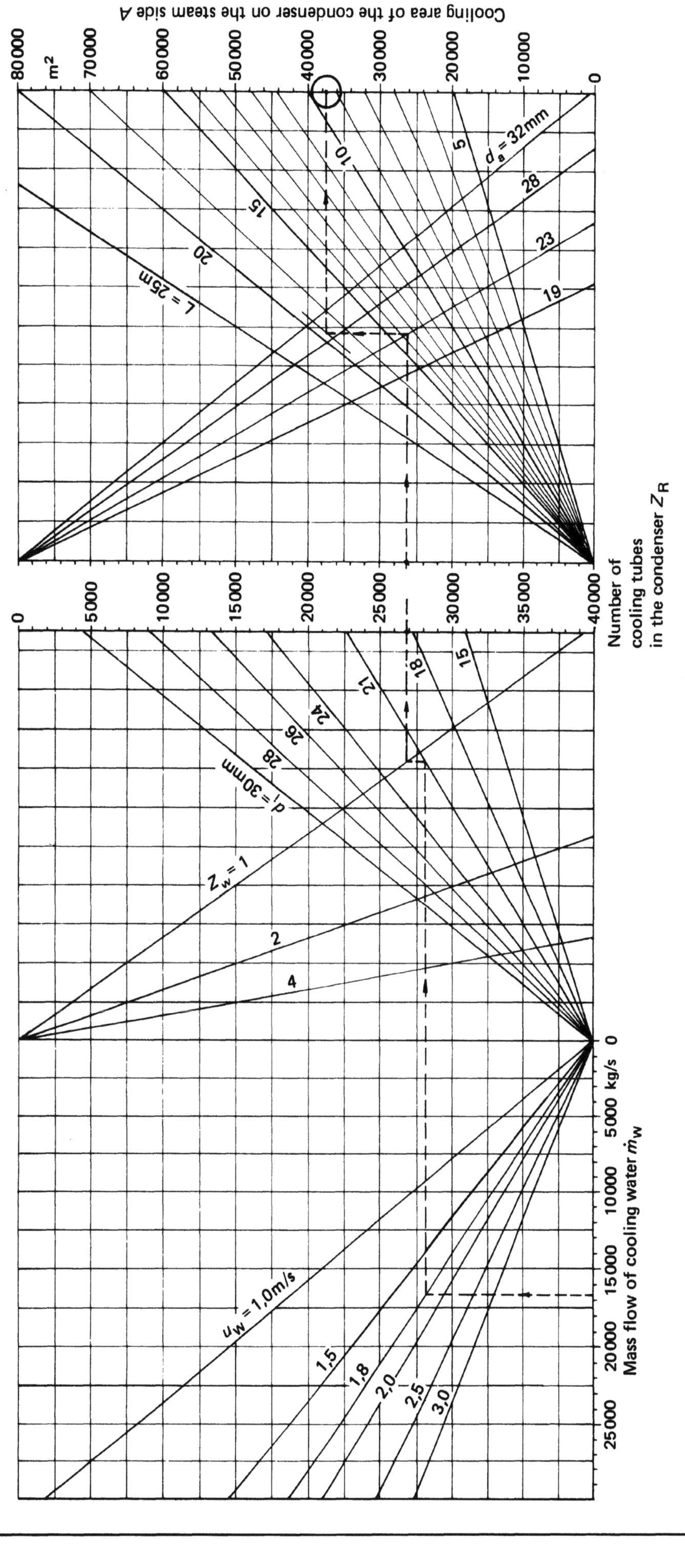

Explanation

With worksheet No 8.3, one can determine the number of condenser tubes Z_R and the cooling area of the condenser A on the steam side.

Definitions:

A in m²   Cooling area of the condenser at the steam side
Z_R in $(-)$ Number of condenser tubes.

Basic equations:

$$Z_R = \frac{Z_W \cdot \dot{m}_W \cdot 4}{\varrho_W \cdot u_W \cdot d_i^2 \cdot \pi} \tag{1}$$

$$A = d_a \cdot \pi \cdot L \cdot Z_R. \tag{2}$$

Example:

Known values:

Density of the cooling water	$\varrho_W = 1\,000$ kg/m³
Mass flow of cooling water	$\dot{m}_W = 16\,750$ kg/s
Effective length of the cooling tubes	$L = 19.1$ m
Velocity of coolant in the condenser tubes	$u_W = 1.8$ m/s
Inner diameter of the tubes	$d_i = 0.021$ m
Outer diameter of the tubes	$d_a = 0.023$ m
Number of water ways	$Z_W = 1$

Results:

Number of condenser tubes	$Z_R = 26\,867$
Cooling area of the condenser	$A = 37\,080$ m².

Explanation

With the calculation scheme in worksheet No 8.8, one can estimate the cold water temperature ϑ_{W2} of every kind of dry cooling tower (direct system (with condensation), non direct system (without condensation), mechanical draught and natural draught), if its operation condition differs from the design condition.

E x a m p l e :

The characteristic diagram of a certain cooling tower with known design data had been calculated. It is not to be transmitted to other conditions. The design condition and an operating point, differing from the design condition, is shown in the chart below. The operating point has the following data:

$\vartheta_{W1,\,ist} = 60.6\,°C$

$\vartheta_{W2,\,ist} = 38.5\,°C$

$\dot{m}_{W,\,ist} = 28\,310\,kg/s$

$\vartheta_{L1,\,ist} = 18.5\,°C$

$p_{L,\,ist} = 980\,mbar$

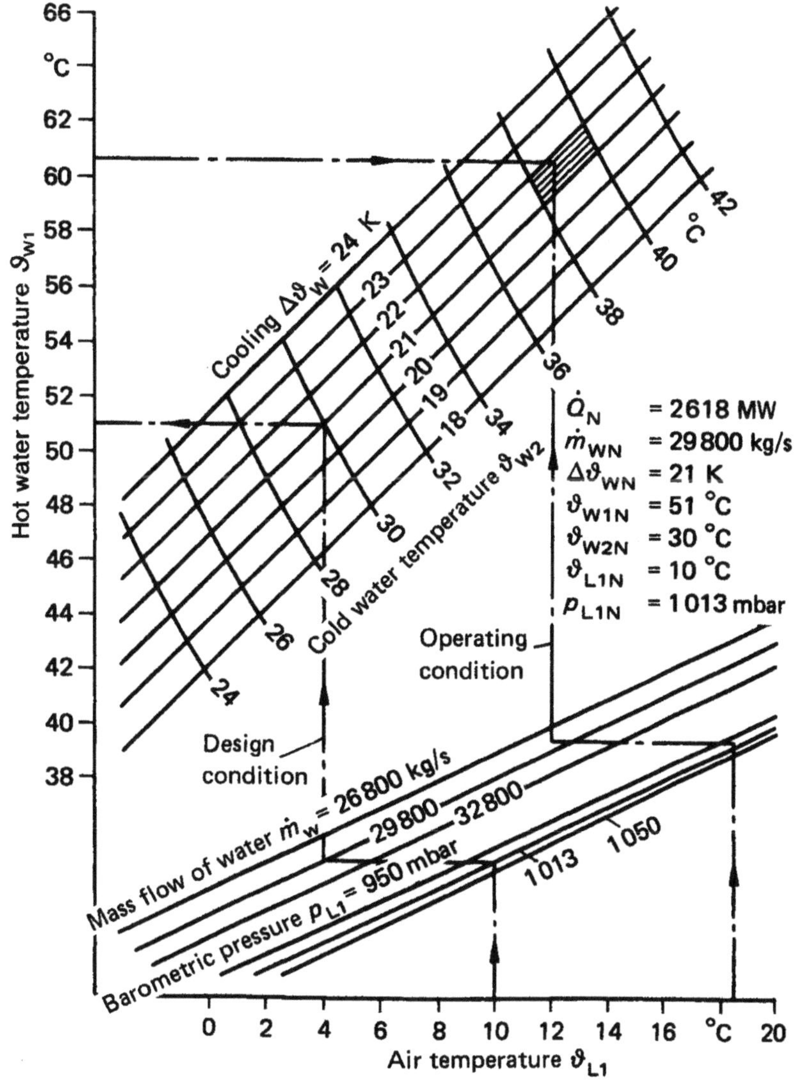

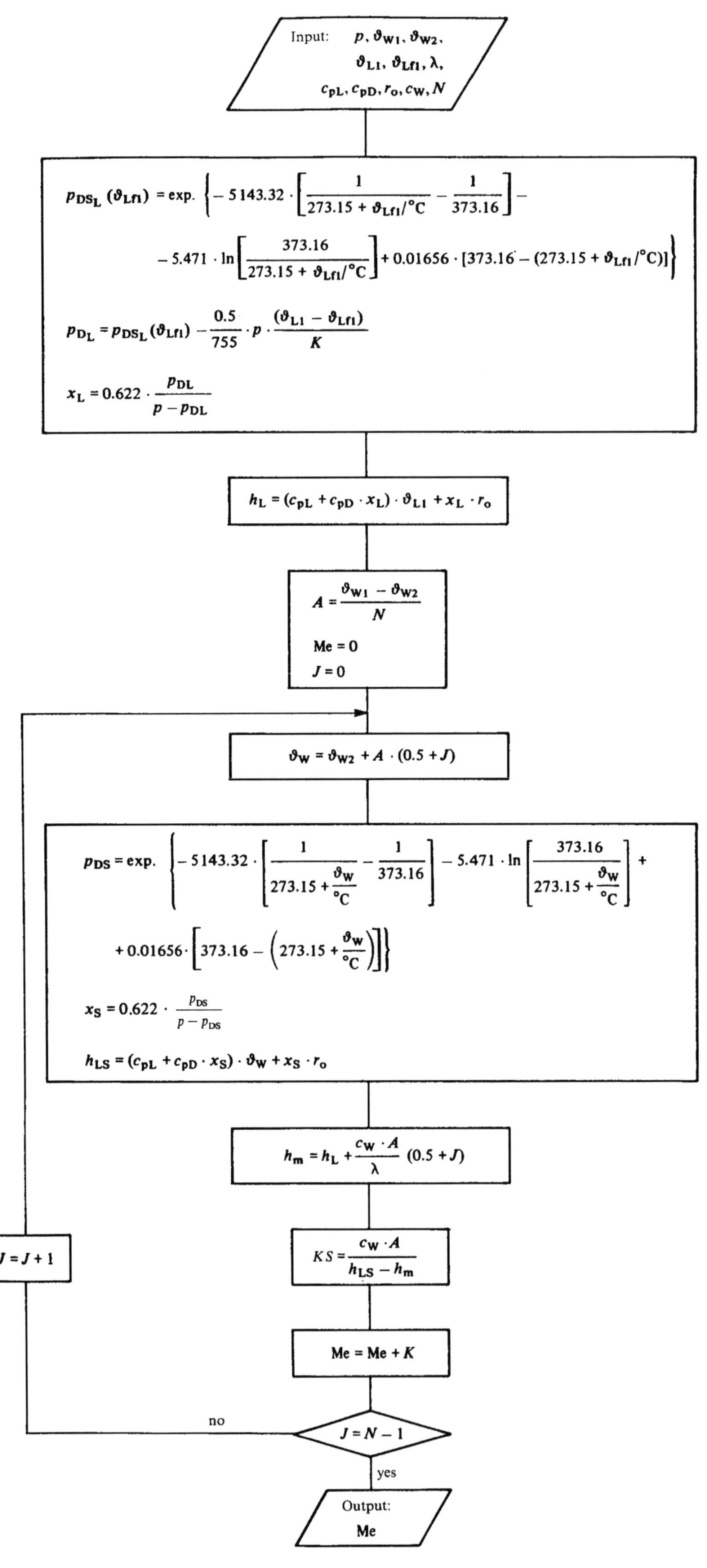

Definitions:

$\dot{m}_{\mathrm{W}}$ in kg/s	Mass flow of water
$\dot{m}_{\mathrm{L}}$ in kg/s	Mass flow of air
$\dot{m}_{\mathrm{D}}$ in kg/s	Mass flow of steam
p in bar	Total pressure
ϑ_{W} in °C	Water temperature
ϑ_{K} in °C	Condensate temperature
ϑ_{L} in °C	Temperature of the air measured at the dry thermometer
ϑ_{Lf} in °C	Temperature of the air measured at the wet thermometer
$\Delta\vartheta_{\mathrm{W}} = \vartheta_{\mathrm{W1}} - \vartheta_{\mathrm{W2}}$	
$\Delta\vartheta_{\mathrm{L}} = \vartheta_{\mathrm{L2}} - \vartheta_{\mathrm{L1}}$	
T in K	Temperature
$h'' - h'$ in kJ/kg	Difference of enthalpy at condensation between steam and condensate
$\dot{Q}$ in kW	Exhaust heat flow
$\dot{C}$ in kW/K	$= \dot{m} \cdot c_{\mathrm{p}}$; Flow of thermal capacity
ω in (−)	Ratio of flow of thermal capacity
$\varkappa$ in (−)	Heat exchanger coefficient
φ in (−)	Operation characteristic of a heat exchanger
B, D	Polynomials

Data depending on construction:

k in W/(m²·K)	Heat transfer coefficient
A in m²	Area of heat exchange
α in W/(m²·K)	Heat transfer coefficient

The values of the exponents, depending on the construction, for the usual baffles in heat exchangers are for

n	between 0.4 and 0.7
m	between −0.2 and −0.4

and the expression

$$\left(\frac{A_{\mathrm{L}} \cdot \alpha_{\mathrm{LN}}}{A_{\mathrm{W}} \cdot \alpha_{\mathrm{WN}}} \right) \quad \text{between 0.1 and 0.3.}$$

Properties of media:

λ in W/(m·K)	Thermal conductivity
η in Pa·s	Dynamic viscosity
Pr in (−)	Prandtl number $= \dfrac{\eta \cdot c_{\mathrm{p}}}{\lambda}$
c_{W} in kJ/(kg·K)	Specific heat capacity of water
c_{pL} in kJ/(kg·K)	Specific heat capacity of dry air
c_{pD} in kJ/(kg·K)	Specific heat capacity of steam
x in kg/kg	Absolute water content of the air

Indexes:

N	Design data
L	Air
W	Water
m	Average value
R	Fins
1	At the entrance
2	At the exit

Bibliography

Buxmann, J.: Berechnung der Kennfelder von Trockenkühltürmen. BWK 26 (1974) No. 10, p. 421/428

Explanation

With the Merkel number Me, one can perform the design of wet cooling towers if the basic values are known. The shown calculation scheme calculates the Merkel numbers dependent on the air number λ.

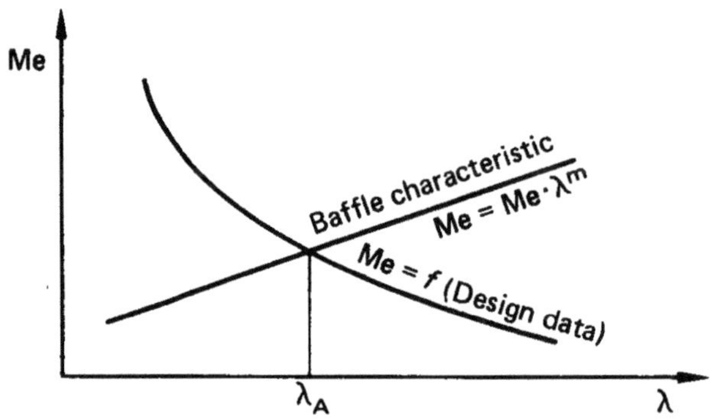

First, one has to determine the point of intersection between the line of baffle characteristics, which depends on producer specifications:

$$Me = Me_0 \cdot \lambda^m \tag{1}$$

and the characteristic curve of the design data:

$$Me = \int_{\vartheta_{W2}}^{\vartheta_{W1}} \frac{c_W \cdot d\vartheta_W}{h_{LS} - h_L} \tag{2}$$

With the air number λ_A and, with known friction factor ξ as a function of the air velocity in the cooling tower u_L, one can determine the sprinkled area in relation to the height of cooling towers with natural draught, as well as the sprinkled area in relation to the power consumption of the fan in cooling towers with mechanical draught.

Definitions:

ϑ_W in °C	Water temperature	h in kJ/kg	Enthalpy
ϑ_L in °C	Temperature of the air measured at the dry thermometer	ξ in $(-)$	Friction factor
		u in m/s	Velocity
ϑ_{Lf} in °C	Temperature of the air measured at the wet thermometer	A	The values precis
		J	Auxiliary figure
p in bar	Ambient pressure	KS	For simplification
p_D in bar	Partial pressure of the steam in the air		
λ in $(-)$	$= \dot{m}_L/\dot{m}_W$ air number		
$\dot{m}_L$ in kg/s	Mass flow of air		
$\dot{m}_W$ in kg/s	Mass flow of water		
c_{pL} in kJ/(kg·K)	Specific heat capacity of dry air		**Indexes:**
c_{pD} in kJ/(kg·K)	Specific heat capacity of steam	0	At 0°C
r_o in kJ/kg	Evaporation enthalpy of water at 0°C	1	At the entrance
x in kg/kg	Absolute water content of the air	2	At the exit
N in $(-)$	Number of steps for iteration	m	Average value
Me in $(-)$	Merkel number	W	Water
c_W in kJ/(kg·K)	Average specific heat capacity of water in the liquid state	L	Air
		S	Saturation
		A	Design

Additional information of this book

(Engineering Reference Book on Energy and Heat; 978-3-642-51125-7;

978-3-642-51125-7_OSFO6) is provided:

http://Extras.Springer.com

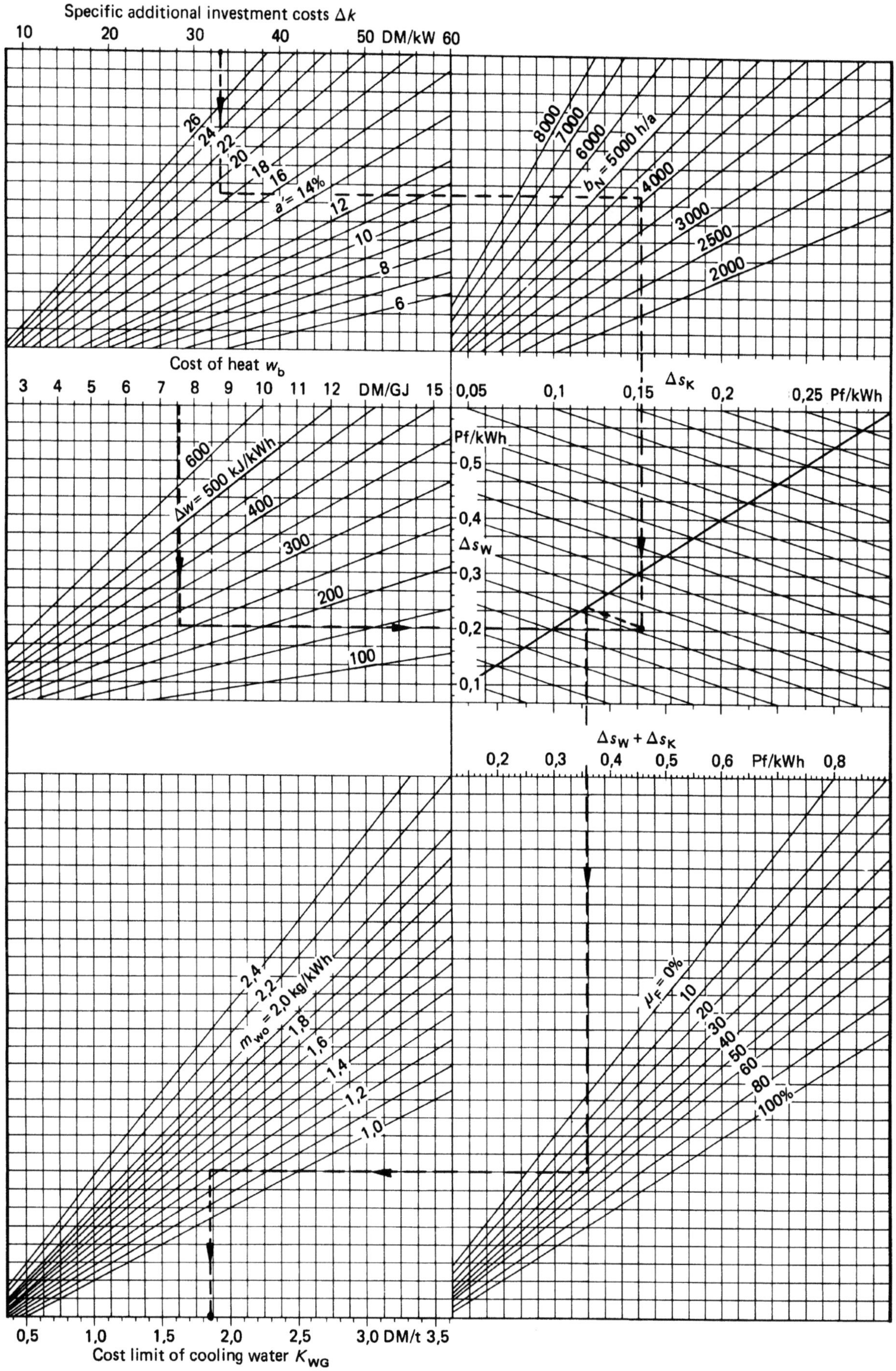

D. Blanck, E. Sauer

Explanation

The cost limit of cooling water K_{WG} are the costs, at which local costs for raw water (incl. cost for water treatment, waste water, fees for waste water etc.) and the higher costs for investment and fuel consumption in case of dry cooling are equivalent to the higher costs of the additional water in case of wet cooling. At known site conditions, the cost limit for cooling water may be a tool in power plant planning.

If the costs for the additional water are lower than K_{WG}, the wet cooling system is more economical; if the costs are higher, than the dry one is preferable.

The values of the example are only valid for special assumptions on the operating conditions for a power station. For further information see bibliography.

Calculation formula:

$$K_{\mathrm{WG}} = \frac{\Delta w \cdot w_{\mathrm{b}} + \Delta k \cdot a'/b_{\mathrm{N}}}{m_{\mathrm{w0}} \cdot (1 + \mu_{\mathrm{F}})}$$

Example:

Known values:

Specific additional investment costs of dry cooling	Δk	$= 33.10 \text{ DM/kW}$
Annuity plus costs for maintenance etc.	a'	$= 0.186/\text{a} \ (18.6\%)$
Hours of nominal output per year	b_{N}	$= 4150 \text{ h/a}$
Cost of heat (fuel costs)	w_{b}	$= 7.56 \text{ DM/GJ}$
Specific *additional* heat consumption (fuel) of dry cooling system versus wet cooling system	Δw	$= 260 \text{ kJ/kWh}$
Blow down in case of wet cooling system	μ_{F}	$= 0.4 \ (40\%)$
Specific evaporation losses (average value per year)	m_{wo}	$= 1.33 \text{ kg/kWh}$

Calculation:

Additional costs of investment	Δs_{k}	$= 0.1484 \text{ Pf/kWh}$
Additional costs of heat consumption	Δs_{W}	$= 0.196 \text{ Pf/kWh}$
Additional costs of electricity in case of dry cooling	$\Delta s_{\mathrm{W}} + \Delta s_{\mathrm{K}}$	$= 0.345 \text{ Pf/kWh}$

Result:

Cost limit for cooling water	K_{WG}	$= 1.85 \text{ DM/t Water}$.

Bibliography

Der Kühlwassergrenzpreis eines Kondensationskraftwerkes, VIK, Essen 1970

$k/d = 0.05$
0,033
0,025
0,0167
0,0125
0,010
0,005
0,0033
0,002
0,001
$5 \cdot 10^{-4}$
$2 \cdot 10^{-4}$
10^{-4}
$5 \cdot 10^{-5}$
$2 \cdot 10^{-5}$
10^{-5}
$k = 0$
Rough tubes
Approximate boundary curve (Re · λ · (k/d) = 200)
Transit region
Turbulent flow in theoretically smooth tubes
Unstable
0,0473
0,0276
Laminar flow $\lambda = 64/Re$
Critical value Re > 2320
The flow is
laminar
turbulent
Reynolds number Re
Tube friction factor λ

K. Wagner

Explanation

Type of flow	Law of flow
laminar	$\lambda = 64/\mathrm{Re}$
turbulent smooth	$1/\sqrt{\lambda} = 2 \cdot \lg\left[(\mathrm{Re} \cdot \sqrt{\lambda})/2.51\right]$
transient region	$1/\sqrt{\lambda} = -2 \cdot \lg\left[2.51/(\mathrm{Re} \cdot \sqrt{\lambda}) + k/(3.71 \cdot d)\right]$
turbulent rough	$1/\sqrt{\lambda} = 2 \cdot \lg\left[(3.71 \cdot d/k)\right]$

Definitions:

d in m	Inner diameter of the tube
k in m	Absolute hydraulic roughness
k/d in $(-)$	Relative hydraulic roughness
w in m/s	Velocity
v in m^2/s	Kinematic viscosity of the fluid
λ in $(-)$	Tube friction factor
Re in $(-)$	$= (w \cdot d)/v$; Reynolds number

Example:

Flow through a newly welded water pipe made of steel

Known values:

Inner diameter	$d = 2.0$ m
Velocity	$w = 3.5$ m/s
Kinematic viscosity (20 °C)	$v(20\,°\mathrm{C}) = 1 \cdot 10^{-6}$ m^2/s
Absolute hydraulic roughness	$k = 10^{-4}$ m (according to sheet 9.1.1, a higher value chosen for safety reasons)

Results:

Reynolds number	$\mathrm{Re} = 7 \cdot 10^6$
Relative hydraulic roughness	$k/d = 5 \cdot 10^{-5}$
Tube friction factor	$\lambda = 0.011$
The chosen factor (including a safety factor for fouling) is	$\lambda = 0.015$ (to 0.020)

Bibliography

Dubbel: Taschenbuch für den Maschinenbau. Berlin, Heidelberg, New York: Springer-Verlag 1981

Eck, B.: Technische Strömungslehre. Vol. 2; 8th edition Berlin, Heidelberg, New York: Springer-Verlag 1981

VDI-Wärmeatlas. 4th edition Düsseldorf: VDI-Verlag 1984

Herning, F.: Stoffströme in Rohrleitungen. 4th edition Düsseldorf: VDI-Verlag 1966

Schmidt, D.: Stahlrohr-Handbuch. 10th edition Aufl. Vulkan-Verlag 1986

Richter, H.: Rohrhydraulik. 5th edition Aufl. Berlin, Heidelberg, New York: Springer-Verlag 1971

Pipe material, manufacturing process	Pipe condition, condition of the pipe wall	Hydraulic roughness k $[10^{-3}\,\mathrm{m}]$
Asbestos cement pressurized pipes	new, nominal diameter 0.050 m 0.100 m 0.150 m 0.200 m 0.300 m used	0.016 0.015 0.013 0.010 0.030 k-value usually lower
Glass	new used	0.001 to 0.002 up to 0.003
Synthetic material	new used	up to 0.002 up to 0.003
Rubber	new, pressurized rubber tube, technical smooth	~ 0.0016
Clay	new, burned drain pipe made of raw clay bricks	~ 0.7 ~ 9.0
Pipes made of brick work	normal flushed joints (channels, too)	1.3 to 3.0

Bibliography

Dubbel: Taschenbuch für den Maschinenbau. Berlin, Heidelberg, New York: Springer-Verlag 1981

Eck, B.: Technische Strömungslehre. Vol. 2; 8th edition Berlin, Heidelberg, New York: Springer-Verlag 1981

VDI-Wärmeatlas. 4th edition Düsseldorf: VDI-Verlag 1984

Herning, F.: Stoffströme in Rohrleitungen. 4th edition Düsseldorf: VDI-Verlag 1966

Schmidt, D.: Stahlrohr-Handbuch. 10th edition Aufl. Vulkan-Verlag 1986

Richter, H.: Rohrhydraulik. 5th edition Aufl. Berlin, Heidelberg, New York: Springer-Verlag 1971

Pipe material, manufacturing process	Pipe condition, condition of the pipe wall	Hydraulic roughness k $[10^{-3}\,\mathrm{m}]$
Steel seamless milled	new, usual rolling scale	0.02 to 0.06
	unscoured	0.02 to 0.06
	scoured	0.03 to 0.04
	narrow pipes	to 0.01
welded	new, usual rolling scale	0.04 to 0.10
coated	new, usually tinned	0.10 to 0.16
	bitumen lined	~0.05
	concrete lined	~0.18
	used pipe (reference values)	
	regular rust spots	~0.15
	moderately rusted, up to lightly encrusted	0.20 to 0.50
	moderately encrusted	~1.5
	strongly encrusted	2.0 to 4.0
	cleaned after long term operation	0.15 to 0.20
	bitumen lined, bitumen partly detached, with rust spots	~0.10
influence of media	water pipe lines, average value	0.4 to 1.2
	encrusted, rust spots	1.5 to 3.0
	steam pipe lines, average value	0.2 to 0.4
	compressed air pipe lines, average value	0.2 to 0.4
	pipe lines for coke oven and town gas with naphta sedimentation, encrusted	1.0 to 3.0
	natural gas pipe lines, average value	0.1 to 0.2
	blast furnace gas pipe lines	0.8 to 0.2
Concrete	new, usually smooth	0.3 to 0.8
	usually medium rough	1.0 to 2.0
	usually rough	2.0 to 3.0
	steel reinforced concrete, carefully smoothed	0.10 to 0.15
	centrifugal cast concrete without roughcast	0.20 to 0.80
	centrifugal cast concrete with smooth roughcast	0.10 to 0.15
	used, water pipe line with smooth roughcast after long term operation	0.2 to 0.3
	pipe lines with sockets, average value	0.2
	pipe lines without sockets, average value	2.0
Cast iron	new, with typical outer crust, not bitumen lined	0.10 to 0.15
	bitumen lined	
	used, with rust spots	0.50 to 1.0
	slightly up to badly encrusted	1.5 to 3.0
	with big rust spots	4.5
	cleaned after long term operation	0.5 to 1.5
	drinking water pipe lines in operation	1.5
	pipe lines for urban sewage removal	1.5 to 3.0
Sheet steel tinned	smooth, for airducts and ventilating pipes	0.07 to 1.20
Non-ferrous heavy metal for example copper, brass, bronze, aluminium etc. (when used as cover metal too)	new, drawn or extruded pipe technical smooth	0.001 to 0.002
	used	to 0.003

K. Wagner

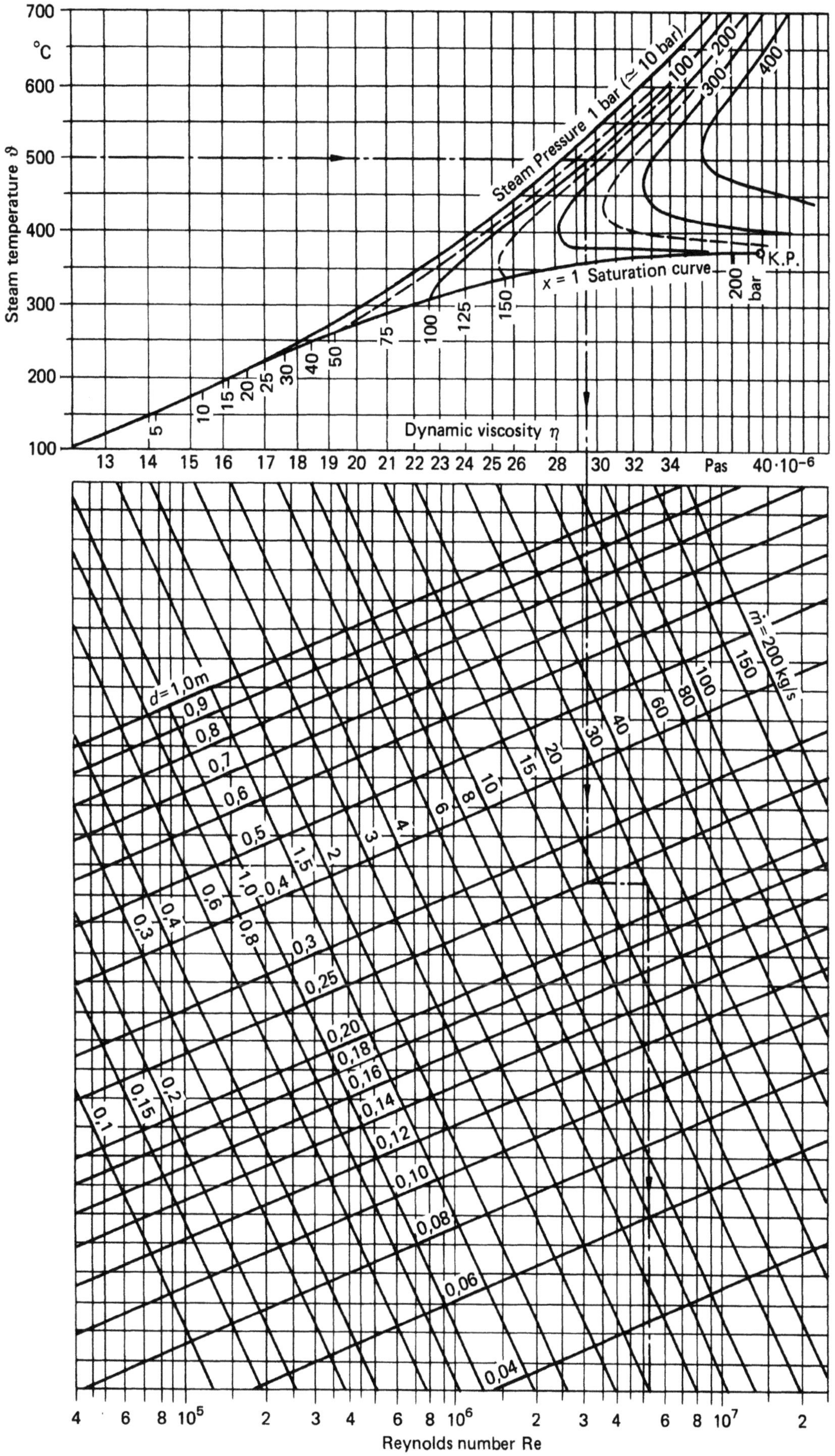

Steam temperature ϑ
°C
700
600
500
400
300
200
100
Steam Pressure 1 bar (≃ 10 bar)
100
200
300
400
K.P.
x = 1 Saturation curve
200 bar
5
10
15
20
25
30
40
50
75
100
125
150
Dynamic viscosity η
13 14 15 16 17 18 19 20 21 22 23 24 25 26 28 30 32 34 Pas 40·10⁻⁶
d = 1,0 m
0,9
0,8
0,7
0,6
0,5
0,4
0,3
0,25
0,20
0,18
0,16
0,14
0,12
0,10
0,08
0,06
0,04
0,3
0,4
0,6
1,0
0,8
1,5
2
3
4
6
8
10
15
20
30
40
60
80
100
150
ṁ = 200 kg/s
0,1
0,15
0,2
4 6 8 10⁵ 2 3 4 6 8 10⁶ 2 3 4 6 8 10⁷ 2
Reynolds number Re

K. Wagner

Explanation

Calculation of the Reynolds number:

$$\mathrm{Re} = (w \cdot d)/v = (4 \cdot \dot{m})/(\pi \cdot d \cdot v \cdot \varrho)$$
$$= (w \cdot d \cdot \varrho)/\eta = (4 \cdot \dot{m})/(\pi \cdot d \cdot \eta)$$

Definitions:

w in m/s	Average flow velocity
d in m	Inner diameter of the tube
v in m^2/s	Kinematic viscosity of the fluid
η in Pa $\cdot$ s	Dynamic viscosity
$\dot{m}$ in kg/s	Mass flow
ϱ in kg/m^3	Density

Example:

Known values:

Steam pressure	p	$= 80$ bar
Steam temperature	ϑ	$= 500\,°\mathrm{C}$
Inner diameter	d	$= 0.25$ m
Mass flow	$\dot{m}$	$= 30$ kg/s

Results:

Dynamic viscosity	η	$= 29.3 \cdot 10^{-3}$ Pa $\cdot$ s
Reynolds number	Re	$= 5.2 \cdot 10^6$

Bibliography

Dubbel: Taschenbuch für den Maschinenbau. Berlin, Heidelberg, New York: Springer-Verlag 1981

Eck, B.: Technische Strömungslehre. Vol. 2; 8th edition Berlin, Heidelberg, New York: Springer-Verlag 1981

VDI-Wärmeatlas. 4th edition Düsseldorf: VDI-Verlag 1984

Herning, F.: Stoffströme in Rohrleitungen. 4th edition Düsseldorf: VDI-Verlag 1966

Schmidt, D.: Stahlrohr-Handbuch. 10th edition Aufl. Vulkan-Verlag 1986

Richter, H.: Rohrhydraulik. 5th edition Aufl. Berlin, Heidelberg, New York: Springer-Verlag 1971

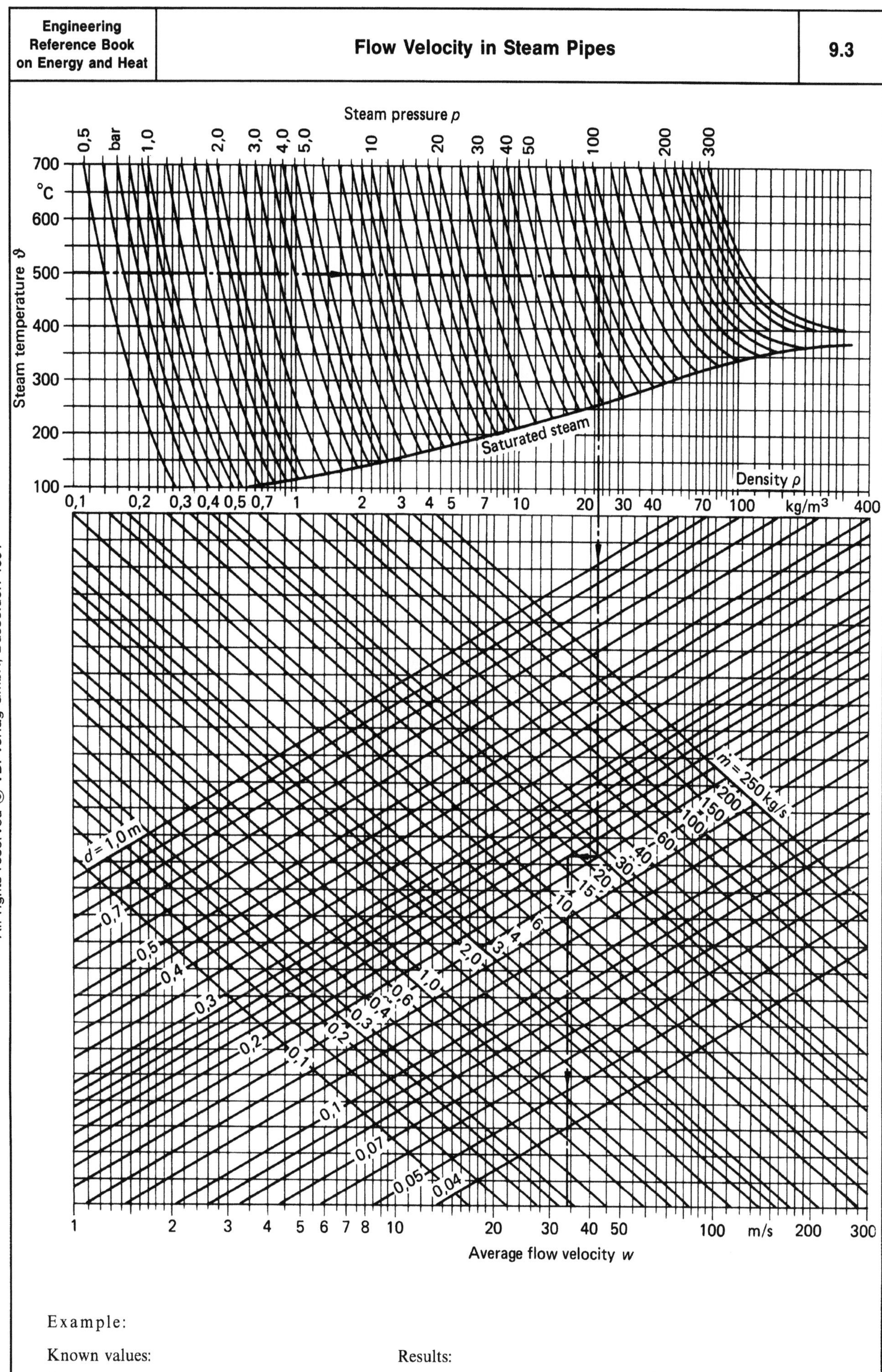

Example:

Known values:		Results:	
Steam temperature	$\vartheta = 500\,°C$	Density	$\varrho = 24$ kg/m³
Steam pressure	$p = 80$ bar	Average flow velocity	$w = 33$ m/s
Mass flow	$\dot{m} = 25$ kg/s		

K. Wagner

Explanation

The velocity of steam flow in pipes can be calculated from:

$$w = \dot{m}/(\varrho \cdot A) = (4 \cdot \dot{m})/(\varrho \cdot \pi \cdot d^2)$$

Definitions:

w in m/s	Average flow velocity
ϱ in kg/m^3	Density of the steam
A in m^2	Cross section of the pipe
d in m	Inner diameter of the pipe
$\dot{m}$ in kg/s	Mass flow

The calculated flow velocity depends on the steam conditions (temperature, pressure). By calculating the average steam conditions, one obtains the average flow velocity in the pipe.

In the practice, the flow velocity at the end of the pipe line (terminal velocity) is of great importance (e.g. concerning the noise level). For these calculations, the real steam conditions at the end of the pipe line (considering changes in temperature and pressure) have to be used.

Bibliography

Dubbel: Taschenbuch für den Maschinenbau. Berlin, Heidelberg, New York: Springer-Verlag 1981

Eck, B.: Technische Strömungslehre. Vol. 2; 8th edition Berlin, Heidelberg, New York: Springer-Verlag 1981

VDI-Wärmeatlas. 4th edition Düsseldorf: VDI-Verlag 1984

Herning, F.: Stoffströme in Rohrleitungen. 4th edition Düsseldorf: VDI-Verlag 1966

Schmidt, D.: Stahlrohr-Handbuch. 10th edition Aufl. Vulkan-Verlag 1986

Richter, H.: Rohrhydraulik. 5th edition Aufl. Berlin, Heidelberg, New York: Springer-Verlag 1971

$(p_1^2 - p_2^2)/l$
10^1 6 4 3 2 10^0 6 4 3 2 10^{-1} 6 4 3 2 10^{-2} 6 bar^2/m
1,0
0,9
0,8
0,7
0,6
0,5
0,4
0,3
0,25
0,2
0,18
0,16
0,14
0,12
$d = 0,1$ m
0,08
0,06
0,04
$\dot{m} = 250$ kg/s
200
150
100
80
60
40
30
20
10
8
6
4
3
2
1,5
1,0
0,8
0,6
0,4
0,3
0,2
0,15
0,1
425
400
380
360
300
100
150
200
450
475
500
550
600
700
800
300 200 100 bar 0
Steam pressure p
$\lambda_R = 0,1$
0,09
0,07
0,05
0,04
0,03
0,02
0,016
0,012
0,010
0,008

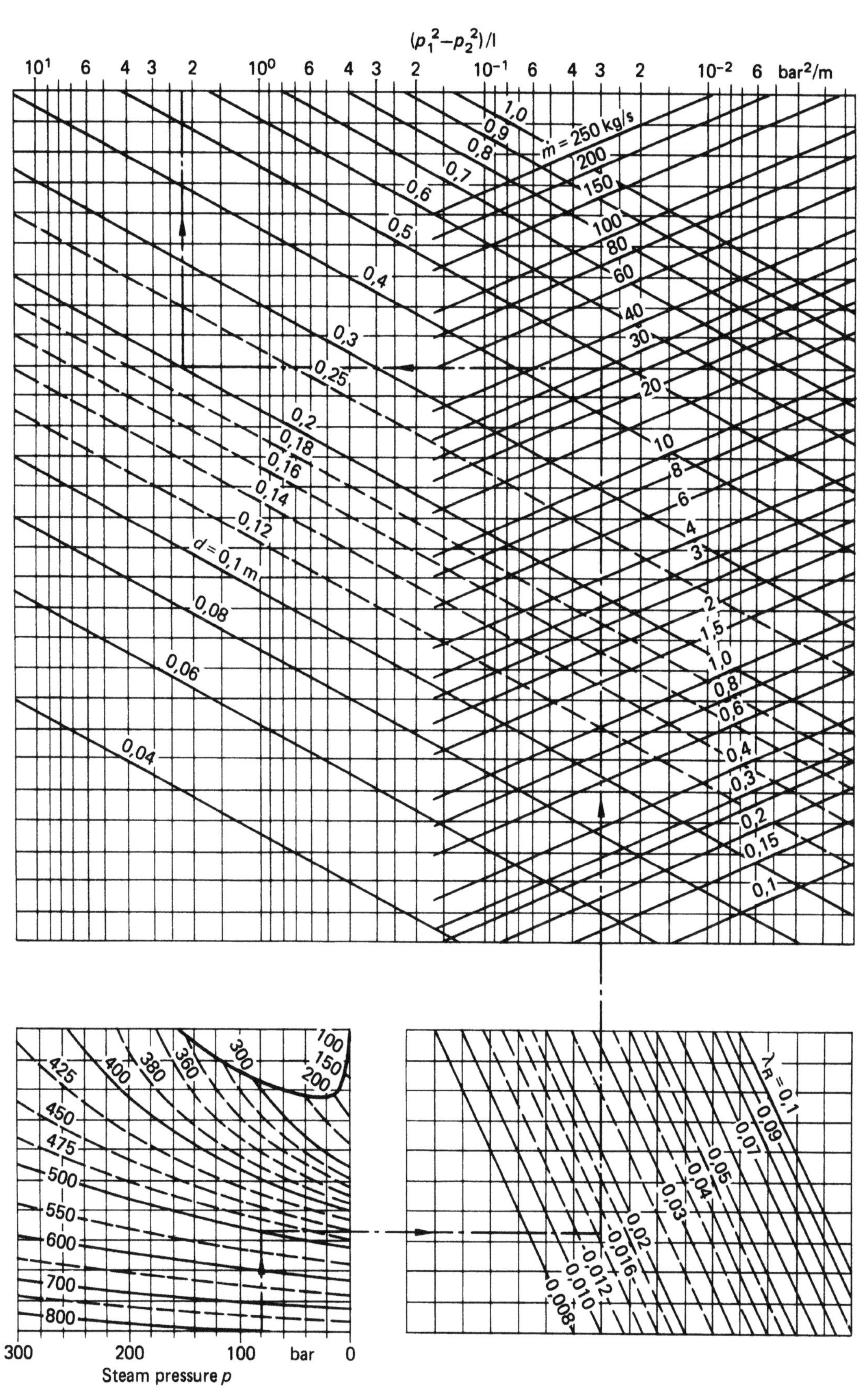

K. Wagner

Explanation

The general form of the equation of pressure drop at isothermal flow is:

$$(p_1^2 - p_2^2)/(2 \cdot p_1) = (\lambda \cdot l \cdot w_1^2 \, \varrho_1)/(2 \cdot d) \quad [\text{Pa}] \, .$$

With the steam pressure p in bar (1 bar $= 10^5$ Pa) and the mass flow $\dot{m}$ in kg/s, one gets the numerical equation:

$$(p_1^2 - p_2^2)/l = 1.62 \cdot 10^{-5} \cdot \lambda \cdot \dot{m}^2 \cdot p_1/(\varrho_1 \cdot d^5) \quad [\text{bar}^2/\text{m}] \, .$$

Definitions:

p_1 in bar   Steam pressure at the entrance of the pipe line
p_2 in bar   Steam pressure at the exit of the pipe line
l in m   Length (i.e. equivalent length) of the pipe line
λ in $(-)$ Friction factor (see worksheet No 9.1)
$\dot{m}$ in kg/s   Mass flow
ϱ in kg/m^3 Density at the entrance of the pipe line
w in m/s   Average flow velocity
d in m   Inner diameter of the pipe

Example:

Known values:

Steam temperature	$\vartheta = 500\,°\text{C}$	Inner diameter	$d = 0.2\,\text{m}$
Steam pressure at the entrance	$p = 80\,\text{bar}$	Density	$\varrho = 24\,\text{kg/m}^3$
Hydraulic roughness	$k = 5 \cdot 10^{-5}\,\text{m}$	Mass flow	$\dot{m} = 30\,\text{kg/s}$

Results:

From worksheet 9.2, one gets:

Dynamic viscosity $\eta \;= 29.4 \cdot 10^{-6}\,\text{Pa} \cdot \text{s}$
Reynolds number $\text{Re} = \;6.5 \cdot 10^6$

With $k = 5 \cdot 10^{-5}\,\text{m}$ and $k/d = 5 \cdot 10^{-5}/0.2 = 2.5 \cdot 10^{-4}$, one obtains from worksheet No 9.1

$$\lambda = 0.0144 \, .$$

thus:

$$(p_1^2 - p_2^2)/l = 1.62 \cdot 10^{-5} \cdot 0.0144 \cdot 30^2 \cdot 80/(24 \cdot 0.2^5) = 2.187\,\text{bar}^2/\text{m} \, .$$

On a pipe line length of $l = 155$ m the pressure drop is:

$$p_1^2 - p_2^2 = 339\,\text{bar}^2$$
$$p_2 = \sqrt{p_1^2 - 339\,\text{bar}^2} = \sqrt{6\,400\,\text{bar}^2 - 339\,\text{bar}^2} = 77.85\,\text{bar}$$
$$p_1 - p_2 = 2.15\,\text{bar}$$

Bibliography

Dubbel: Taschenbuch für den Maschinenbau. Berlin, Heidelberg, New York: Springer-Verlag 1981

Eck, B.: Technische Strömungslehre. Vol. 2; 8th edition Berlin, Heidelberg, New York: Springer-Verlag 1981

VDI-Wärmeatlas. 4th edition Düsseldorf: VDI-Verlag 1984

Herning, F.: Stoffströme in Rohrleitungen. 4th edition Düsseldorf: VDI-Verlag 1966

Schmidt, D.: Stahlrohr-Handbuch. 10th edition Aufl. Vulkan-Verlag 1986

Richter, H.: Rohrhydraulik. 5th edition Aufl. Berlin, Heidelberg, New York: Springer-Verlag 1971

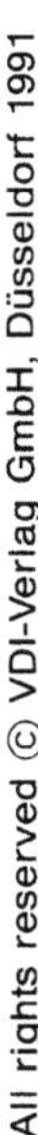

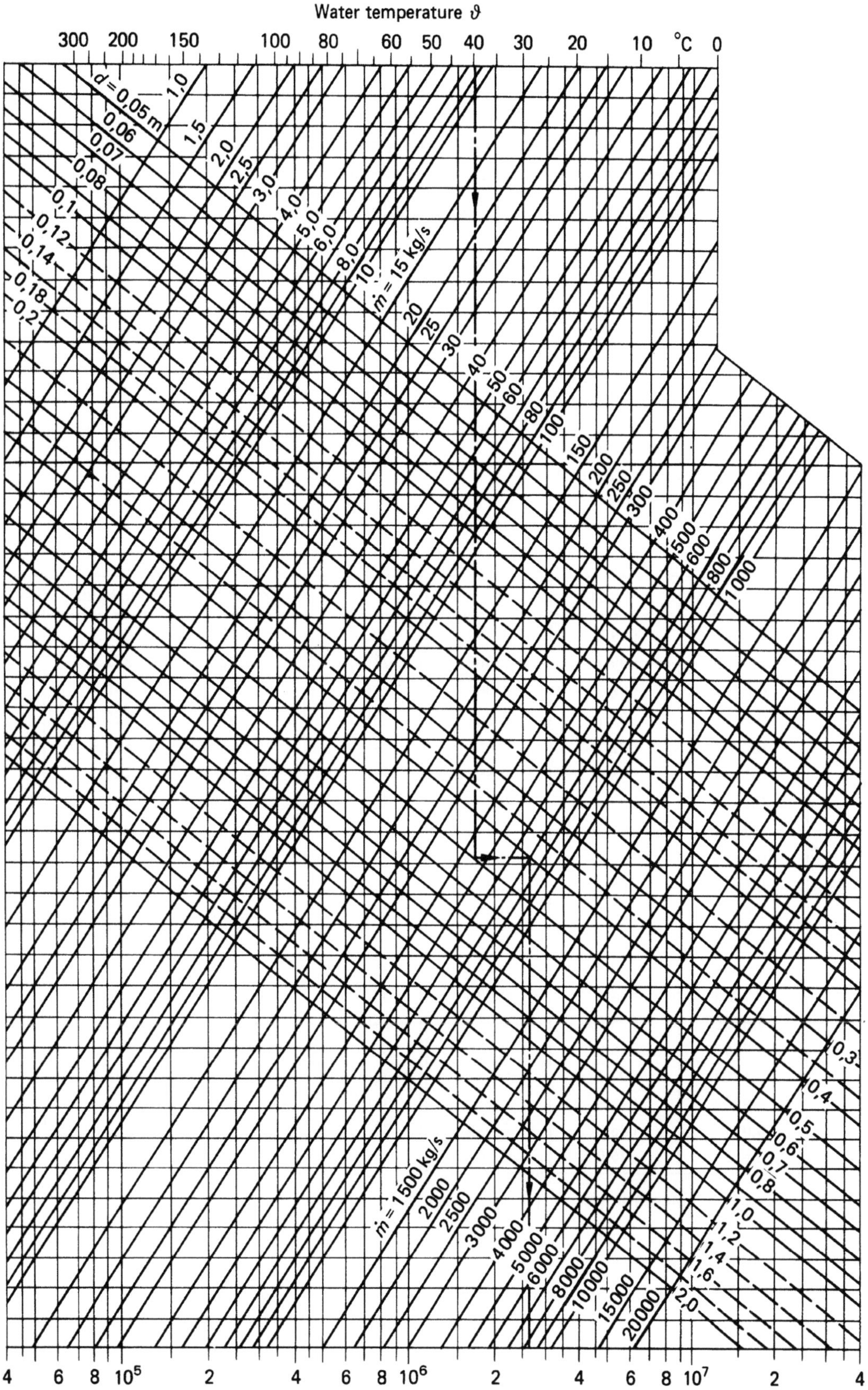

Example:

Known values:

Water temperature ϑ = 40 °C
Inner diameter d = 0.5 m
Mass flow $\dot{m}$ = 700 kg/s

Result:

Reynolds number Re = $2.7 \cdot 10^6$

K. Wagner

Explanation

Reynolds number $Re = \dfrac{w \cdot d}{v}$.

In practice, if the medium is water, the volume flow $\dot{v}$ in l/s is equal to the mass flow $\dot{m}$ in kg/s. At barometric pressure, up to a temperature of 45 °C, the difference is $< 1\%$ and at a temperature of 65 °C $< 2\%$. In high pressure systems (for example in feed water systems of steam generators), the difference will even be lower.

Worksheet 9.5 is based on the following numerical equation for water:

$$Re = 1.274 \cdot \frac{\dot{m}}{d \cdot v \cdot \varrho}$$

Definitions:

$\dot{m}$ in kg/s	Mass flow
$\dot{v}$ in l/s	Volume flow
v in m^2/s	Kinematic viscosity of water
d in m	Inner diameter of the tube
ϱ in kg/m^3	Density of the water

ϑ in °C	p in bar	ρ in kg/m^3	v in 10^{-6} m^2/s	ϑ in °C	p in bar	ρ in kg/m^3	v in 10^{-6} m^2/s
0.01	0.0061	999.8	1.792	200	15.551	864.7	0.154
10	0.0123	999.7	1.308	210	19.080	852.8	0.149
20	0.0234	998.3	1.004	220	23.201	840.4	0.144
30	0.0424	995.7	0.801	230	27.979	827.3	0.140
40	0.0737	992.2	0.658	240	33.480	813.6	0.136
50	0.1234	988.0	0.554	250	39.776	799.2	0.132
60	0.1992	983.1	0.475	260	46.940	783.9	0.129
70	0.3115	977.7	0.414	270	55.051	767.8	0.127
80	0.4736	971.6	0.365	280	64.191	750.5	0.124
90	0.7011	965.1	0.326	290	74.448	732.1	0.122
100	1.0133	958.1	0.294	300	85.917	712.2	0.120
110	1.4326	950.7	0.268	310	98.697	690.6	0.119
120	1.9854	942.8	0.246	320	112.900	666.9	0.117
130	2.7012	934.6	0.228	330	128.646	640.4	0.116
140	3.6136	925.9	0.212	340	146.079	610.2	0.115
150	4.7597	916.8	0.198	350	165.367	574.5	0.114
160	6.1804	907.3	0.187	360	186.737	528.3	0.114
170	7.9202	897.3	0.174	370	210.528	448.3	0.115
180	10.027	886.9	0.168	373.99	220.67	322.0	0.112
190	12.552	876.0	0.161				

Table 9.5.1: Density ϱ, kinematic viscosity v and steam pressure of water at the saturation curve

Bibliography

Dubbel: Taschenbuch für den Maschinenbau. Berlin, Heidelberg, New York: Springer-Verlag 1981

Eck, B.: Technische Strömungslehre. Vol. 2; 8th edition Berlin, Heidelberg, New York: Springer-Verlag 1981

VDI-Wärmeatlas. 4th edition Düsseldorf: VDI-Verlag 1984

Herning, F.: Stoffströme in Rohrleitungen. 4th edition Düsseldorf: VDI-Verlag 1966

Schmidt, D.: Stahlrohr-Handbuch. 10th edition Aufl. Vulkan-Verlag 1986

Richter, H.: Rohrhydraulik. 5th edition Aufl. Berlin, Heidelberg, New York: Springer-Verlag 1971

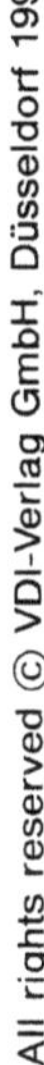

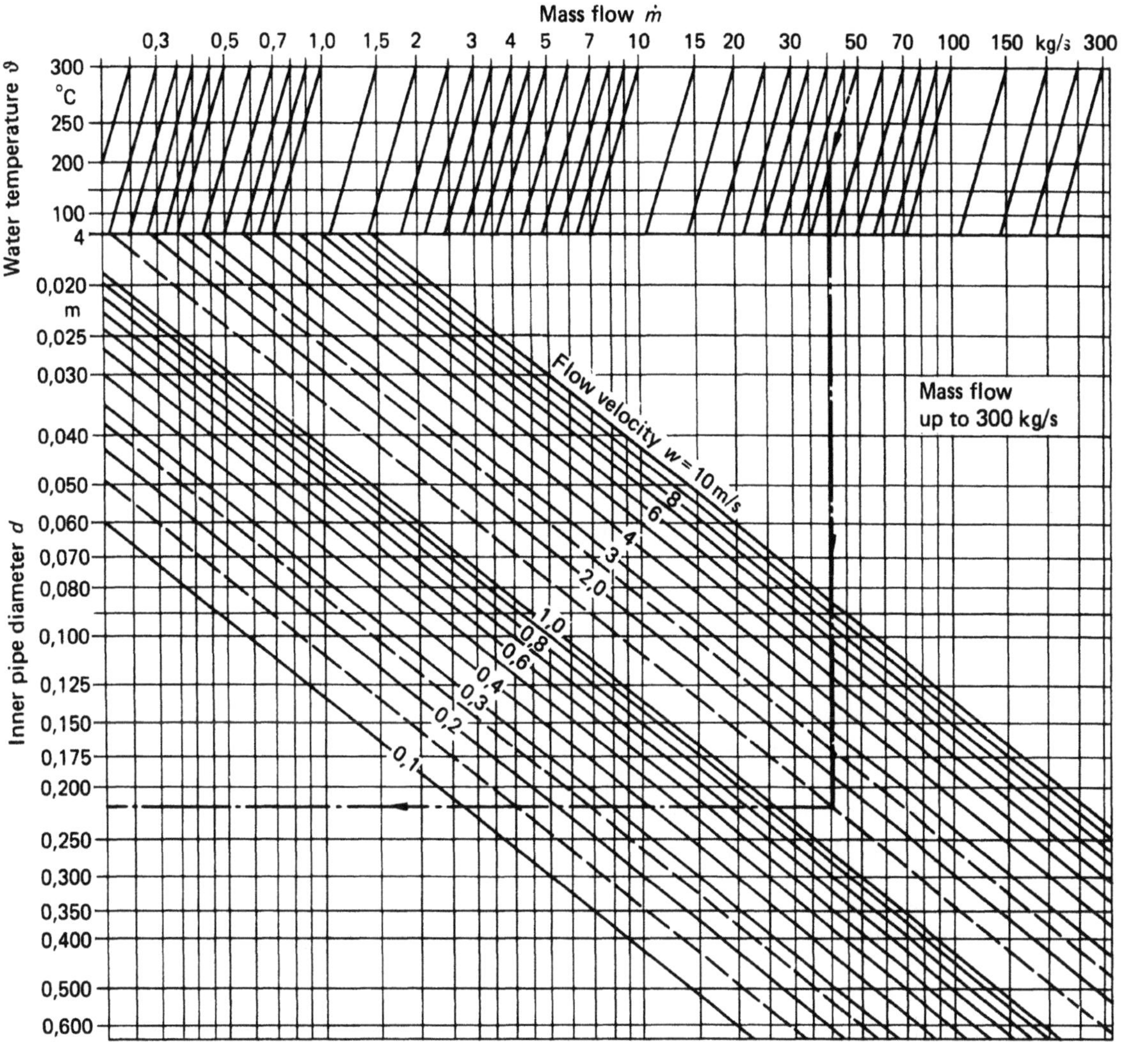
Mass flow ṁ
0,3 0,5 0,7 1,0 1,5 2 3 4 5 7 10 15 20 30 50 70 100 150 kg/s 300
Water temperature ϑ
300 °C 250 200 100 4
Inner pipe diameter d
0,020 m 0,025 0,030 0,040 0,050 0,060 0,070 0,080 0,100 0,125 0,150 0,175 0,200 0,250 0,300 0,350 0,400 0,500 0,600
Flow velocity w = 10 m/s
8 6 4 3 2,0 1,0 0,8 0,6 0,4 0,3 0,2 0,1
Mass flow
up to 300 kg/s

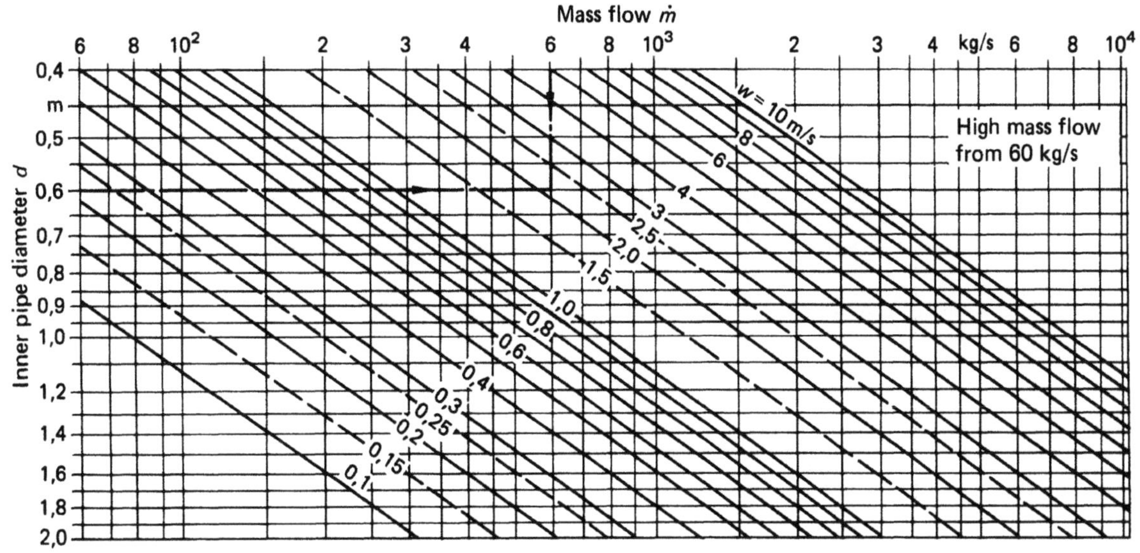
Mass flow ṁ
6 8 10^2 2 3 4 6 8 10^3 2 3 4 kg/s 6 8 10^4
Inner pipe diameter d
0,4 m 0,5 0,6 0,7 0,8 0,9 1,0 1,2 1,4 1,6 1,8 2,0
w = 10 m/s
8 6 4 3 2,5 2,0 1,5 1,0 0,8 0,6 0,4 0,3 0,25 0,2 0,15 0,1
High mass flow
from 60 kg/s

K. Wagner

Explanation

The upper part of the worksheet is valid for a water mass flow up to 300 kg/s. A correction for higher water temperatures (saturation condition) has been included.

The lower part of the worksheet is valid for a mass flow up to 10 000 kg/s.

$$\dot{m} = 0.785 \cdot d^2 \cdot w \cdot \varrho \quad [\text{kg/s}],$$

$$w = 1.274 \cdot \frac{\dot{m}}{d^2 \cdot \varrho} \quad [\text{m/s}].$$

Definitions:

$\dot{m}$ in kg/s	Mass flow
w in m/s	Flow velocity
d in m	Inner diameter of the tube
ϱ in kg/m^3	Density of the water
ϑ in °C	Water temperature

The volume flow $\dot{v}$ in l/s is practically equal to the mass flow $\dot{m}$ in kg/s at a temperature of 45 °C with an error of $<1\%$, and at a temperature of 65 °C $< 2\%$. The values for the density of the water ϱ have been listed in table 9.5.1.

Example:

1. Upper diagram

   Known values:

Water temperature	$\vartheta = 200\,°\text{C}$
Density	$\varrho = 864.68\ \text{kg/m}^3$
Mass flow	$\dot{m} = 50\ \text{kg/s}$
Flow velocity	$w = 1.5\ \text{m/s}$

   Result:

Inner diameter	$d = 0.22\ \text{m}$

2. Lower diagram

   Known values:

Water temperature	$\vartheta = 10\,°\text{C}$
Density	$\varrho \sim 1\,000\ \text{kg/m}^3$
Mass flow	$\dot{m} = 600\ \text{kg/s}$
Inner diameter	$d = 0.6\ \text{m}$

   Result:

Flow velocity	$w = 2.12\ \text{m/s}$

If in example No 1 the nominal diameter of the pipe will be chosen according to DIN 2448 to $D = 0.267$ m and the inner diameter to $d = 0.2544$ m, the flow velocity will be $w = 1.14$ m/s.

Bibliography

Dubbel: Taschenbuch für den Maschinenbau. Berlin, Heidelberg, New York: Springer-Verlag 1981

Eck, B.: Technische Strömungslehre. Vol. 2; 8th edition Berlin, Heidelberg, New York: Springer-Verlag 1981

VDI-Wärmeatlas. 4th edition Düsseldorf: VDI-Verlag 1984

Herning, F.: Stoffströme in Rohrleitungen. 4th edition Düsseldorf: VDI-Verlag 1966

Schmidt, D.: Stahlrohr-Handbuch. 10th edition Aufl. Vulkan-Verlag 1986

Richter, H.: Rohrhydraulik. 5th edition Aufl. Berlin, Heidelberg, New York: Springer-Verlag 1971

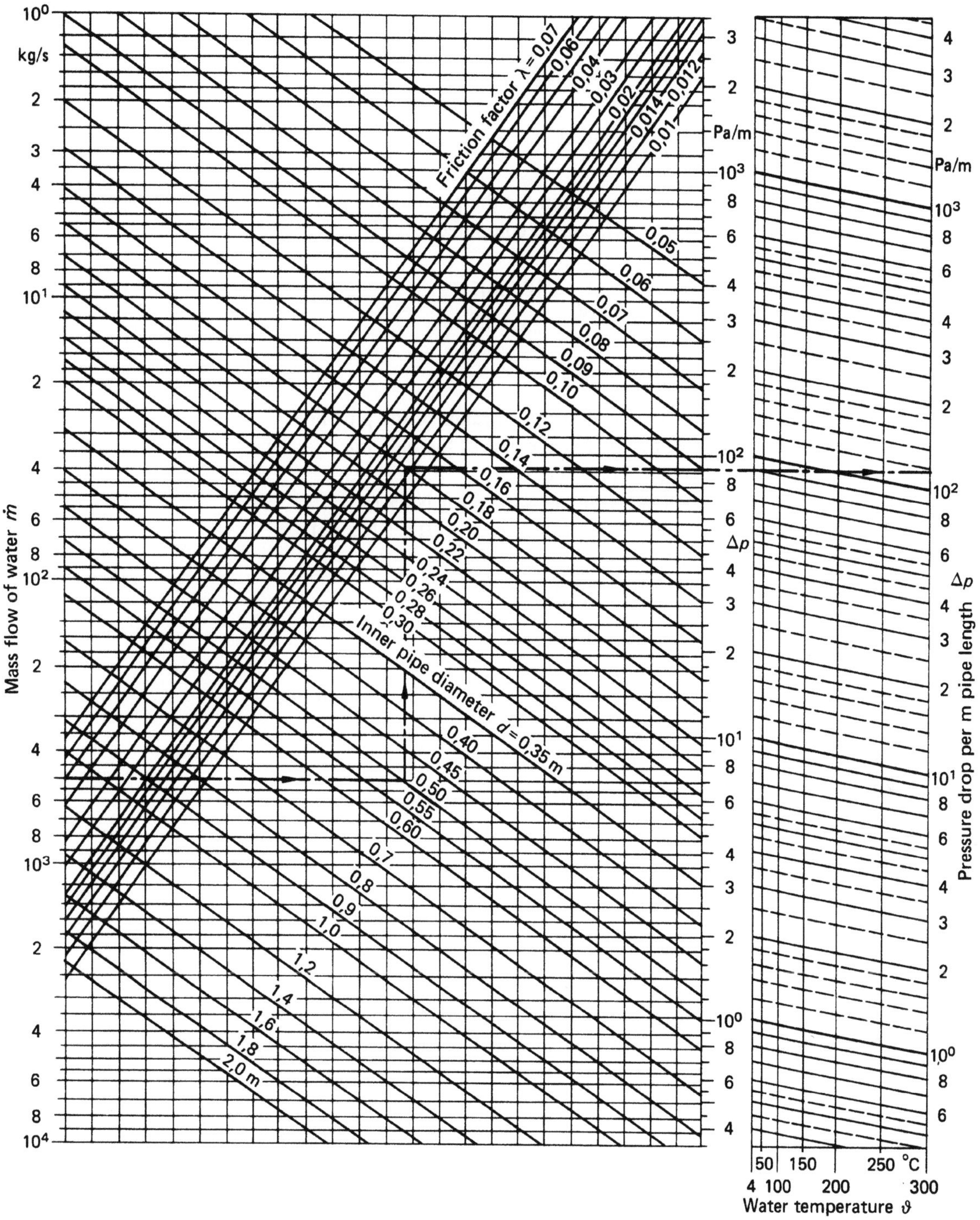

Example:

Known values:

Mass flow	$\dot m$	$= 500\ \text{kg/s}$
Inner diameter	d	$= 0.5\ \text{m}$
Water temperature	ϑ	$= 300\,^{\circ}\text{C}\ (\varrho = 712.2\ \text{kg/m}^3)$
Friction factor	λ	$= 0.013$

(from worksheet No 9.1)

Result:

$$\frac{p_1 - p_2}{l} = 118\ \text{Pa/m}$$

for a pipe length of

$l = 280\ \text{m}$ is $p_1 - p_2 = 33\,040\ \text{PA} \sim 0.33\ \text{bar}$

K. Wagner

Explanation

The pressure drop, which occurs at the flow of incompressible media through pipes, can be calculated by the following numerical equation:

$$p_1 - p_2 = \lambda \cdot \varrho \, \frac{l \cdot w^2}{2 \cdot d} \quad [\text{Pa}].$$

According to worksheet No 9.6 is

$$w = 1.274 \cdot \frac{\dot{m}}{d^2 \cdot \varrho} \quad [\text{m/s}].$$

From this, the numerical equation for the pressure drop per meter length of pipe, shown in worksheet No 9.7, can be computed:

$$\frac{p_1 - p_2}{l} = 0.811 \cdot \lambda \cdot \frac{\dot{m}^2}{d^5 \cdot \varrho} \quad [\text{Pa/m}].$$

Definitions:

p_1 in Pa	Pressure head at the entrance of the pipe line
p_2 in Pa	Pressure head at the exit of the pipe line
l in m	Length of pipe line
λ in $(-)$	Friction factor (see worksheet No 9.1)
ϱ in kg/m^3	Density of the water (see worksheet No 9.5)
$\dot{m}$ in kg/s	Mass flow
d in m	Inner diameter of the tube

Example:

Known values:

			Result:
Mass flow	$\dot{m}$	$= 500\ \text{kg/s}$	$\dfrac{p_1 - p_2}{l} = 118\ \text{Pa/m}$
Inner diameter	d	$= 0.5\ \text{m}$	
Water temperature	ϑ	$= 300\,°\text{C}$	
Density	ϱ	$= 712.2\ \text{kg/m}^3$	
Friction factor	λ	$= 0.013$	
(from worksheet No 9.1)			

For a pipe length of $l = 280$ m one gets a total pressure drop of $p_1 - p_2 = 118\ \text{Pa/m} \cdot 280\ \text{m} = 33\,040\ \text{Pa}$.

For the calculation of pipe networks, nowadays extensive proved computer programs exist. They are partly combined with plot programs, which plot graphs of the computed data.

Bibliography

Dubbel: Taschenbuch für den Maschinenbau. Berlin, Heidelberg, New York: Springer-Verlag 1981

Eck, B.: Technische Strömungslehre. Vol. 2; 8th edition Berlin, Heidelberg, New York: Springer-Verlag 1981

VDI-Wärmeatlas. 4th edition Düsseldorf: VDI-Verlag 1984

Herning, F.: Stoffströme in Rohrleitungen. 4th edition Düsseldorf: VDI-Verlag 1966

Schmidt, D.: Stahlrohr-Handbuch. 10th edition Aufl. Vulkan-Verlag 1986

Richter, H.: Rohrhydraulik. 5th edition Aufl. Berlin, Heidelberg, New York: Springer-Verlag 1971

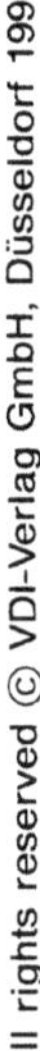

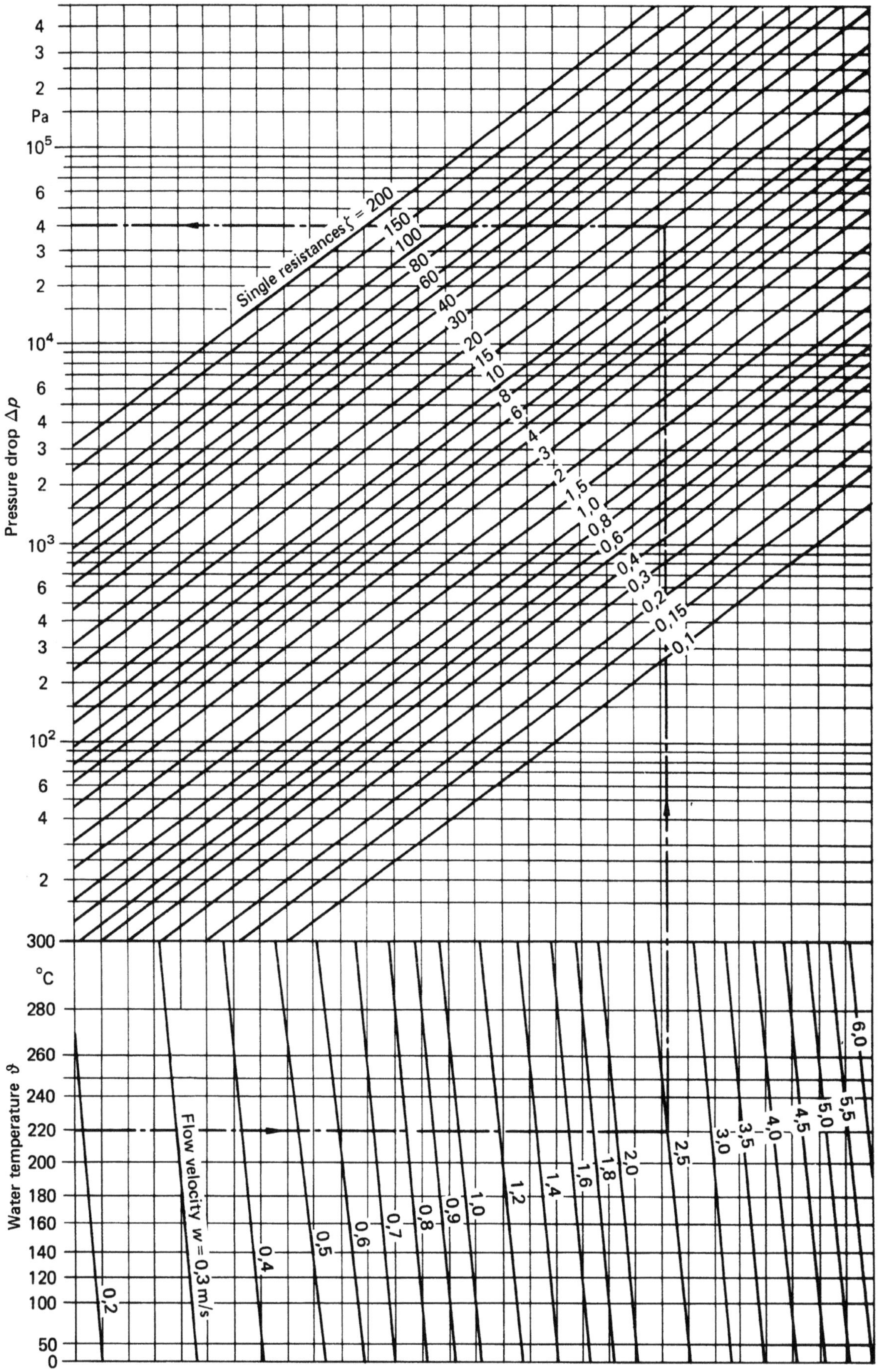
Pressure drop Δp
Pa
Single resistances ζ = 200
150
100
80
60
40
30
20
15
10
8
6
4
3
2
1,5
1,0
0,8
0,6
0,4
0,3
0,2
0,15
0,1
Water temperature ϑ
°C
Flow velocity w = 0,3 m/s
0,2
0,4
0,5
0,6
0,7
0,8
0,9
1,0
1,2
1,4
1,6
1,8
2,0
2,5
3,0
3,5
4,0
4,5
5,0
5,5
6,0

K. Wagner

Explanation

In water pipes, additional pressure drop is caused by changes of direction, cross section, fittings and other installations. In the calculation, they are considered by the resistance coefficient ζ. The value of ζ depends on the kind of the resistance. The most common resistance factors have been listed in the worksheets No 9.10 up to 9.13.

If there are several different installations in a pipe line, all pressure drops have to be added.

$$\Delta p = \Sigma \left(\zeta_i \cdot \frac{\varrho_i}{2} \cdot w_i^2 \right)$$

Definitions:

Δp in Pa	Pressure by single resistance
ζ_i in $(-)$	Resistance coefficient
ϱ_i in kg/m^3	Density of the water $(\varrho = f(\vartheta))$
w_i in m/s	Flow velocity in the installation (from worksheet No 9.7 or 9.18)

Example:

Known values:

			Result:
Resistance factor	ζ	$= 15$	Pressure drop $\quad \Delta p = 40\,000$ Pa
Temperature	ϑ	$= 220\,°C$	
Flow velocity	w	$= 2.5$ m/s	
Density	ϱ	$= 840.3$ kg/m^3	

Bibliography

Dubbel: Taschenbuch für den Maschinenbau. Berlin, Heidelberg, New York: Springer-Verlag 1981

Eck, B.: Technische Strömungslehre. Vol. 2; 8th edition Berlin, Heidelberg, New York: Springer-Verlag 1981

VDI-Wärmeatlas. 4th edition Düsseldorf: VDI-Verlag 1984

Herning, F.: Stoffströme in Rohrleitungen. 4th edition Düsseldorf: VDI-Verlag 1966

Schmidt, D.: Stahlrohr-Handbuch. 10th edition Aufl. Vulkan-Verlag 1986

Richter, H.: Rohrhydraulik. 5th edition Aufl. Berlin, Heidelberg, New York: Springer-Verlag 1971

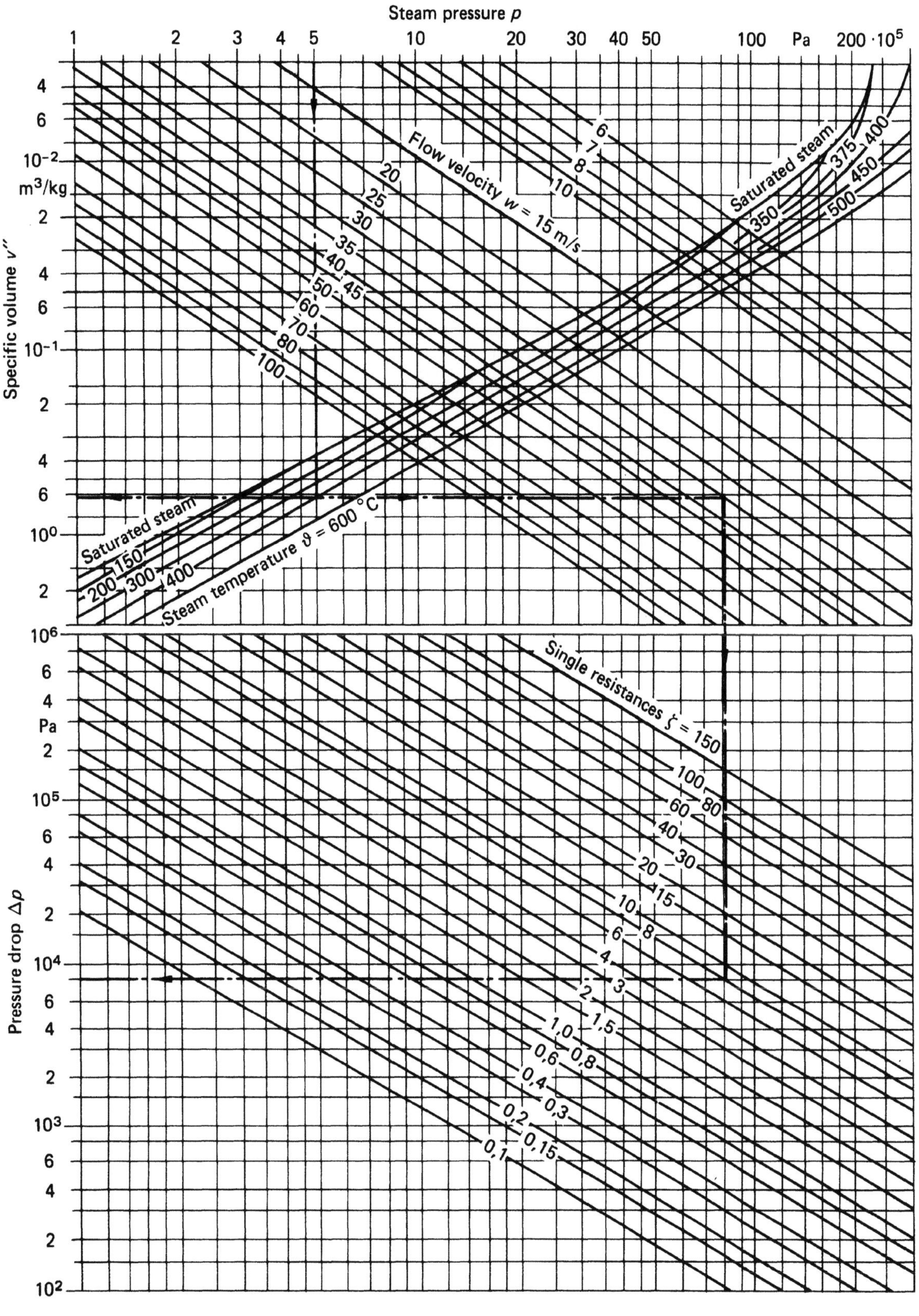

Example:

Known values:

Pressure	p	$= 5 \cdot 10^5$ Pa
Temperature	ϑ	$= 400\,°C$
Flow velocity	w	$= 35$ m/s
Resistance factor	ζ	$= 8$

Result:

Specific volume	v''	$= 0.617\ \mathrm{m^3/kg}$
Density	ϱ	$= 1.621\ \mathrm{kg/m^3}$
Pressure drop	Δp	$= 7\,950$ Pa

K. Wagner

Explanation

In steam pipes, additional pressure drop is caused by changes of direction, cross section, fittings and other installations. In the calculation, they are considered by the resistance coefficient ζ. The value of ζ depends on the kind of the resistance. The most common resistance factors have been listed in the worksheets No 9.10 up to 9.13.

If there are several different installations in a pipe line, all pressure drops have to be added.

$$\Delta p = \sum \left(\zeta_i \cdot \frac{\varrho_i}{2} \cdot w_i^2 \right) \quad [\text{Pa}] .$$

Definitions:

Δp in Pa	Pressure by single resistance
ζ_i in $(-)$	Resistance coefficient
ϱ_i in kg/m^3	Density of the steam (calculated from pressure and temperature $\varrho = 1/v''$)
w_i in m/s	Flow velocity (from worksheet No 9.3)

Bibliography

Dubbel: Taschenbuch für den Maschinenbau. Berlin, Heidelberg, New York: Springer-Verlag 1981

Eck, B.: Technische Strömungslehre. Vol. 2; 8th edition Berlin, Heidelberg, New York: Springer-Verlag 1981

VDI-Wärmeatlas. 4th edition Düsseldorf: VDI-Verlag 1984

Herning, F.: Stoffströme in Rohrleitungen. 4th edition Düsseldorf: VDI-Verlag 1966

Schmidt, D.: Stahlrohr-Handbuch. 10th edition Aufl. Vulkan-Verlag 1986

Richter, H.: Rohrhydraulik. 5th edition Aufl. Berlin, Heidelberg, New York: Springer-Verlag 1971

Plain bends

angle	δ	11.25°	22.5°	30°	45°	60°	90°
				ζ			
smooth	$R/d = 1$	0.03	0.045	0.05	0.14	0.19	0.21
	2	0.03	0.045	0.05	0.09	0.12	0.14
	4	0.03	0.045	0.05	0.08	0.10	0.11
	6	0.03	0.045	0.045	0.075	0.09	0.09
	10	0.03	0.045	0.045	0.07	0.07	0.11
rough	$R/d = 1$	0.07	0.13	0.22	0.30	0.38	0.51
	2	0.06	0.11	0.13	0.18	0.26	0.30
	4	0.05	0.09	0.10	0.17	0.21	0.23
	6	0.05	0.09	0.09	0.15	0.18	0.18
	10	0.05	0.08	0.08	0.13	0.15	0.20

Bends, welded in segments

angle	δ	15°	22.5°	30°	45°	60°	90°
number of circumferential seams		1	1	2	2	3	3
				ζ			
	$a/d = 1.5$	0.06	0.07	0.10	0.13	0.19	0.24
	2	0.06	0.08	0.10	0.15	0.20	0.26
	4	0.07	0.09	0.11	0.16	0.22	0.28
	6	0.07	0.09	0.11	0.17	0.23	0.29

Elbows (circular area)

angle	δ	15°	22.5°	30°	45°	60°	90°
				ζ			
smooth		0.04	0.07	0.11	0.24	0.47	1.13
rough		0.06	0.11	0.17	0.32	0.68	1.27

Elbows

	a/d	1	1.5	2	3	4
				ζ		
smooth			0.16	0.14	0.15	0.15	0.17
rough			0.31	0.28	0.26	0.25	0.24

	a/d	1	1.5	2	3	4
smooth			0.17	0.16	0.15	0.15	0.16
rough			0.32	0.29	0.27	0.25	0.25

	l/d	0.8	1.5	2	3	4	5	6
					ζ			
smooth		0.45	0.28	0.30	0.36	0.38	0.39	0.40
rough		0.47	0.38	0.40	0.43	0.44	0.44	0.45

	l/d	0.8	1.5	2	3	4	5	6
smooth		0.19	0.18	0.16	0.17	0.19	0.19	0.20
rough		0.40	0.32	0.32	0.35	0.36	0.36	0.36

K. Wagner, K. Beck

Cast iron bends 90°

n.d.	50	100	200	300	400	500
ζ	1.3	1.5	1.8	2.1	2.2	2.2

Lateral, straight (the ζ values are related to the entrance velocity)

δ	15°	22.5°	30°	45°	60°	90°
ζ	0.15	0.23	0.30	0.7	1.0	1.4

Lateral, curved (the ζ values are related to the entrance velocity)

R/d	0.5	0.75	1	1.5	2
ζ	1.1	0.6	0.4	0.25	0.2

Built-up bends (2 · 90°)

$\zeta = 1.3 \cdot \zeta_{90°}$ $\zeta = 1.7 \cdot \zeta_{90°}$ $\zeta = 1.7 \cdot \zeta_{90°}$ $\zeta_{90°}$ from table plain bends

Bellow bend 90°

$\zeta = 0.4$

Composite specials

$\zeta = 2.0$ to 2.5 $\zeta = 3$ $\zeta = 4$ to 5

Strainers	Tubes (for example for compressed air)	Circumferential seams

with inlet valve
$\zeta = 2.2$ to 2.5
	Tube joints $\zeta = 0.5$ to 1.0
	Screw joints $\zeta = 1.5$ to 2.0
	Tube clutches
	  with metal sleeves $\zeta = 1.9$ to 2.0
	  with rubber seals $\zeta = 2.0$ to 3.0
		$\zeta = 0.02$ to 0.05

Compensators

Shaft compensator $\zeta = 0.2$ per shaft

Lyra bend
smooth tube $\zeta = 0.7$
bellow tube $\zeta = 1.4$

Metal tube compensator
with inner spiral $\zeta = 0.5$ to 1.0
without inner spiral $\zeta = 1$ to 2

Water separator (the ζ values are related to the entrance velocity)

Entrance:
axial: $\zeta = 2.5$

tangential: $\zeta = 3$
normal: $\zeta = 5$ to 8
with scrubber baffles: $\zeta = 4$ to 7

Water flow meter

Principle of measurement

Torque: $\zeta = 6$
Disc principle: $\zeta = 8$
Piston principle: $\zeta = 12$
Nozzle, orifice disc: see worksheet No 9.11

Electrical induction

without reduction: ζ-values are the same as for a straight pipe of the same dimensions

with reduction: ζ-values are the same as for a gate valve with the same diameter ratio

Flow separation:

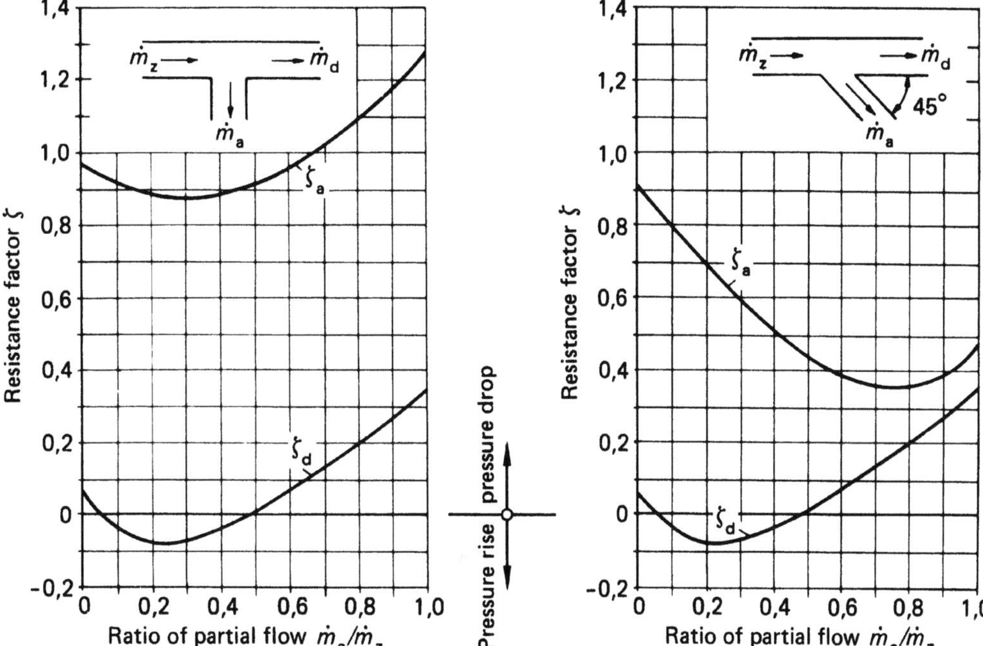

Inflow:

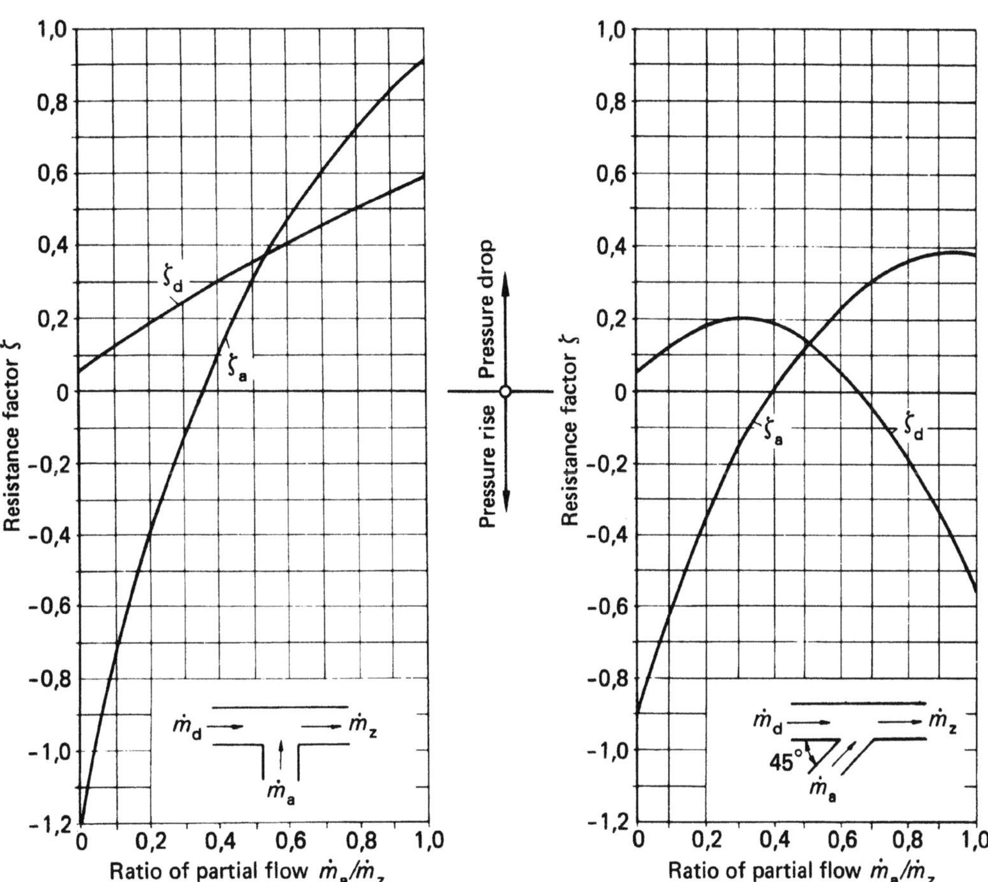

Explanation:

$\dot{m}_z$ total flow
$\dot{m}_a$ off-flow resp. in-flow
$\dot{m}_d$ straight flow

$$\Delta p_a = \zeta_a \cdot \frac{\varrho \cdot w_z^2}{2}$$

$$\Delta p_d = \zeta_d \cdot \frac{\varrho \cdot w_z^2}{2}$$

K. Wagner, K. Beck

Pipe inlets

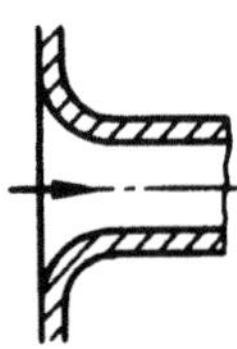

Well rounded
$\zeta_E = 0.005$ up to 0.06
depending on the
smoothness

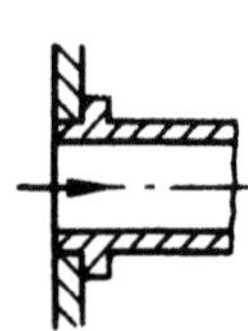

Sharp-edged
Sharp: $\zeta_E = 0.5$
Slightly broken:
$\zeta_E = 0.25$

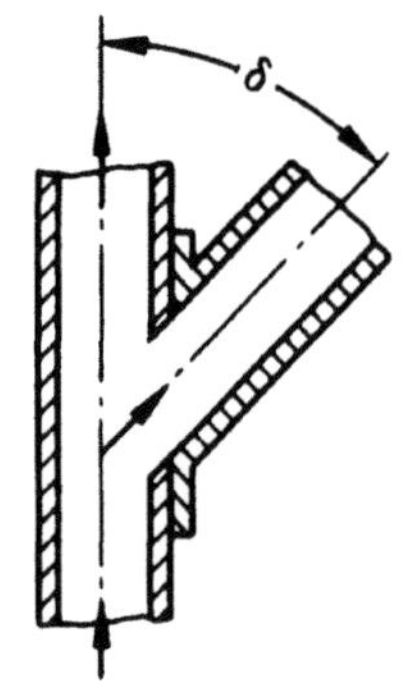

Oblique-angled, sharp-edged
$\zeta_E = 0.5 + 0.3 \cos\delta + 0.2 \cos^2\delta$

δ	22.5°	30°	45°	60°	90°
ζ_E	0.95	0.9	0.8	0.7	0.5

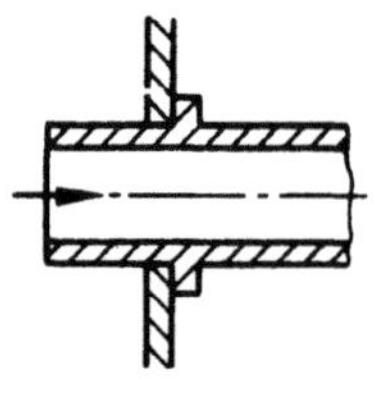

Immersed, with edges
Sharp-edged: $\zeta_E = 3$
Slightly broken: $\zeta_E = 0.6$

Orifices and Venturi meters
(according to DIN 1952)

Ratio of cross sectional areas

$$m = \frac{A_B}{A_1} = \frac{d_B^2}{d_1^2}$$

For orifices and Venturi meters the ζ-values are related to the inlet velocity w_1.

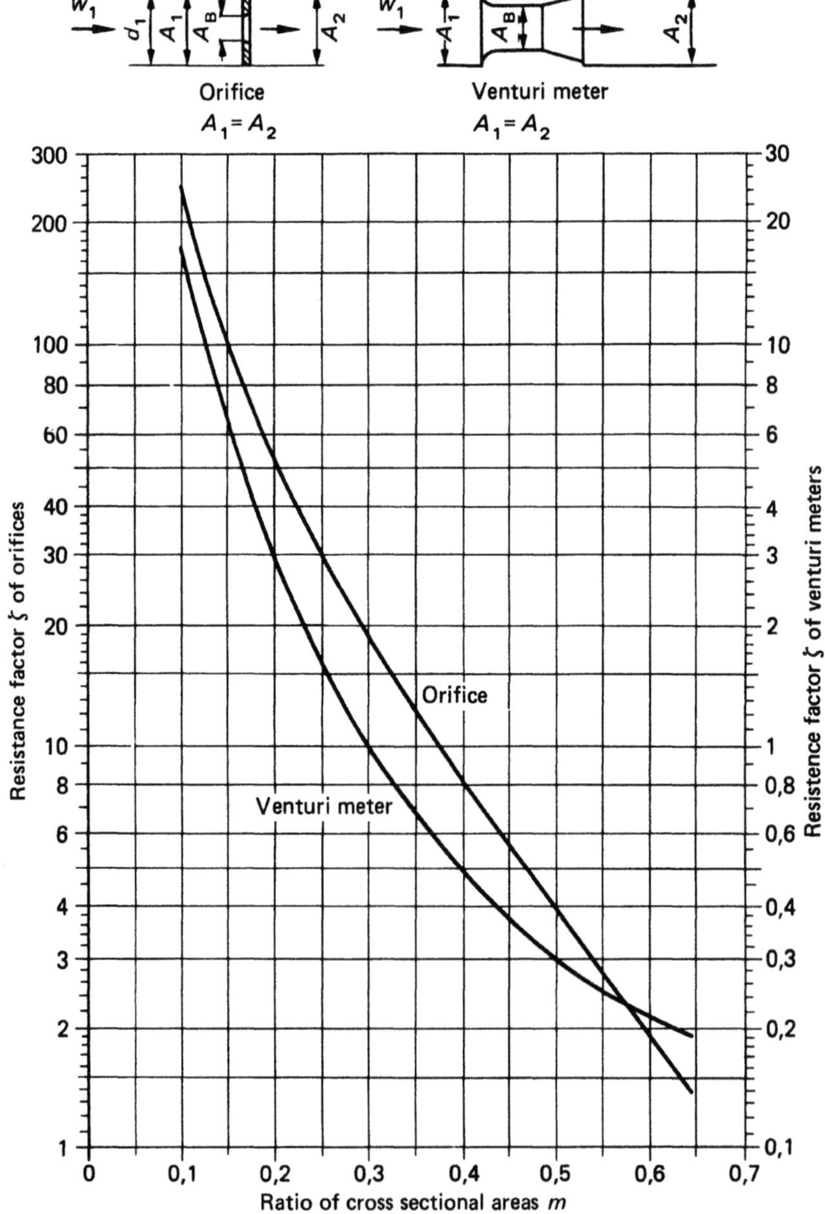

Continuous enlargement in cross-section

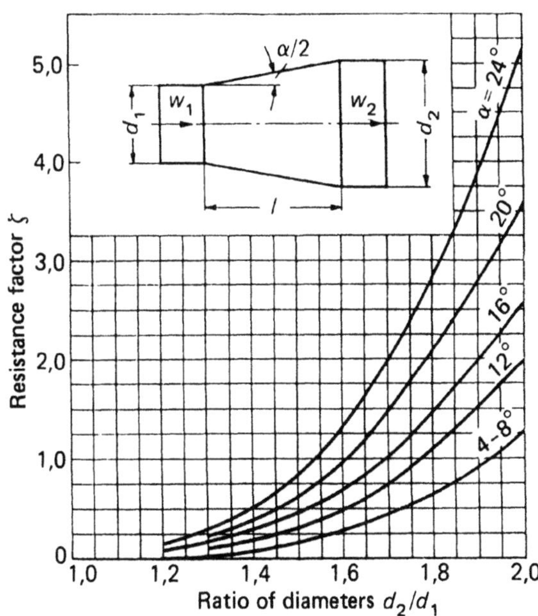

$$\Delta p = \zeta \cdot \frac{\varrho \cdot w_2^2}{2}$$

The minimum losses occur at a small incline in diameter. Here, the absolute hydraulic roughness of the pipe wall has a strong influence. Thus, for example, on a smooth surface ($k = 2 \cdot 10^{-5}$ m) an angle of 7° and on a rough surface ($k = 4 \cdot 10^{-5}$ m) an angle of 9° are of advantage.

The losses at constant reduction in cross-section are very low.

$$\Delta p = \zeta \cdot \frac{\varrho \cdot w_2^2}{2}$$

Constant reduction in cross-section

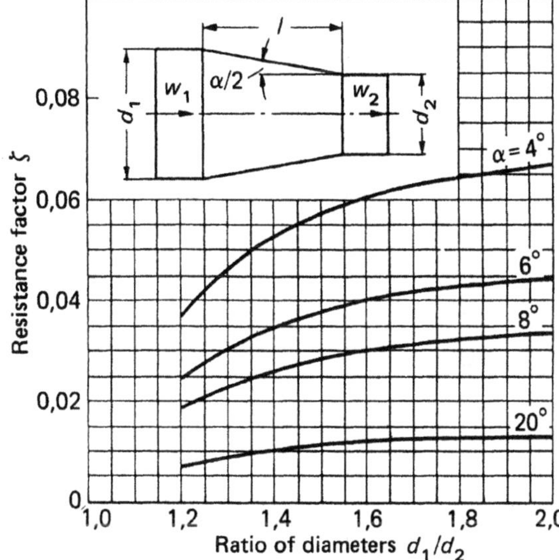

Sudden reduction in cross-section

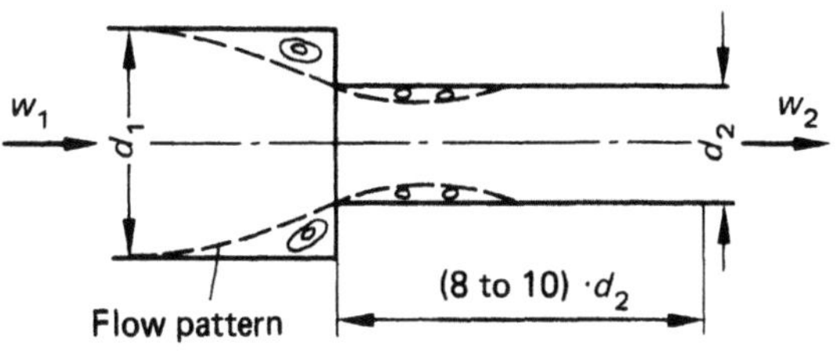

$$\Delta p = \zeta_E \cdot \frac{\varrho \cdot w_2^2}{2}$$

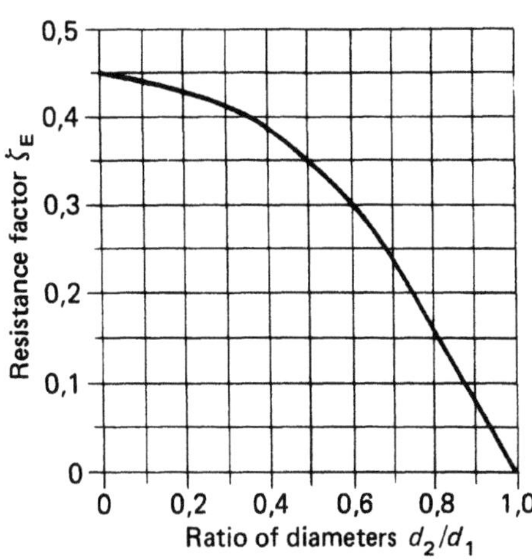

K. Wagner, K. Beck

Sudden enlargement in cross-section:

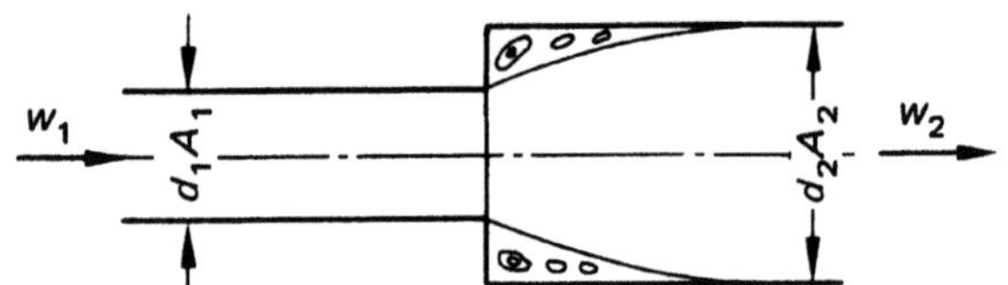

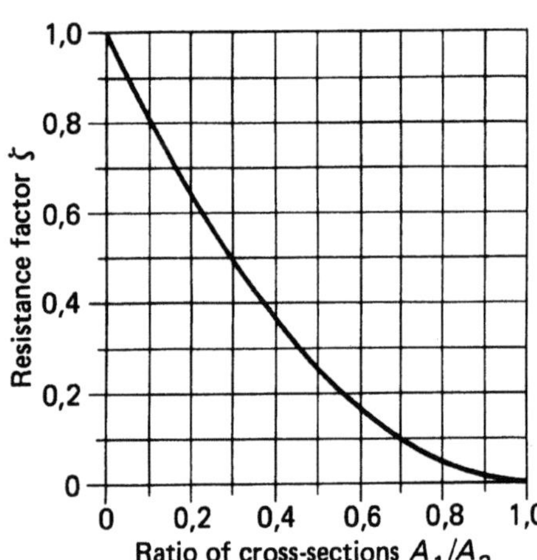

In the case of sudden enlargement in cross-section, there occurs an expanding jet with strong a vortex, which, after about $(8 \text{ to } 10) \cdot d_2$, touches the wall again.

$$\Delta p = \left(1 - \frac{A_1}{A_2}\right)^2 \cdot \frac{\varrho \cdot w_1^2}{2}$$

$$\zeta = \left(1 - \frac{A_1}{A_2}\right)^2$$

Resistance factor ζ vs. Ratio of cross-sections A_1/A_2

T-pieces of different shape:
(at flow separation)

a)

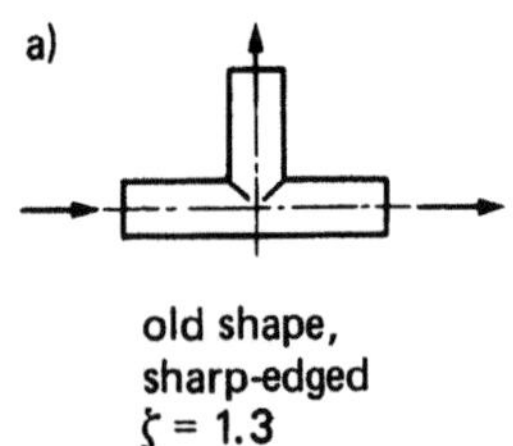

old shape,
sharp-edged
$\zeta = 1.3$

b)

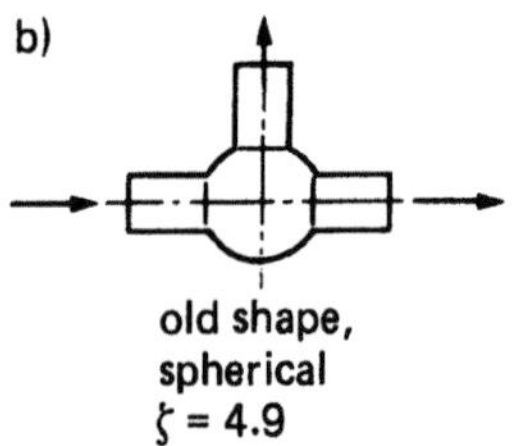

old shape,
spherical
$\zeta = 4.9$

c)

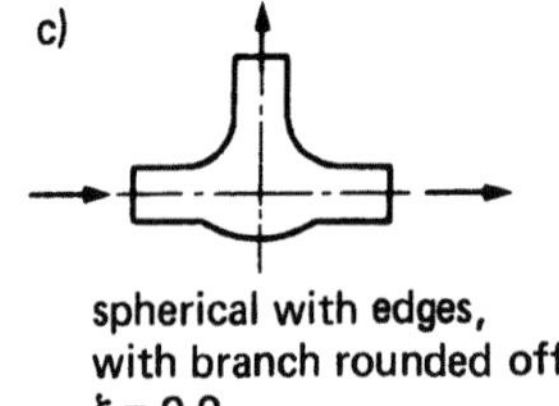

spherical with edges,
with branch rounded off
$\zeta = 0.9$

d)

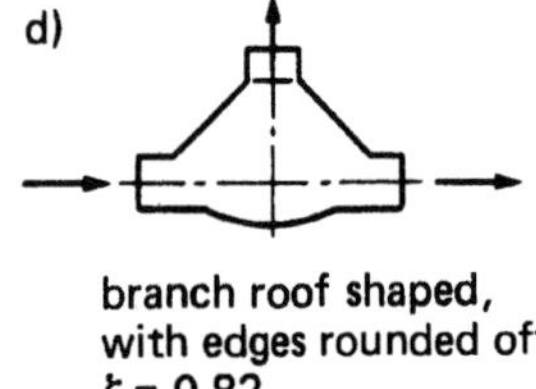

branch roof shaped,
with edges rounded off
$\zeta = 0.82$

e)

plain tube main, with edges
of branch rounded off
$\zeta = 0.73$

f)

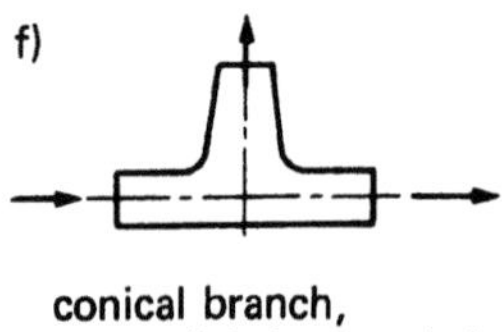

conical branch,
edges slightly rounded
$\zeta = 0.75$

The ζ-values are related to the flow velocity of the total flow.

Bibliography

Dubbel: Taschenbuch für den Maschinenbau. Berlin, Heidelberg, New York: Springer-Verlag 1981

Eck, B.: Technische Strömungslehre. Vol. 2; 8th edition Berlin, Heidelberg, New York: Springer-Verlag 1981

VDI-Wärmeatlas. 4th edition Düsseldorf: VDI-Verlag 1984

Herning, F.: Stoffströme in Rohrleitungen. 4th edition Düsseldorf: VDI-Verlag 1966

Schmidt, D.: Stahlrohr-Handbuch. 10th edition Aufl. Vulkan-Verlag 1986

Richter, H.: Rohrhydraulik. 5th edition Aufl. Berlin, Heidelberg, New York: Springer-Verlag 1971

The pressure drop of fittings can be calculated by

$$\Delta p = \zeta_v \cdot \frac{\varrho \cdot w^2}{2};$$

with the flow velocity w in the cross-section of the connected pipe with the nominal diameter. The following data are suggested values with the presupposition of a straight inlet line of at least $12 \cdot d$ length. (Otherwise higher ζ-values have to be used for calculation.) Additionally, the ζ-values of the fittings depend on the type of construction, the operating pressure, the operating conditions and the flow velocity.

Gate valve (with reduction in cross-section):

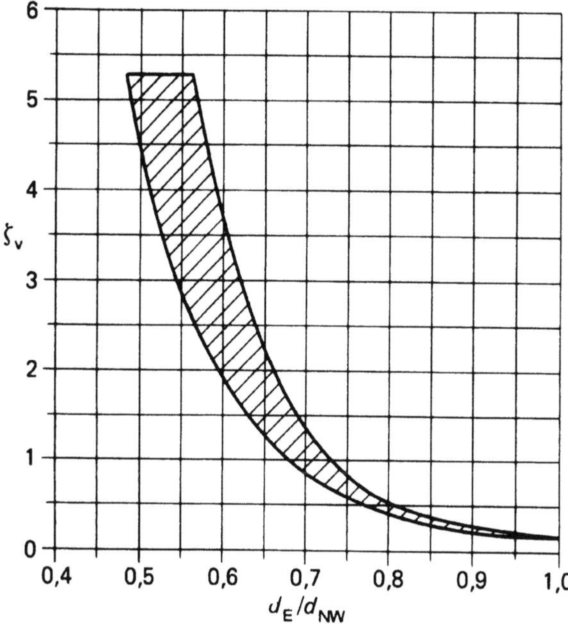

d_E diameter of the reduced area
d_{NW} nominal diameter
ζ_v values for the valve entirely open

Parallel slide valve:

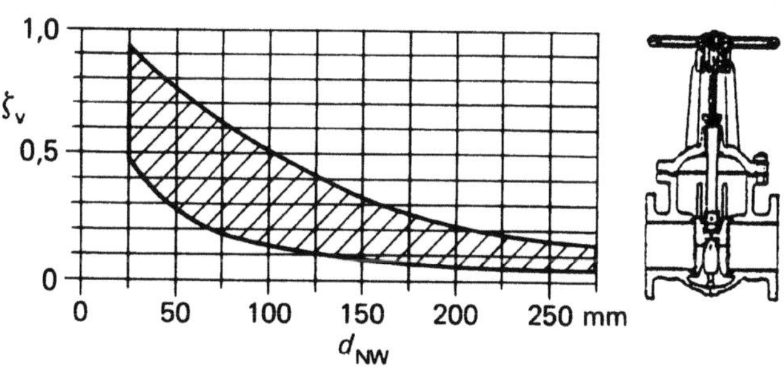

ζ_v-values for the valve entirely open

Valves:

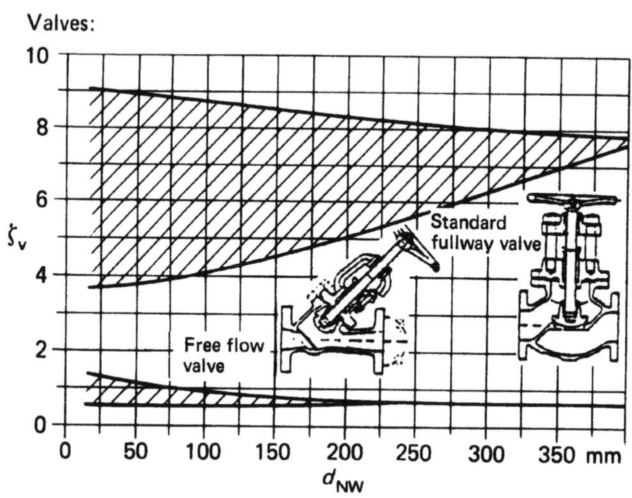

The diagram shows the ζ_v-values for the entirely open valve. The higher ζ_v-values of the standard fullway valves are due to the bad flow pattern in the valve.

K. Wagner, K. Beck

Resistance factors for different types of construction of valves of 100 mm nominal diameter:

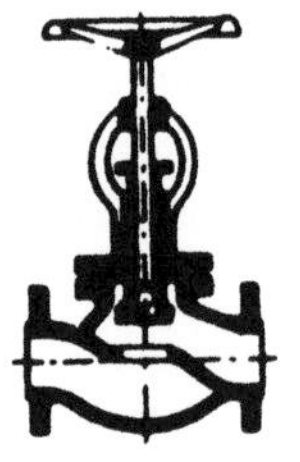
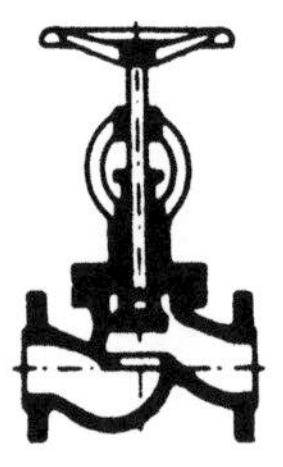
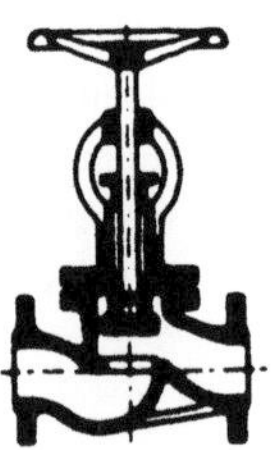

Corner valves:

Nominal diameter [mm]	50	100	200	300	400
ζ_v	3.3	4.1	5.3	6.2	6.6

Non-return valves:

Nominal diameter [mm]	50	100	125	150	200
ζ_v	5.5	4.6	4.8	4.8	4.8

Non-return flaps:

Nominal diameter [mm]	50	100	150	200	500
ζ_v	1.4	1.2	1.0	1.0	1.0

Cocks

For cocks with constant or enlarged cross-section, the ζ_v-values vary from 0.6 up to 1.0. If there are reduction or impact zones, the resistance factor can rise up to the value $\zeta_v = 4$.

Bibliography

Dubbel: Taschenbuch für den Maschinenbau. Berlin, Heidelberg, New York: Springer-Verlag 1981

Eck, B.: Technische Strömungslehre. Vol. 2; 8th edition Berlin, Heidelberg, New York: Springer-Verlag 1981

VDI-Wärmeatlas. 4th edition Düsseldorf: VDI-Verlag 1984

Herning, F.: Stoffströme in Rohrleitungen. 4th edition Düsseldorf: VDI-Verlag 1966

Schmidt, D.: Stahlrohr-Handbuch. 10th edition Aufl. Vulkan-Verlag 1986

Richter, H.: Rohrhydraulik. 5th edition Aufl. Berlin, Heidelberg, New York: Springer-Verlag 1971

Explanation

For the convective heat transfer to an uninsulated horizontal pipe, which is in unstirred air, the equation of *Churchill* and *Chu* (VDI-Wärmeatlas, 4th edition, Page Fa 5) is valid:

$$\mathrm{Nu} = \left\{ A + \frac{0.387 \cdot \mathrm{Ra}^{1/6}}{\left[1 + \left(\dfrac{B}{\mathrm{Pr}} \right)^{9/16} \right]^{8/27}} \right\}^2 \qquad (1).$$

Dimensionless references numbers

$$\mathrm{Nu} = \frac{\alpha_k \cdot l}{\lambda}; \quad \mathrm{Ra} = \mathrm{Gr} \cdot \mathrm{Pr};$$

$$\mathrm{Gr} = \frac{g \cdot l^3 \cdot \beta \cdot (\vartheta_0 - \vartheta_L)}{v^2}; \quad \mathrm{Pr} = \frac{v}{\alpha} = \frac{v \cdot \varrho \cdot c_p}{\lambda}.$$

Definitions:

A, B in $(-)$	constants
α_k in W/(m² · K)	heat transfer coefficient
l in m	characteristic length
g in m/s²	$= 9.81$ m/s²; acceleration due to gravity
ϑ_0, ϑ_L in °C	temperatures at the pipe surface resp. of the air
β in 1/K	volumetric expansion coefficient

Properties of the air (to be calculated at $\vartheta_m = (\vartheta_0 + \vartheta_L)/2$, for example according to VDI-Wärmeatlas, Page Db 8):

λ in W/(m² · K)	thermal conductivity
v in m²/s	kinematic viscosity
ϱ in kg/m³	density
c_p in J/(kg · K)	specific heat capacity at constant pressure

Range of validity and characteristic length:

$$\mathrm{Ra} < 10^3: \quad A = 0.6; \quad B = 0.559; \quad l = d_a;$$
$$d_a: \text{ outer diameter of the pipe}$$
$$\mathrm{Ra} > 10^3: \quad A = 0.825; \quad B = 0.492; \quad l = d_a \, \pi/2.$$

The heat loss per m length of pipe due to natural convection is:

$$(\phi/l)_k = \pi \, d_a \, \alpha_k (\vartheta_0 - \vartheta_L) \qquad (2).$$

This equation is shown in diagram 9.14.1 a for $\vartheta_L = 20\,°\mathrm{C}$.

The heat loss per m length of pipe due to radiation is:

$$(\phi/l)_s = \pi \, d_a \, C \left[\left(\frac{\vartheta_0 + 273}{100} \right)^4 - \left(\frac{\vartheta_L + 273}{100} \right)^4 \right] \qquad (3).$$

This equation is shown in diagram 9.14.1 b for $\vartheta_L = 20\,°\mathrm{C}$ and the radiation coefficient $C = C_1 = 4.65$ W/(m²K⁴).

Correction factors:

If the air temperature differs from the value of $20\,°\mathrm{C}$, then the equations

$$(\phi/l)_k = f_{T_k} (\phi/l)_{k, N}$$

and

$$(\phi/l)_s = f_{T_s} (\phi/l)_{s, N}$$

are valid.

The correction factors f_{T_k} and f_{T_s} can be calculated from the diagrams 9.14.1 c and 9.14.1 d. If the radiation coefficient differs from the value $C_1 = 4.65$ W/(m²K⁴), $(\phi/l)_s$, then it has to be multiplied by the factor C/C_1.

Here C is the real radiation coefficient.

Thus, the total heat loss per m uninsulated pipe in unstirred air is:

$$\phi/l = f_{T_k} (\phi/l)_{k, N} + f_{T_s} (C/C_1) (\phi/l)_{s, N} \qquad (4).$$

$(\phi/l)_{k, N}$ in W/m		specific heat loss by convection in unstirred air at $\vartheta_L = 20\,°\mathrm{C}$ (from diagram 9.14.1 a)
$(\phi/l)_{s, N}$ in W/m		specific heat loss by radiation at $C = C_1$ and at an air temperature of $\vartheta_L = 20\,°\mathrm{C}$ (from diagram 9.14.1 a)
$f_{T_k}; f_{T_s}$ in $(-)$		correction factor at an air temperature different from $20\,°\mathrm{C}$ (from diagram 9.14.1 c or 9.14.1 d)
C in W/(m²K⁴)		real radiation coefficient (for example from table 11, VDI-guideline No. 2055)
C_1 in W/(m²K⁴)	$= 4.65$	

Example:

Known values:

Temperature at the pipe suface:	$\vartheta_0 = 225\,°\mathrm{C}$
Unstirred air (i.e. heat transfer by natural convection)	
Temperature of the air:	$\vartheta_L = 10\,°\mathrm{C}$
Outer diameter of the pipe:	$d_a = 0.35$ m
Radiation coefficient:	$C = 3.5$ W/(m²K⁴)

Solution:

from diagram 9.14.1 a:	$(\phi/l)_{k, N} = 1\,615$ W/m
from diagram 9.14.1 b:	$(\phi/l)_{s, N} = 2\,771$ W/m
from diagram 9.14.1 c:	$f_{T_k} = 1.07$
from diagram 9.14.1 d:	$f_{T_s} = 1.02$

Correction of the radiation coefficient:

$$C/C_1 = 3.5/4.65 = 0.75 \, .$$

Specific heat loss according to equation No. (4)

$$\phi/l = (1.07 \cdot 1\,615 + 1.02 \cdot 0.75 \cdot 2\,771) \text{ W/m} =$$
$$= 3\,848 \text{ W/m} \, .$$

Additional information of this book

(Engineering Reference Book on Energy and Heat; 978-3-642-51125-7; 978-3-642-51125-7_OSFO7) is provided:

http://Extras.Springer.com

5
m/s
4
3
2
1
0
Air velocity w
d_a = 1 m
0,9
0,8
0,7
0,6
0,5
0,4
0,3
0,2
0,1
0,01
ϑ_m = 25 °C
50
100
150
200
300
400
$\vartheta_0 - \vartheta_L$ = 150 °C
50
100
200
250
300
350
400
450
500
$\vartheta_m = \frac{\vartheta_0 + \vartheta_L}{2}$
0
5000
10000
15000
20000
25000
30000 W/m 35000
Heat loss per m length of pipe $(\phi/l)_{k,W}$

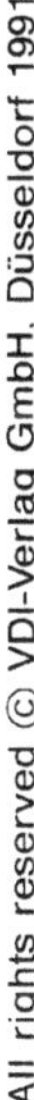

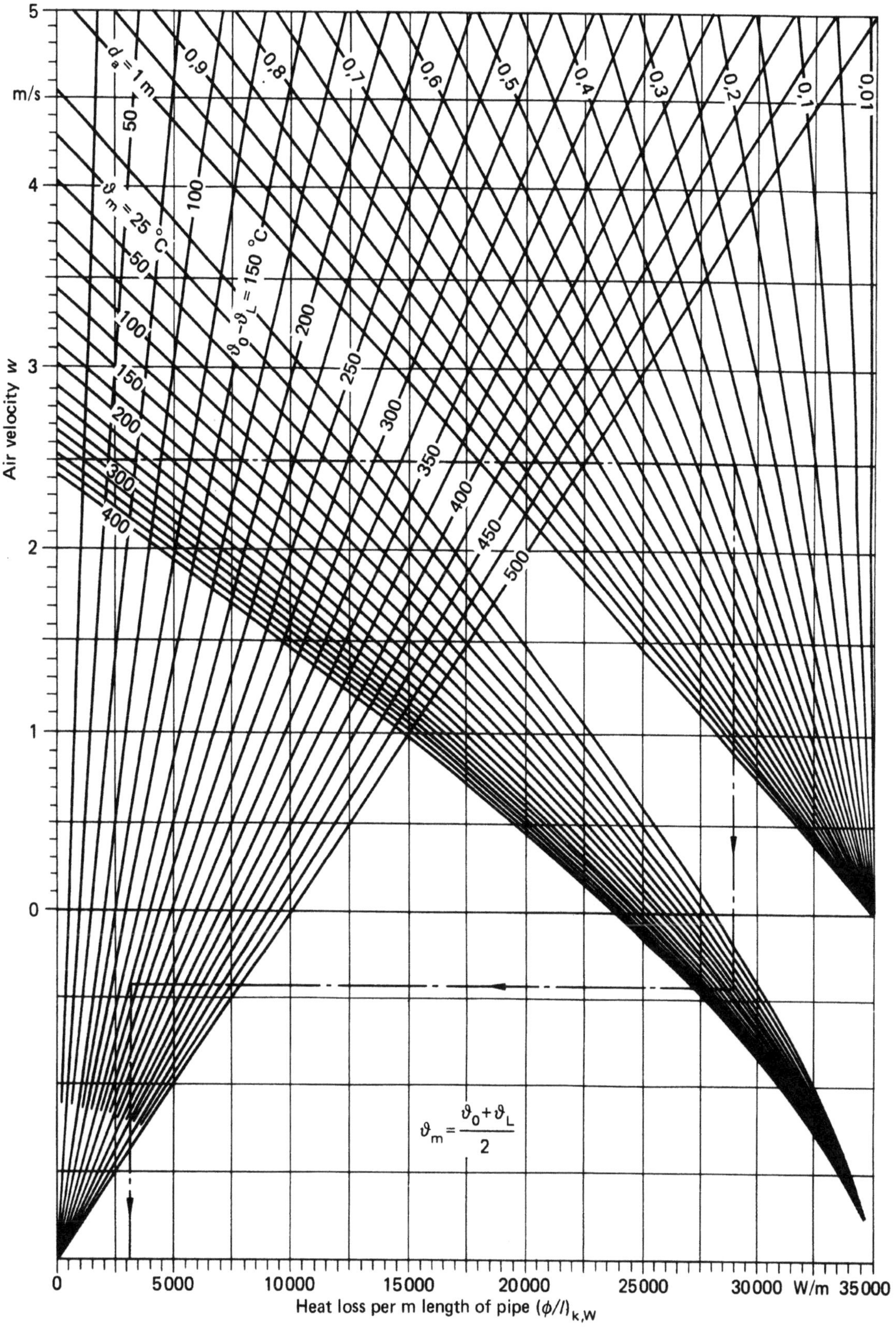

H. Auracher

Explanation

If there is a cross flow of air vertical to the axis of the pipe, then the convective part of the heat transfer will change, whilst the radiation part will not. Here the first part of equation No (4) has to be replaced by $(\phi/l)_{k,w}$. This value can be determined from worksheet No 9.14.2 as a function of the wind velocity w, the outer diameter of the pipe, the temperature difference $\vartheta_O - \vartheta_L$ and the average temperature $\vartheta_m = (\vartheta_O - \vartheta_L)/2$. The nomograph is based on the equations shown in the VDI-Wärmeatlas (4th edition 1984, section Ge).

The total heat loss per m uninsulated pipe with forced convection is:

$$\phi/l = (\phi/l)_{k,w} + f_{T_s}(C/C_1)(\phi/l)_{S,N} \tag{5}$$

$(\phi/l)_{k,w}$ is the specific heat loss of a pipe with forced convection (worksheet No 9.14.2).

The other signs have been already defined for equation No (4).

At small air velocities, $(\phi/l)_{k,w}$ may become smaller than $f_{T_k}(\phi/l)_{k,N}$. In this case, the higher value has to be chosen for calculation.

Example:

Known values:

Temperature at the pipe surface	$\vartheta_O = 225\,°C$
Temperature of the air	$\vartheta_L = 10\,°C$
Velocity of the air	$w = 2.5\,m/s$
Outer diameter of the pipe	$d_a = 0.35\,m$
Radiation coefficient	$C = 3.5\,W/(m^2K^4)$

Solution:

$$\vartheta_m = \frac{\vartheta_O - \vartheta_L}{2} = \frac{(225 + 10)\,°C}{2} = 117.5\,°C.$$

$\vartheta_O - \vartheta_L = 215\,°C.$

From worksheet No 9.14.2: $(\phi/l)_{k,w} = 3\,362\,W/m.$

The other values see example No 1.

Specific heat loss according to equation No (5)

$$\phi/l = (3\,362 + 1.02 \cdot 0.75 \cdot 2\,771)\,W/m = 5\,482\,W/m.$$

Heat Loss and Surface Temperature of Insulated Pipes (Heat Resistance Factor)

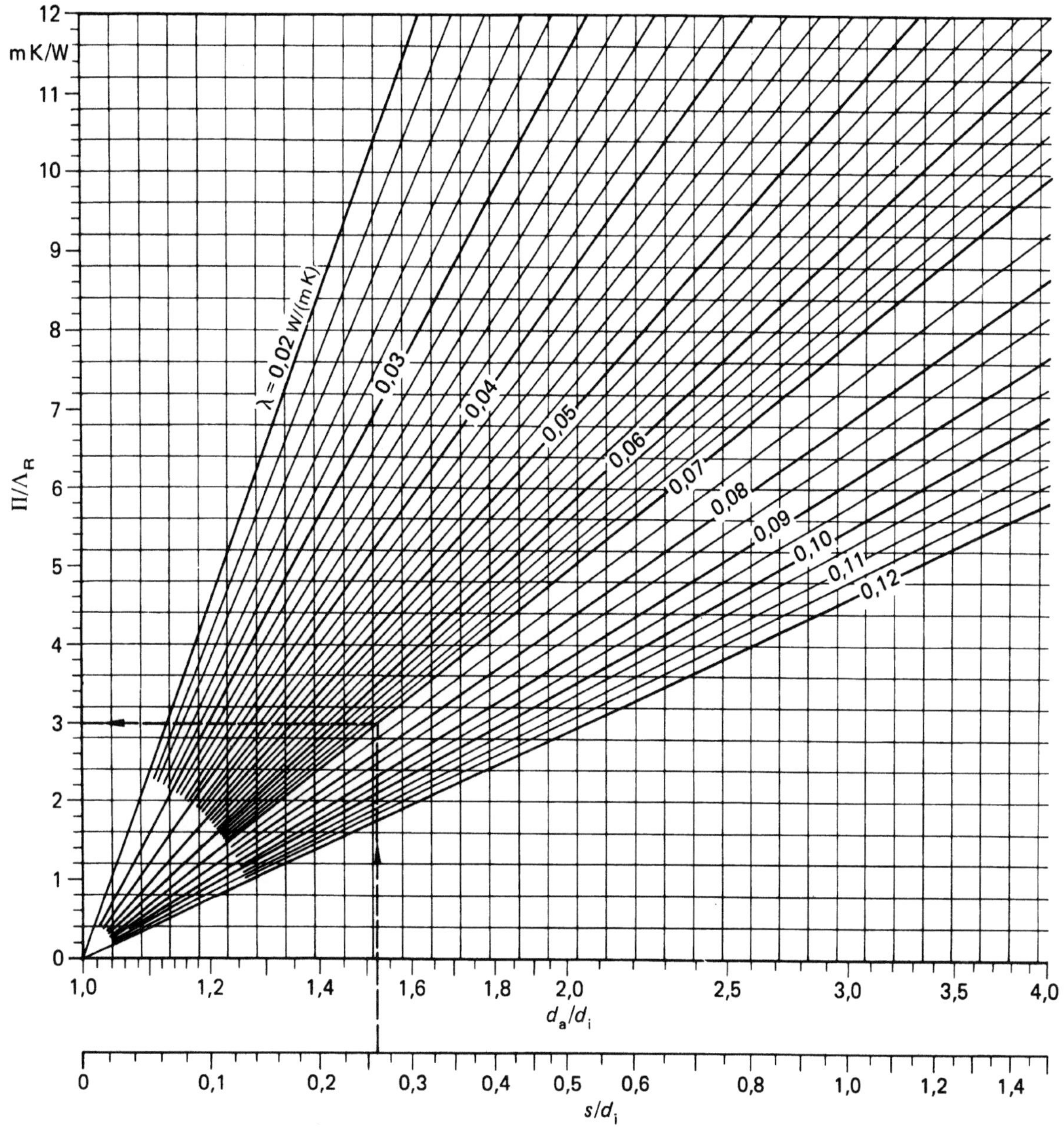

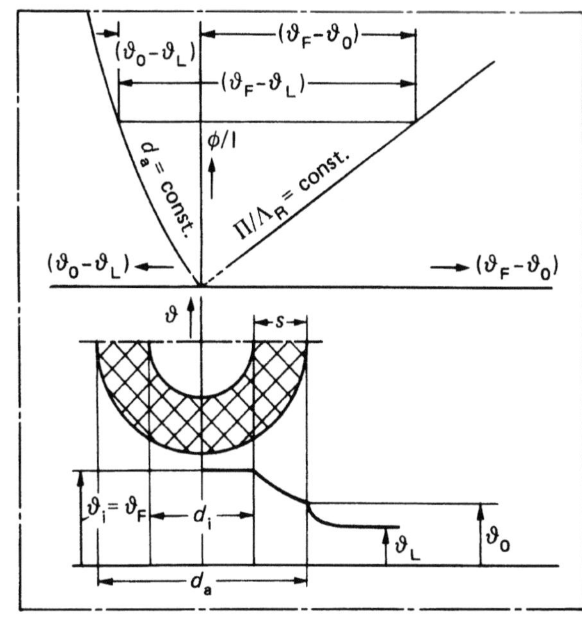

Schematic calculation scheme

U. Grigull, H. Auracher

Explanation

The difficulties in the determination of the heat loss of an insulated pipe are the dependencies of the outer heat transfer coefficient on the flow conditions around the pipe (unstirred air or wind) and on the exchange of radiation with the surroundings, as well as on the surface temperature of the insulation material, which is an unknown factor at the beginning. In the worksheets No 9.15.1 to 9.15.4 a graphic method is shown, which is based on the fact, that under steady conditions the heat flow through the insulation is equal to the heat flow from the surface of the insulation material to the surroundings.

Fundamentals of calculation:

Neglecting the thermal resistance between fluid and pipe, as well as the wall thickness of the pipe, the heat flow from the fluid to the surface of the insulating material per m pipe length is:

$$\phi/l = \frac{\vartheta_F - \vartheta_O}{1/\varLambda_R} \qquad (1).$$

The heat resistance coefficient of the pipe is:

$$\frac{1}{\varLambda_R} = \frac{\ln(d_a/d_i)}{2\pi\lambda} = \frac{\ln(1 + 2\,s/d_i)}{2\pi\lambda} \qquad (2).$$

Definitions:

d_i in m	inner diameter of the pipe
d_a in m	outer diameter of the insulation layer
s in m	thickness of the insulation layer
λ in W/(m · K)	thermal conductivity of the insulation layer
ϑ_F in °C	temperature of the fluid
ϑ_O in °C	surface temperature of the insulation layer

The specific heat flow according to equation No (1) has to be equal to the heat flow from the surface of the insulating layer to the surroundings. This heat flow has a convection $((\phi/l)_k)$ and a radiation $((\phi/l)_s)$ part. Thus, it is:

$$\phi/l = (\phi/l)_k + (\phi/l)_s \qquad (3).$$

Here, there is no need for equations to calculate these parts, because they have been considered in the worksheets No 9.15.2 to 9.15.4.

As shown in the schematic calculation, the correlations (1) and (3) for the specific heat loss have been shown in the diagram in that way, that the specific heat loss can be calculated from the common ordinate. Parameter on the one side is the heat transmission coefficient $\pi/\varLambda_R$ (eq. No (2)) multiplied by π and, on the other side, the outer diameter of the pipe d_a (independent variable in eq. No (3)).

Although the value of ϑ_O is unknown, there does exist only one value of (ϕ/l), at which the total difference of temperature between the fluid and the surroundings $(\vartheta_F - \vartheta_L)$ fills the space between the curves of the known values of d_a and $\pi/\varLambda_R$. If the value of (ϕ/l) will be determined for a known difference of temperature $(\vartheta_F - \vartheta_L)$ (from sheet No 9.15.2 to 9.15.4), the difference of temperature and, thus, the temperature of the surface may be determined from the abscissa.

Procedure of calculation:

(1) Determination of the value of $\pi/\varLambda_R$ for known values of λ, s and d_i (from the diagram; see the following example)

(2) Determination of the specific heat loss (ϕ/l) and the surface temperature of the insulated pipe at a known difference of temperature $(\vartheta_F - \vartheta_L)$ (from sheet No 9.15.2 to 9.15.4 regarding the influence of wind and radiation).

Example:

Known values:

Inner diameter of the pipe	$d_i = 0.267$ m
Outer diameter of the insulating layer	$d_a = 0.407$ m
Heat transmission coefficient of the insulating layer	$\lambda = 0.07$ W/(mK)

Solution:

$s = (d_a - d_i)/2 = 0.07$ m; $\;d_a/d_i = 1.524$; $\;s/d_i = 0.262$.

From the diagram $\pi/\varLambda_R = 3$ mK/W may be determined.

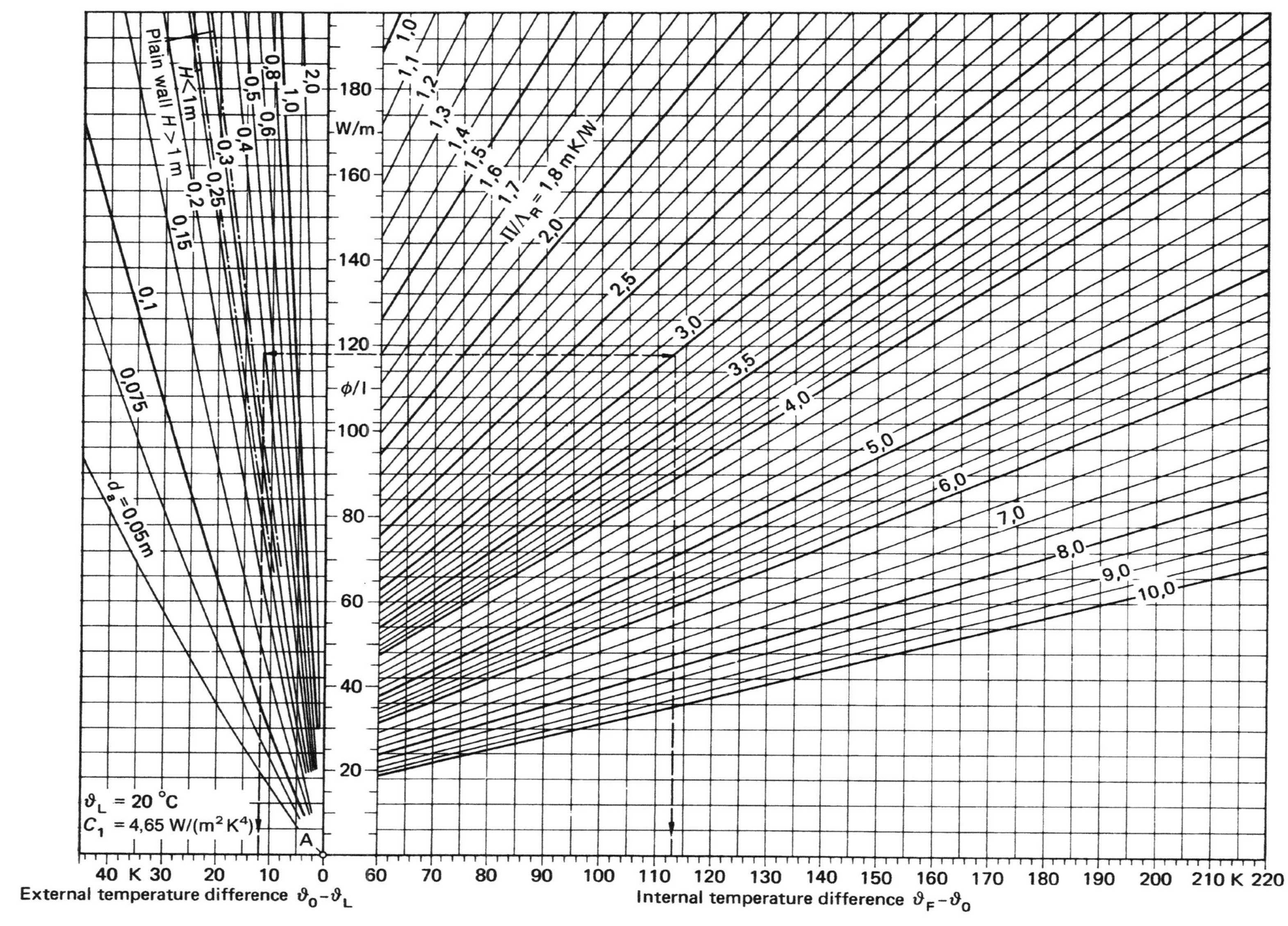

U. Grigull, H. Auracher

Explanation

The diagram is based on the following assumptions:

Air temperature $\qquad\vartheta_L = 20\,°C$

Air velocity (heat transfer by natural convection) $\qquad w = 0\ \text{m/s}$

Radiation coefficient $\qquad C_1 = 4.65\ \text{W/(m}^2\text{K}^4)$

The value of π/Λ_R, drawn out in the right part of the diagram, has to be determined from worksheet No 9.15.1.

The heat loss ϕ/l may be determined from the ordinate of a horizontal line between the terminal points on the curves of π/Λ_R and d_a. The length of the line is equal to the differences of temperature $(\vartheta_F - \vartheta_O)$ (right side of the abscissa) and $(\vartheta_O - \vartheta_L)$ (left side of the abscissa), so that the addition of these values is equal to the known difference of temperature $\vartheta_F - \vartheta_L$.

For example, to determine this, from the point A, the line with the value $\vartheta_F - \vartheta_L$ on the right side can be measured by a compass. Now the left point of the compass has to be moved along the curve for the known d_a, until the right point of the compass touches the curve of the known value π/Λ_R. Doing this, the line has to be parallel to the abscissa. ϕ/l can now be determined on the abscissa by adding $\vartheta_F - \vartheta_O$ and $\vartheta_O - \vartheta_L$.

Example:

Known values:

Difference of temperature $\qquad \vartheta_F - \vartheta_L = 125\ \text{K}$

π/Λ_R according to worksheet No 9.15.1 $\qquad \pi/\Lambda_R = 3.0\ \text{mK/W}$

Outer diameter of the insulating layer $\qquad d_a = 0.407\ \text{m}$

Solution:

Specific heat loss $\qquad \phi/l = 118\ \text{W/m}$

Difference of temperature between inner and outer temperature of the insulation $\qquad \vartheta_F - \vartheta_O = 113\ \text{K}$

Difference of temperature between surface temperature of the insulating layer and ambient temperature $\qquad \vartheta_O - \vartheta_L = 12\ \text{K}$

Surface temperature of the insulating layer $\qquad \vartheta_O = 32\,°C$

If the ambient temperature and the radiation coefficient differ from the assumptions, then the left diagram is not valid any more. In this case, it is easy to determine the values of ϕ/l and ϑ_O. Normally the outer difference of temperature $\vartheta_O - \vartheta_L$ is small compared with the inner one. Therefore, these calculation does not have to be very accurate.

Procedure of calculation:

The outer heat transfer by natural convection is:

$$(\phi/l)_k = \pi \cdot d_a \cdot \alpha_k (\vartheta_O - \vartheta_L) \qquad (1).$$

The heat transfer coefficient α_k can be approximated by (VDI-Guideline No 2055):

$$\alpha_k = 1.35 \sqrt[4]{\frac{\vartheta_O - \vartheta_L}{d_a}}\ \text{W/(m}^2\text{K)}; \quad (d_a \text{ in m}) \qquad (2).$$

With C as the radiation coefficient of the insulating layer (for example according to VDI-Guideline No 2055), table 11), the heat transfer by radiation is:

$$(\phi/l)_S = \pi \cdot d_a \cdot C \left[\left(\frac{\vartheta_O + 273}{100}\right)^4 - \left(\frac{\vartheta_L + 273}{100}\right)^4\right] \qquad (3).$$

By means of equation (1) to (3), now, with the usually known values of d_a and ϑ_L, the total heat flux

$$\phi/l = (\phi/l)_k + (\phi/l)_S \qquad (4)$$

can be calculated. At first, the unknown surface temperature is the parameter. Now, the values of ϕ/l and of $\vartheta_O - \vartheta_L$ have to be marked in the left part of the diagram. One obtains a curve for the known value of d_a. With this curve, it is possible to determine – with the above mentioned method – the value of ϕ/l and ϑ_O. Usually it is sufficient to determine some few values of the curve, as the unknown values can be estimated by the diagram.

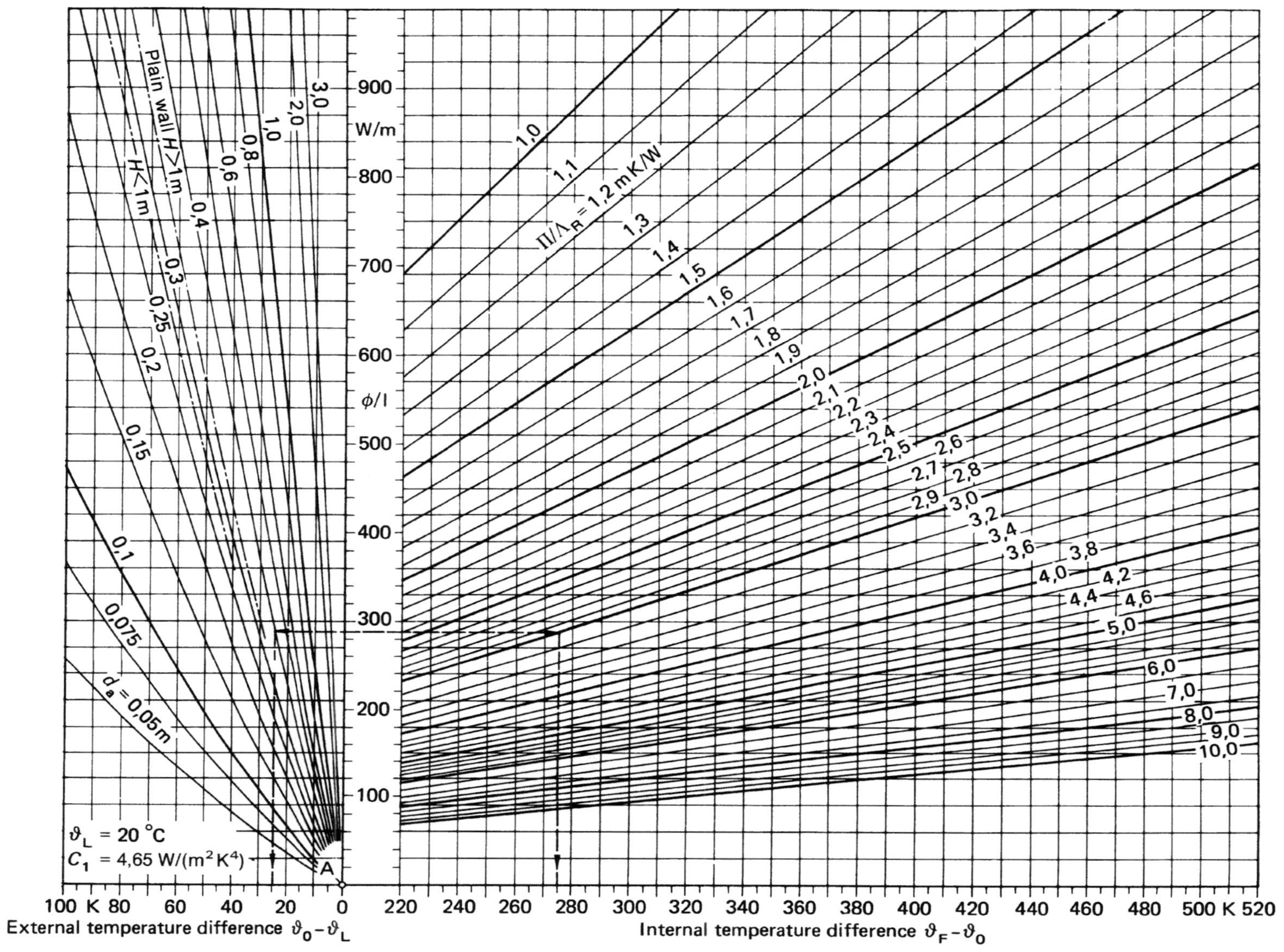

U. Grigull, H. Auracher

Explanation

The diagram is based on the following assumptions:

Air temperature $\qquad\qquad\qquad\qquad\qquad$ $\vartheta_L = 20\,°C$

Air velocity
(heat transfer by natural convection) $\quad w = 0\ \text{m/s}$

Radiation coefficient $\qquad\qquad\qquad\quad C_1 = 4.65\ \text{W/(m}^2\text{K}^4)$

The value of π/Λ_R, drawn out in the right part of the diagram, has to be determined from worksheet No 9.15.1.

The heat loss ϕ/l may be determined from the ordinate of a horizontal line between the terminal points on the curves of π/Λ_R and d_a. The length of the line is equal to the differences of temperature $(\vartheta_F - \vartheta_O)$ (right side of the abscissa) and $(\vartheta_O - \vartheta_L)$ (left side of the abscissa), so that the addition of these values is equal to the known difference of temperature $\vartheta_F - \vartheta_L$.

For example, to determine this, from the point A, the line with the value $\vartheta_F - \vartheta_L$ (on the right side) can be measured by a compass. Now the left point of the compass has to be moved along the curve for the known d_a, until the right point of the compass touches the curve of the known value π/Λ_R. In doing this, the line has to be parallel to the abscissa. ϕ/l can now be determined on the abscissa by adding $\vartheta_F - \vartheta_O$ and $\vartheta_O - \vartheta_L$.

Example:

Known values:

Difference of temperature $\qquad\qquad\qquad\qquad\quad \vartheta_F - \vartheta_L = 300\ \text{K}$

π/Λ_R according to worksheet No 9.15.1 $\qquad\quad \pi/\Lambda_R \quad = 3.0\ \text{mK/W}$

Outer diameter of the insulating layer $\qquad\quad d_a \qquad = 0.407\ \text{m}$

Solution:

Specific heat loss $\qquad\qquad\qquad\qquad\qquad\quad \phi/l \qquad = 288\ \text{W/m}$

Difference of temperature between inner
and outer temperature of the insulation $\qquad \vartheta_F - \vartheta_O = 275\ \text{K}$

Difference of temperature between surface
temperature of the insulating layer and
ambient temperature $\qquad\qquad\qquad\qquad\qquad \vartheta_O - \vartheta_L = 25\ \text{K}$

Surface temperature of the insulating layer $\quad \vartheta_O \qquad = 45\,°C$

If the ambient temperature and the radiation coefficient differ from the assumptions, then the left diagram is not valid any more. A method to determine the values of φ/l and ϑ_O has already been described in worksheet 9.15.2. It can be used without changes for worksheet 9.15.3.

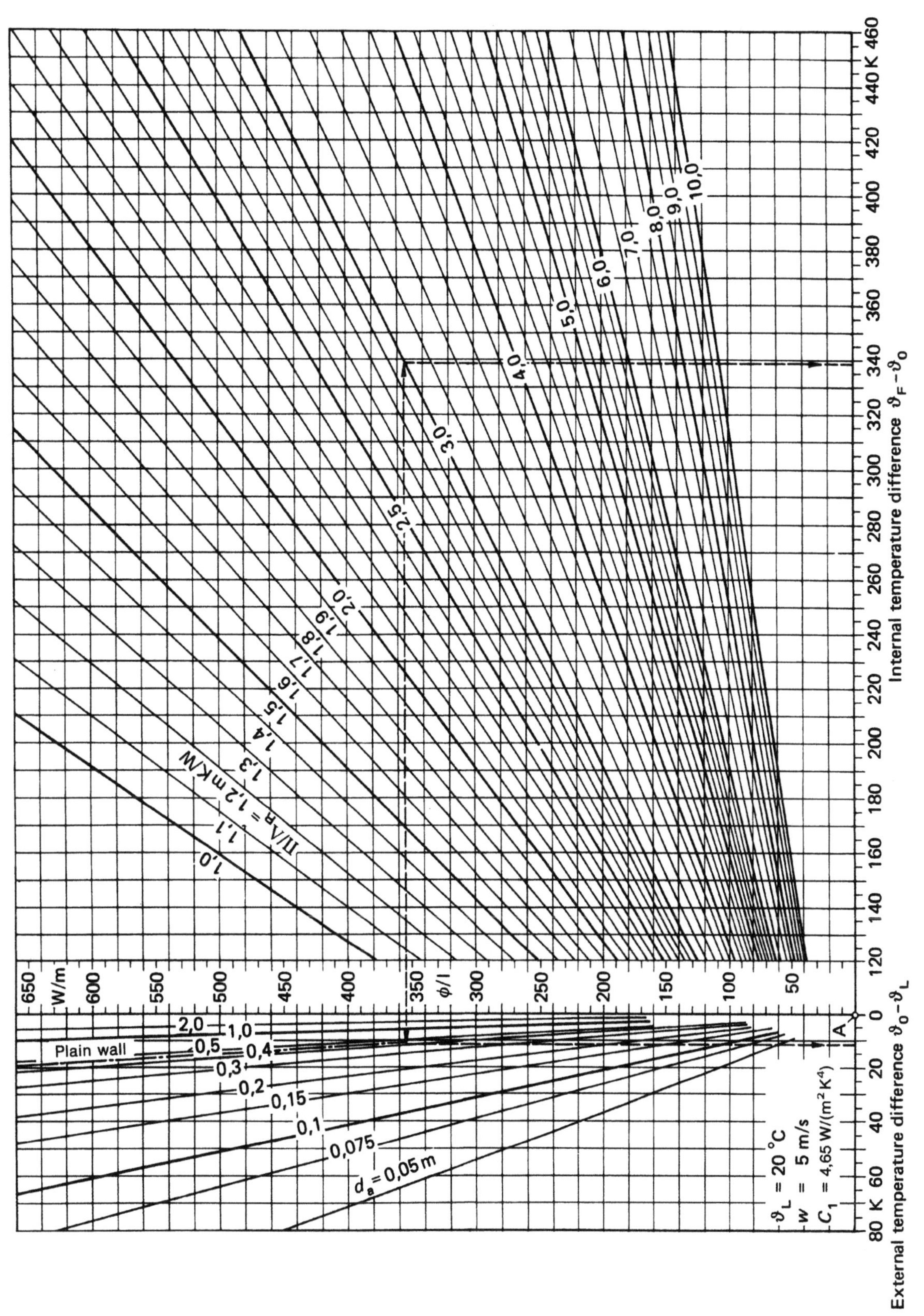

U. Grigull, H. Auracher

Explanation

The diagram is based on the following assumptions:

Air temperature	$\vartheta_L = 0\,°C$
Wind velocity	$w = 5\ m/s$
Radiation coefficient	$C_1 = 4.65\ W/(m^2K^4)$

The value of π/Λ_R, drawn out in the right part of the diagram, has to be determined from worksheet No 9.15.1.

The heat loss ϕ/l may be determined from the ordinate of a horizontal line between the terminal points on the curves of π/Λ_R and d_a. The length of the line is equal to the differences of temperature $(\vartheta_F - \vartheta_O)$ (right side of the abscissa) and $(\vartheta_O - \vartheta_L)$ (left side of the abscissa), so that the addition of these values is equal to the known difference of temperature $\vartheta_F - \vartheta_L$.

For example, to determine this, from the point A, the line with the value $\vartheta_F - \vartheta_L$ (on the right side) can be measured by a compass. Now the left point of the compass has to be moved along the curve for the known d_a, until the right point of the compass touches the curve of the known value π/Λ_R. In doing this, the line has to be parallel to the abscissa. ϕ/l can now be determined on the abscissa by adding $\vartheta_F - \vartheta_O$ and $\vartheta_O - \vartheta_L$.

Example:

Known values:

Difference of temperature	$\vartheta_F - \vartheta_L$	$= 350\ K$
π/Λ_R according to worksheet No 9.15.1	π/Λ_R	$= 3.0\ mK/W$
Outer diameter of the insulating layer	d_a	$= 0.407\ m$

Solution:

Specific heat loss	ϕ/l	$= 355\ W/m$
Difference of temperature between inner and outer temperature of the insulation	$\vartheta_F - \vartheta_O$	$= 339\ K$
Difference of temperature between surface temperature of the insulating layer and ambient temperature	$\vartheta_O - \vartheta_L$	$= 11\ K$
Surface temperature of the insulating layer	ϑ_O	$= 11\,°C$

If the ambient temperature and the radiation coefficient differ from the assumptions, then the left diagram is not valid any more. A method to determine the values of φ/l and ϑ_O has been already described in worksheet 9.15.2. It can be used for worksheet 9.15.4 too. There is only one change: instead of equation No (2) in worksheet 9.15.2 the approximation for the heat transfer coefficient at forced convection (VDI-guideline No 2055):

$$\alpha_k = 4.13 \cdot \frac{w^{0.8}}{d_a^{0.2}}\ W/(m^2K)$$

with w in m/s and d_a in m has to be used.

Fittings, for example valves and gate valves, built into the insulation layer material, for example spacer, supports, tube hanger and flanges cause additional heat loss, which has to be added to the heat loss of normal insulated pipe (worksheet 9.15). It is not possible, to calculate exactly these additional losses. Below, rough standard values for these corrections have been listed.

a) Valves and gate valves

The influence of valves, gate valves and flanges can be considered by adding a fictitious equivalent pipe length Δl to the existing one l. The total heat loss has to be calculated from the specific heat loss of the pipe (worksheet No 9.15) multiplied by $(l + \Delta l)$.

Standard values for the equivalent length Δl of an insulated pipe without flanges (see b), with the following assumptions, have been listed in the table below:

Usual thickness of the insulating layer.

Thermal conductivity λ of the insulating material at 100 °C: 0.08 W/(mK)

Thermal conductivity λ of the insulating material at 400 °C: 0.1 W/(mK)

Inner diameter of the pipe d_i	Location of the pipe line:	Open area			Indoors		
in m	temperature of the pipe in °C	100	200	300	400	500	600
		equivalent pipe length Δl					
	insulating condition	m	m	m	m	m	m
0.1	3/4 insulated	4.5	5.0	6.0	2.5	3.5	5.0
	2/3 insulated	6.0	6.5	8.0	3.0	4.0	6.0
	not insulated	15.0	17.5	22.0	6.0	10.0	16.0
0.2	3/4 insulated	5.5	7.0	8.0	2.8	4.0	6.0
	2/3 insulated	7.0	8.0	10.0	4.0	5.5	8.0
	not insulated	18.0	22.0	27.0	7.8	12.0	21.0
0.5	3/4 insulated	6.0	7.0	8.5	3.0	5.0	7.5
	2/3 insulated	7.0	8.5	11.0	4.0	6.0	10.0
	not insulated	19.0	24.0	32.0	9.0	15.0	26.0

b) Pair of flanges

The additional heat loss of the pair of flanges belonging to the insulated pipe has to be considered by the following equivalent length:

Not insulated:

The equivalent length is a third of the equivalent length of an uninsulated valve of the same nominal diameter.

Insulated:

If the total length of the pipe includes the flange length, then no additional length has to be considered. For pipes with double-flanged butterfly valve, 1 m equivalent length has to be added.

c) Pipe supports and pipe hangers

The additional heat loss has to be considered by a percentage addition to the loss of the normal insulated pipe:

Location	Addition
Indoors	15%
Open area without wind	20%
Open area with wind	25%

d) Material built in into the insulating layer

The additional heat loss, caused by built in material into the insulating layer, for example spacer and supports, has to be considered by adding a value $\Delta\lambda$ to the thermal conductivity λ of the insulating material. The specific heat loss of the pipe, according to worksheet No 9.15, has to be determined with an equivalent thermal conductivity $\lambda + \Delta\lambda$.

Spacer or support material	Addition
Steel	0.010 W/(mK)
Ceramics	0.003 W/(mK)

Bibliography

VDI-Guideline 2055

U. Grigull, H. Auracher

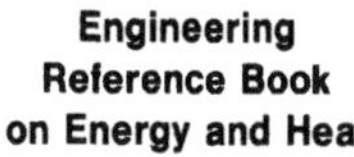

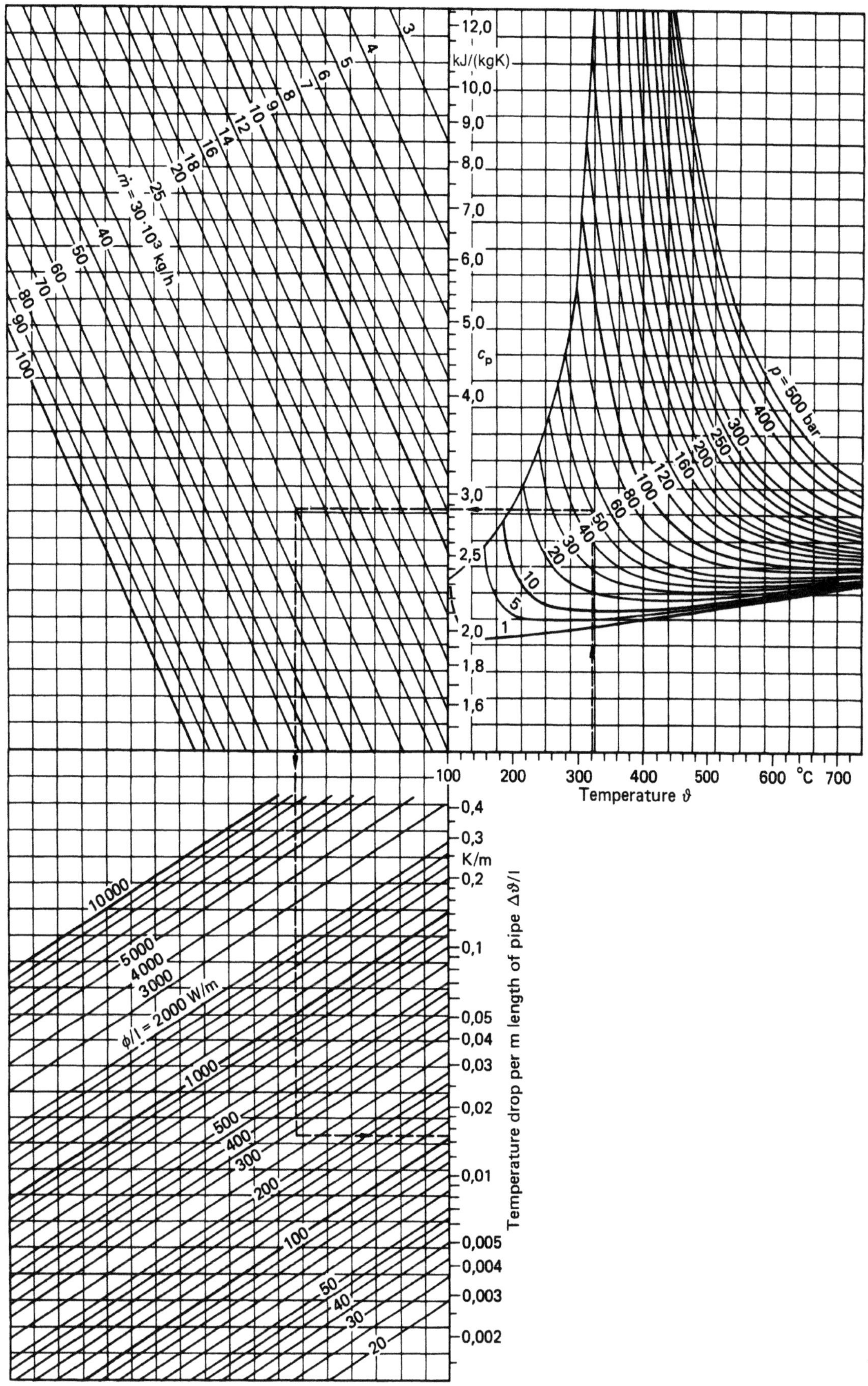

U. Grigull, W. Goldstern, H. Auracher

Explanation

The specific temperature drop $\Delta\vartheta/l$ of a fluid flowing through a pipe is:

$$\frac{\Delta\vartheta}{l} = \frac{\phi/l}{\dot{m}\,c}.$$

ϕ/l in W/m   specific heat loss of the pipe (see worksheet No 9.14, 9.15 and 9.16)

$\dot{m}$ in kg/s   mass flow

c in kJ/(kgK)   specific heat capacity of the fluid (the value may be found from a collection of the properties of media; the specific heat capacity at constant pressure c_p can be used instead of c for an approximation)

The diagram at the upper right side shows the specific heat capacity of steam. For other media the equivalent c_p-value on the axis has to be used.

The difference of temperature between the fluid and the surroundings is a decrease at constant ambient conditions, if the fluid is cooled in the direction of flow. Thus, the specific heat loss ϕ/l decreases, too. Therefore the above mentioned equation resp. the nomograph is valid only for short pipe sections. Thus, the total temperature drop of the fluid in a pipe has to be calculated section by section. The length of a section has to be chosen in such a way, that the temperature drop in this section $\Delta\vartheta$ is small compared with the difference of temperature between the fluid and the ambient air $[\Delta\vartheta \ll (\vartheta_F - \vartheta_L)]$. Then the specific heat loss of this section can be assumed to be constant with a good approximation. Frequently, the total temperature drop of the fluid in the pipe is small enough to dispense with sectional calculation.

The above shown procedure of calculation resp. the nomograph is valid only, if the fluid in the pipe does not change the state of aggregate (condensation, evaporation).

Another commonly used assumption for this calculation is the disregard of the cooling process of the pipe and insulating material.

Example:

Steam with a mass flow of $\dot{m} = 6.95$ kg/s and a pressure of $p = 50$ bar flows through an insulated $l = 8$ m long pipe. The temperature at pipe entrance is $\vartheta_1 = 320\,°$C. The exit temperature ϑ_2 has to be determined.

Solution:

The specific heat capacity for these steam conditions (from the nomograph or the steam tables) is $c_p = 2.87$ kJ/(kgK). The specific heat loss at $\vartheta_1 = \vartheta_F = 320\,°$C is $\phi/l = 288$ W/m (from worksheets Nos 9.14, 9.15 and 9.16). The specific drop of temperature is:

$$\frac{\Delta\vartheta}{l} = \frac{288 \text{ W/m}}{6.95 \text{ kg/s} \cdot 2.87 \text{ kJ/(kgK)}} = 0.0144 \text{ K/m}.$$

The temperature at the exit of the pipe is:

$$\vartheta_2 = \vartheta_1 - 8 \text{ m}\,(\Delta\vartheta/l) = (320 - 8 \cdot 0.0144)\,°\text{C} = 319.89\,°\text{C}.$$

There is no need for a sectional calculation, because the temperature of the fluid in the pipe has not changed much.

Economic thickness of insulation s_w
1000 700 500 400 300 200 100 70 50 40 30 20 10
1,5
1,0
0,8
0,6
20
B = 10
6
4
2
0,4
1
0,6
0,3
0,4
0,2
0,2
0,1
0,15
0,06
0,1
0,04
0,1
0,02
0,15
0,2
0,08
0,3
0,06
0,4
0,5
0,04
0,6
0,7
0,03
d_i = 1,0m
0,8
Cost coefficient K
20 10 5 2 1 0,5 0,2 0,05
0,02
0,015
Coefficient of thermal protection $\sigma = s_w/d_i$
K_0 = 200 DM/m²
150
100
80
60
40
30
20
15
10
K' = 10 DM/(m²cm)
5
4
3
2
1
0,5
0,01 0,02 0,03 0,04 0,05 0,07 0,1 0,2 0,3 0,4 0,5 0,7 1,0 2,0 m 3,0
Inner diameter d_i

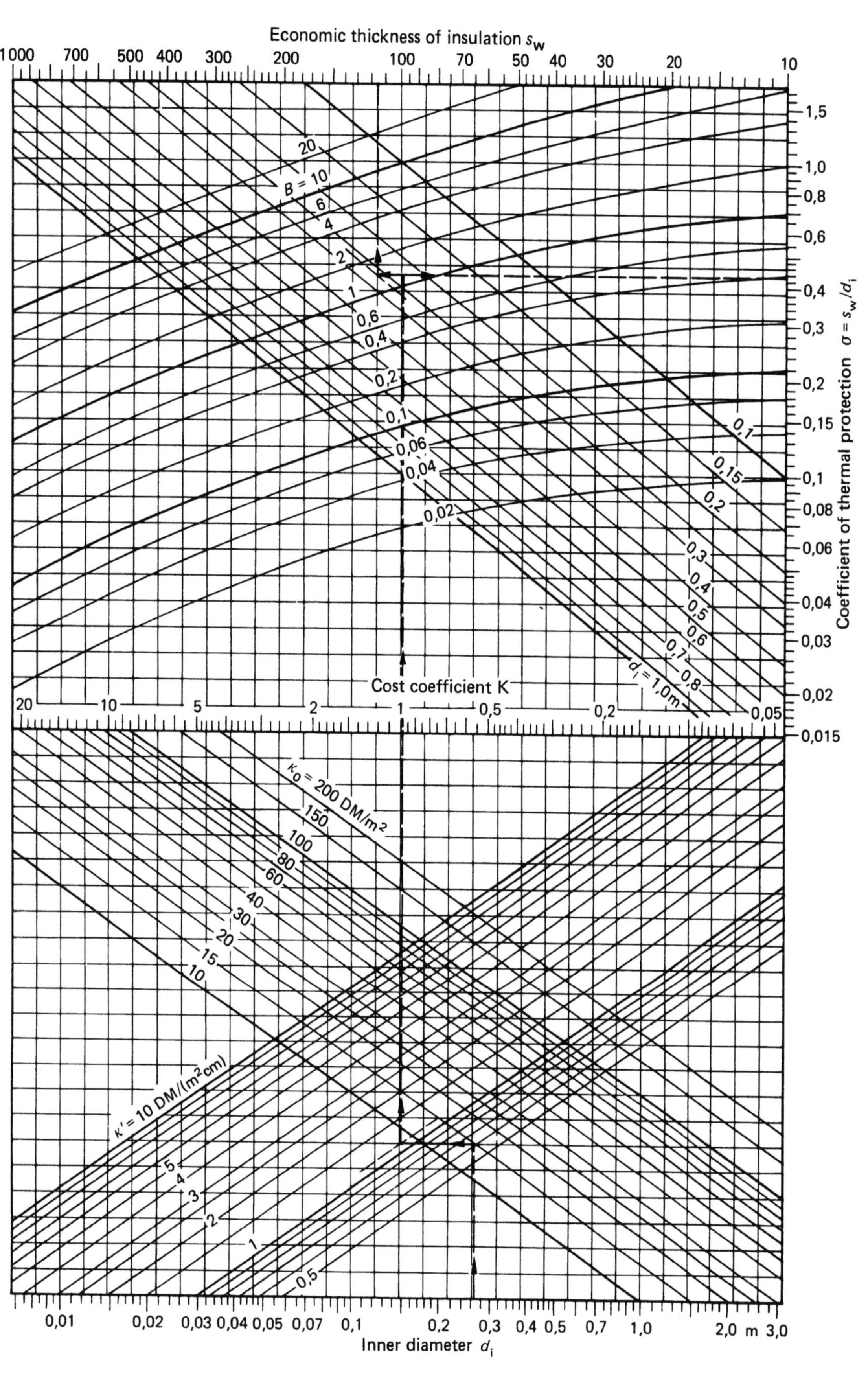

H. Auracher

Explanation

The cost for thermal protection rises with increasing thickness of the insulating layer. Simultaneously, the heat losses and the related costs decrease. The total cost may be calculated:

$$K_{\text{ges}} = \phi/l \cdot W \cdot \beta \cdot 3.6 \cdot 10^{-6} + b \cdot \varkappa \cdot \pi \cdot d_{\text{a}}\,.$$

At the so called "economic thickness of the insulating layer" s_{W}, K_{ges} has a minimum. According to *Grigull*, with the cost coefficient

$$K_{\text{D}} = 50\,\frac{d_{\text{i}}\,\varkappa'}{\varkappa_{\text{o}}}$$

and the operating factor

$$B = \frac{\lambda\,(\vartheta_{\text{i}} - \vartheta_{\text{o}})\,\beta\,W}{d_{\text{i}}\,b\,\varkappa_{\text{o}}} \cdot 3.6 \cdot 10^{-6}$$

the thickness of the layer, s_{W}, can be determined from the worksheets Nos 9.18.1 and 9.18.2 (see VDI-guideline No 2055).

Definitions:

$\vartheta_{\text{i}} - \vartheta_{\text{o}}$: Difference of temperature between temperature of the pipe surface and the temperature of the insulating layer surface.

ϑ_{i} is nearly equal to the temperature of the fluid, because the thermal resistance between the fluid and the pipe, as well as the pipe wall, can normally be neglected. Normally, the difference of temperature, $\vartheta_{\text{i}} - \vartheta_{\text{o}}$, is somewhat smaller than the usually known difference of temperature, based on the ambient temperature $\vartheta_{\text{i}} - \vartheta_{\text{L}}$. However, with a rough estimation of s_{W}, the unknown difference of temperature $\vartheta_{\text{i}} - \vartheta_{\text{o}}$ can be determined from worksheet No 9.15.

$\varkappa, \varkappa_{\text{o}}, \varkappa'$: Specific costs of thermal insulation

These are the costs per square meter surface of thermal insulation, including material (insulation material, surface protection, spacer, screws), assembly and auxiliary costs as, for example, additional costs for scaffolding, pipe ducts and storage. $\varkappa_{\text{o}}$ are the fictitious costs of a thermal insulation with $s = 0$ in DM/m^2 and $\varkappa'$ the increase in costs on a growing thickness of the insulating layer s of 1 cm. Thus, $\varkappa'$ has the dimension of DM/(m$^2 \cdot$ cm). Here the simplification has been made so that the specific costs $\varkappa$ of thermal insulation is a linear function of the layer thickness: $\varkappa = \varkappa_{\text{o}} + \varkappa'\,s$.

β: Time of operation

This is the number of hours per year, at which the plant works at design temperature. There is no dependency on the output. If the plant works at a temperature below design temperature, the number of hours has to be reduced accordingly.

W: Costs for heat produced

These are combined costs from fixed and operating costs. E.g. the additional costs for the boiler or for personnel, and the operating costs are fuel costs (see VDI-guidelines Nos 2055 and 2067). If there is additional heat loss, caused by fittings and pipe supports (worksheet No 9.16), the costs caused by heat loss will increase. In this case, it has to be considered by a higher estimation of W.

b: Cost annuity

It is the yearly amortization rate related to the investment costs, which has to be paid at the end of each year for depreciation and interest. Empirical data for maintenance, repair and auxiliary costs are additionally included (average value: $b = 0.2$ per year).

ϕ/l: Heat loss per m pipe length in W/m (see worksheet No 9.15).

λ: Thermal conductivity of the insulating material. In this case, the factors according to worksheet No 9.16 have to be considered.

d_{i}: Inner diameter of the pipe.

Example:
(shown in the worksheets Nos 9.18.1 and 9.18.2)

Known values:

Difference of temperature between the temperature at the pipe surface and the temperature at the surface of the insulating layer	$\vartheta_{\text{i}} - \vartheta_{\text{o}} = 250\ \text{K}$
Specific costs of the thermal insulation	$\varkappa_{\text{o}} = 15.-\ \text{DM/m}^2$
Increase in the specific costs	$\varkappa' = 1.10\ \text{DM/(m}^2\,\text{cm})$
Time of operation	$\beta = 6\,800\ \text{h/a}$
Costs for heat produced	$W = 12.-\ \text{DM/GJ}$
Cost annuity	$b = 0.2$
Thermal conductivity of the insulating material	$\lambda = 0.014\ \text{W/(mK)}$
Inner diameter of the pipe	$d_{\text{i}} = 0.27\ \text{m}$

Solution:

From worksheet No 9.18.2, one obtains the operating factor $B = 1.27$.

Now, from worksheet No 9.18.1, the cost coefficient $K_{\text{D}} = 0.99$, the coefficient of thermal protection $\sigma = 0.43$ and finally the economic thickness of the insulating layer $s_{\text{W}} = 116\ \text{mm}$ can be determined.

Bibliography

Grigull, U.: Die Ermittlung der wirtschaftlichen Isolierstärke. BWK 2 (1950) No 5, p. 125/127

Operating Factor for Pipe Insulation (with Worksheet No 9.18.1)

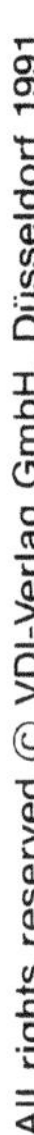

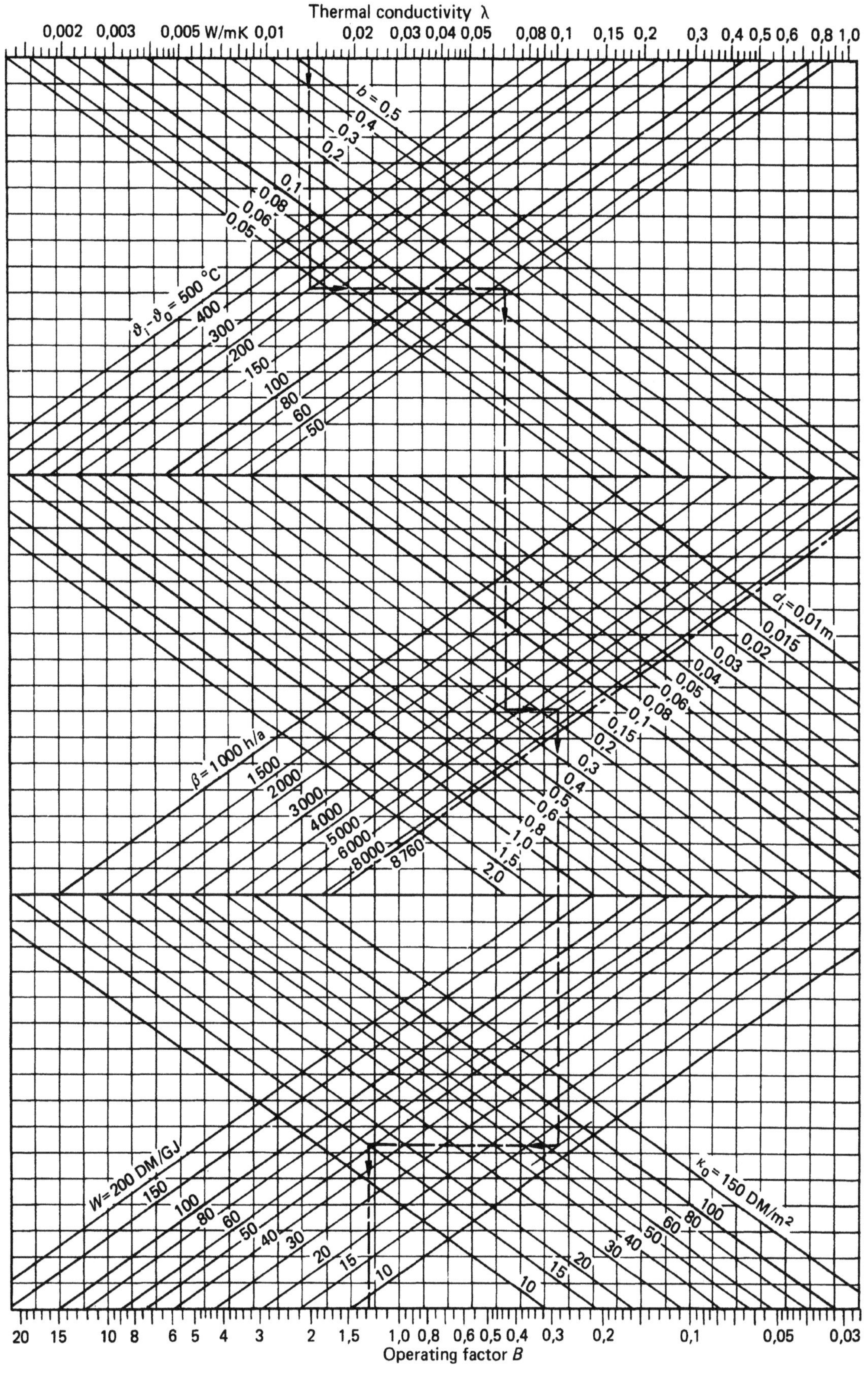
Thermal conductivity λ
0,002 0,003 0,005 W/mK 0,01 0,02 0,03 0,04 0,05 0,08 0,1 0,15 0,2 0,3 0,4 0,5 0,6 0,8 1,0
b = 0,5
0,4
0,3
0,2
0,1
0,08
0,06
0,05
ϑi - ϑo = 500 °C
400
300
200
150
100
80
60
50
di = 0,01 m
0,015
0,02
0,03
0,04
0,05
0,06
0,08
0,1
0,15
0,2
0,3
0,4
0,5
0,6
0,8
1,0
1,5
2,0
β = 1000 h/a
1500
2000
3000
4000
5000
6000
8000
8760
W = 200 DM/GJ
150
100
80
60
50
40
30
20
15
10
κo = 150 DM/m²
100
80
60
50
40
30
20
15
10
20 15 10 8 6 5 4 3 2 1,5 1,0 0,8 0,6 0,5 0,4 0,3 0,2 0,1 0,05 0,03
Operating factor B

H. Auracher

Reynolds number Re
Turbulent | Laminar
Pipe diameter d = 0.02 m
0.03
0.04
0.06
0.08
0.1
0.15
0.2
0.3
0.4
0.5
3
1.5 2
0.8 1.0
0.4 0.6
0.2 0.3
Flow velocity w = 0.1 m/s
Re = 2320
Temperature ϑ
°C
120
100
80
60
40
20
0
Very heavy fuel oil (type bunker c)
Heavy fuel oil (type bunker b)
Semi heavy fuel oil
Tar oil
Diesel oil
Fuel oil type L
Gas oil
Kinematic viscosity ν
10·10⁻⁶ m²/s

K. Beck

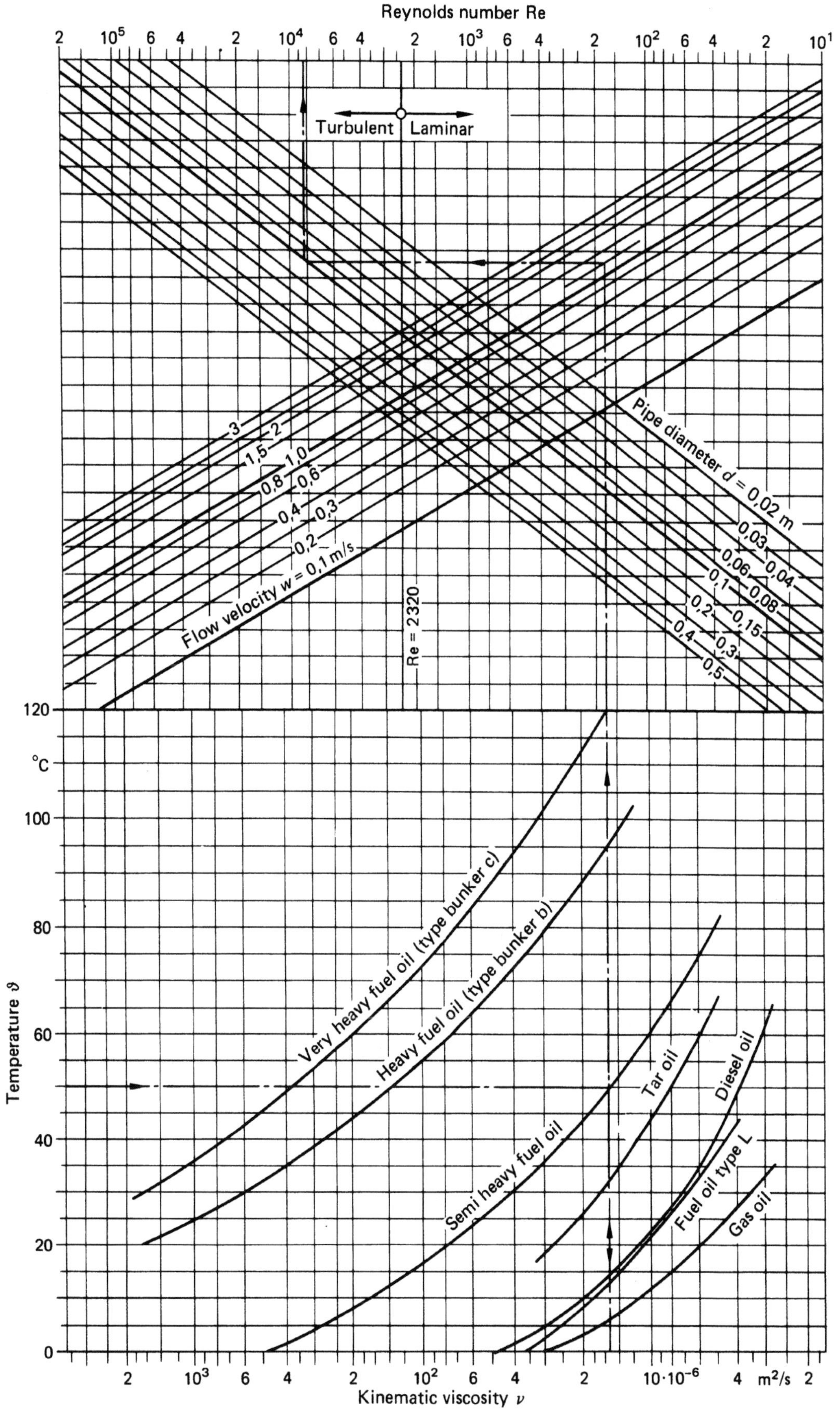

Explanation

$$\text{Reynolds number: } Re = \frac{w \cdot d}{v}$$

Definitions:

w in m/s   Average flow velocity
d in m   Inner diameter of the pipe
v in m²/s   Kinematic viscosity of the fluid

The value of v does very much depend on the temperature of the medium. Some approximate values for different types of fuel oils have been shown in the lower part of the diagram.

At a Reynolds number of 2 320, the flow through a pipe changes from laminar to turbulent conditions. If, for example, there are temperature changes in the neighbourhood of this Reynolds number, the pattern of flow easily changes from laminar to turbulent conditions (resp. vice versa). The pressure drop is strongly influenced by this. Between these two flow conditions, there exists a transit region, which is added to the turbulent conditions.

Depending on Re and the hydraulic roughness of the pipe wall, from worksheet No 9.1, the friction factor λ can be determined. At laminar flow, the friction factor λ is nearly independent of the pipe wall conditions and is in a reverse proportion to the Reynolds number:

$$\lambda = \frac{64}{Re} \quad \text{(worksheet No 9.1)}$$

Also for turbulent flow conditions, the friction factor can be determined from worksheet No 9.1.

Example:

Semi heavy fuel oil at a temperature of 50 °C.

Known values:
Kinematic viscosity $v = 15.4 \cdot 10^{-6}\,\text{m}^2/\text{s}$
Flow velocity $w = 1.3\,\text{m/s}$
Inner diameter $d = 100\,\text{mm}$

Results:
Re = 8 442

It is a turbulent flow.

Bibliography

Dubbel: Taschenbuch für den Maschinenbau. Berlin, Heidelberg, New York: Springer-Verlag 1981

Eck, B.: Technische Strömungslehre. Vol. 2; 8th edition Berlin, Heidelberg, New York: Springer-Verlag 1981

VDI-Wärmeatlas. 4th edition Düsseldorf: VDI-Verlag 1984

Herning, F.: Stoffströme in Rohrleitungen. 4th edition, Düsseldorf: VDI-Verlag 1966

Schmidt, D.: Stahlrohr-Handbuch. 10th edition, Vulkan-Verlag 1986

Richter, H.: Rohrhydraulik. 5th edition, Berlin, Heidelberg, New York: Springer-Verlag 1971

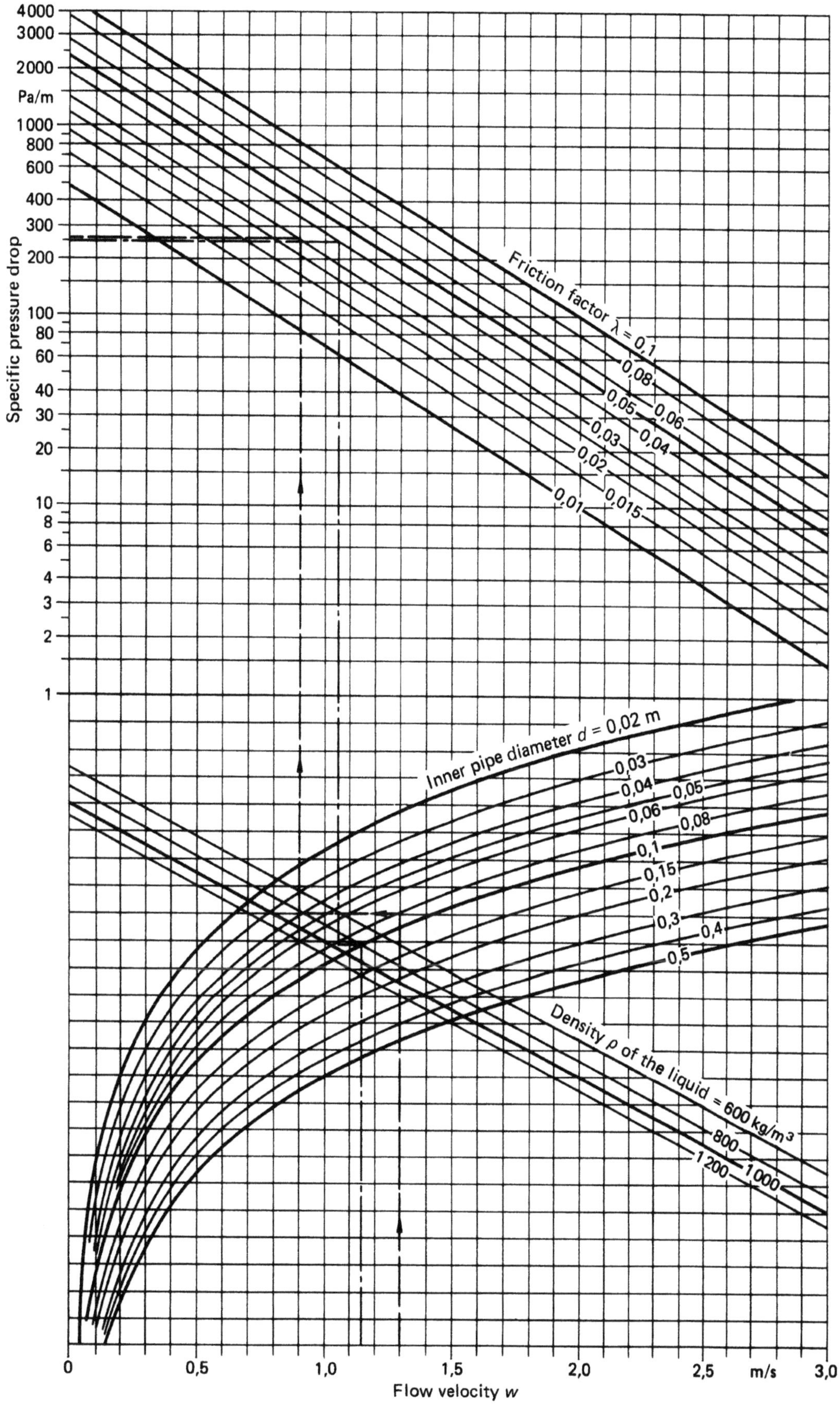

K. Beck, K. Wagner

Explanation

The specific pressure drop is:

$$\frac{p_1 - p_2}{l} = \lambda \cdot \frac{\varrho \cdot w^2}{2d}$$

p_1 in Pa   pressure head at the pipe line entrance
p_2 in Pa   pressure head at the pipe line exit
l in m   length of the pipe line
w in m/s   flow velocity
ϱ in kg/m^3 density of the fluid at the temperature ϑ
d in m   inner diameter of the pipe

The density ϱ and the kinematic viscosity v can be determined from the diagram below. The Reynolds number can be determined from worksheet No 9.19. If Re is $< 2\,320$, then the flow is laminar.

Dependent on the hydraulic roughness k, the inner diameter of the pipe d and the Reynolds number Re, the friction factor λ can be determined from worksheet No 9.1.

Example 2: Fuel oil type L		Example 1: Heavy fuel oil	
Known values:		**Known values:**	
Temperature	$\vartheta = 40\,°\mathrm{C}$	Temperature	$\vartheta = 60\,°\mathrm{C}$
Inner diameter	$d = 0.08\ \mathrm{m}$	Inner diameter	$d = 0.1\ \mathrm{m}$
Velocity	$w = 1.3\ \mathrm{m/s}$	Velocity	$w = 115\ \mathrm{m/s}$
Dynamic viscosity	v (from the diagram) $= 4.6 \cdot 10^{-6}\,\mathrm{m^2/s}$	Dynamic viscosity	v (from the diagram) $= 72 \cdot 10^{-6}\,\mathrm{m^2/s}$
Density	ϱ (from the diagram) $= 816\ \mathrm{kg/m^3}$	Density	ϱ (from the diagram) $= 943\ \mathrm{kg/m^3}$
Reynolds number	Re (from worksheet No 9.19) $= 22\,610$	Reynolds number	Re (from worksheet No 9.19) $= 1\,597$
Friction factor	λ (from worksheet No 9.1) $= 0.03$	Friction factor	λ (from worksheet No 9.1) $= 0.04$

There is turbulent flow.

Result: $\Delta p/l = 260\ \mathrm{Pa/m}$

There is laminar flow.

Result: $\Delta p/l = 250\ \mathrm{Pa/m}$

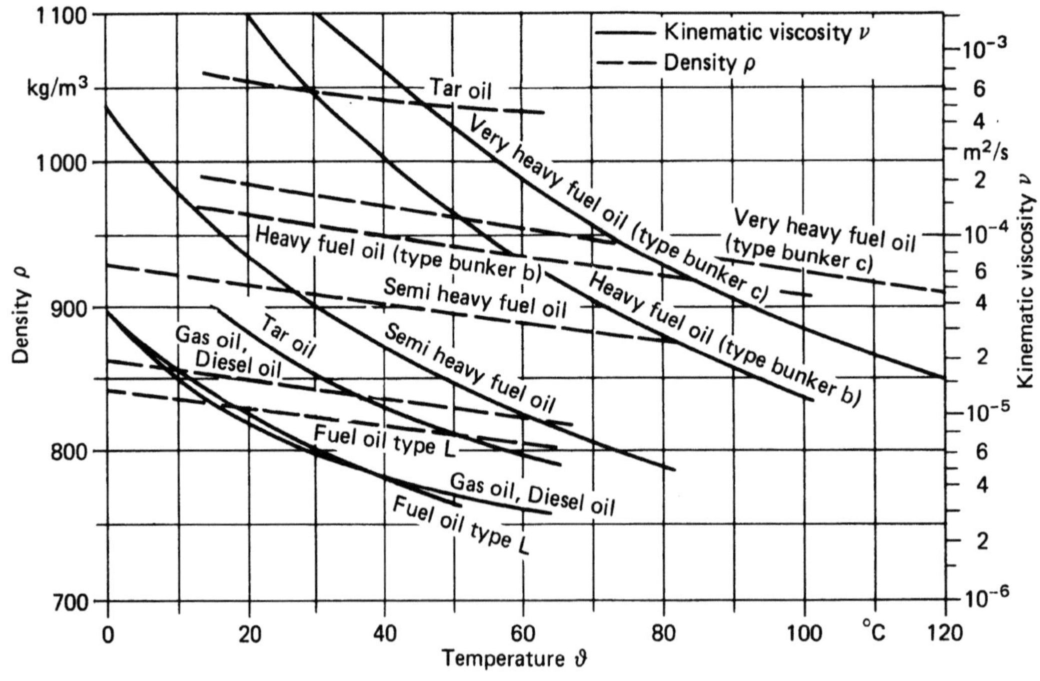

Bibliography

Dubbel: Taschenbuch für den Maschinenbau. Berlin, Heidelberg, New York: Springer-Verlag 1981
Eck, B.: Technische Strömungslehre. Vol. 2; 8th edition Berlin, Heidelberg, New York: Springer-Verlag 1981
VDI-Wärmeatlas. 4th edition Düsseldorf: VDI-Verlag 1984
Herning, F.: Stoffströme in Rohrleitungen. 4th edition, Düsseldorf: VDI-Verlag 1966
Schmidt, D.: Stahlrohr-Handbuch. 10th edition, Vulkan-Verlag 1986
Richter, H.: Rohrhydraulik. 5th edition, Berlin, Heidelberg, New York: Springer-Verlag 1971

Volume Related Storage Capacity of a Drop Accumulator

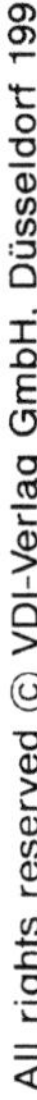

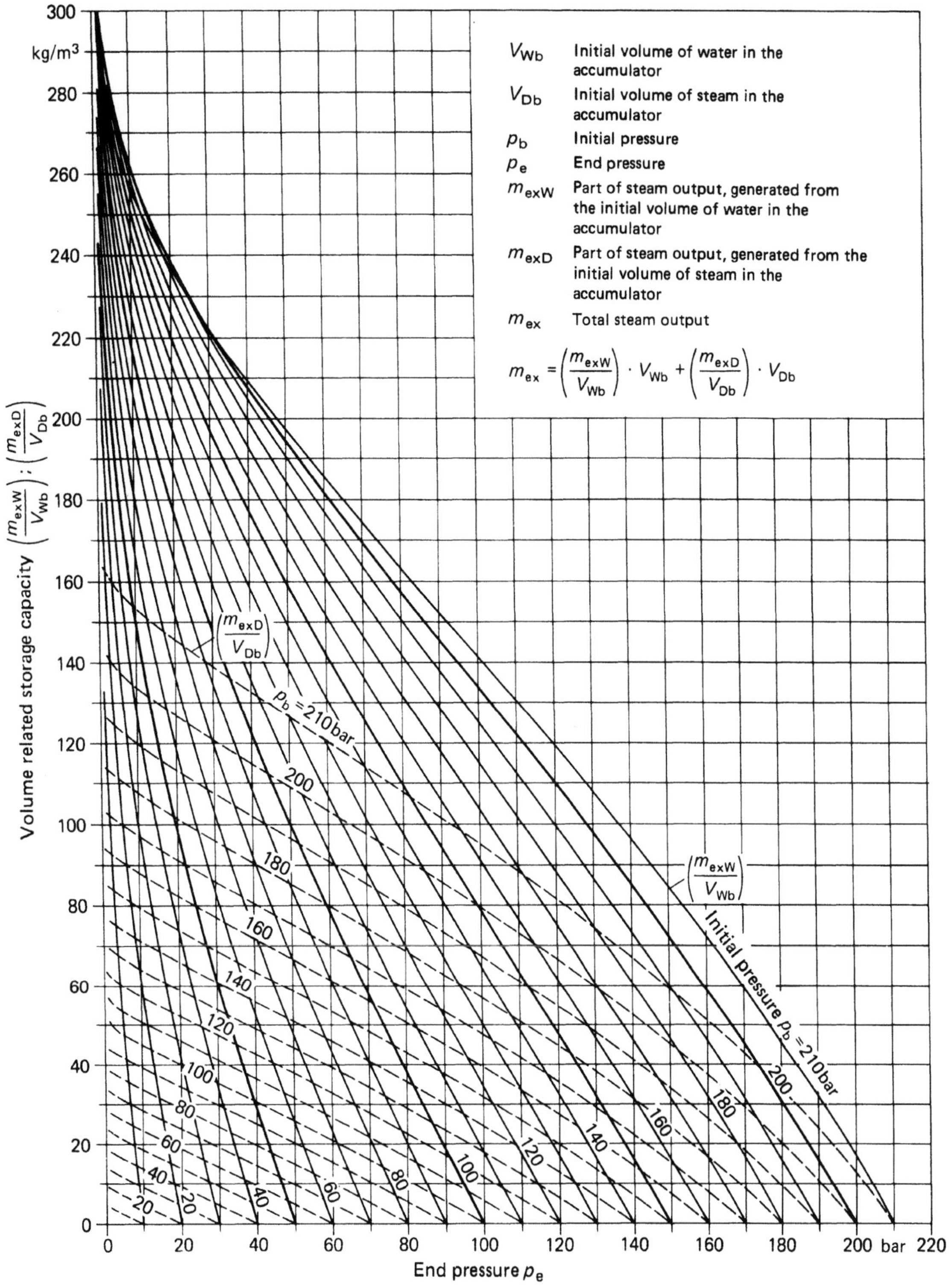

W. Bitterlich

Explanation

With the worksheets Nos 10.1 and 10.2, one can determine graphically the steam output of a drop accumulator, if the pressure at the beginning and at the end are known. The nomograph was calculated according to results of the exact numeric calculation of the complete equation system of a pressure drop accumulator.

The following assumptions are valid for the nomographs Nos 10.1 and 10.2:

- adiabatic pressure vessel
- no thermal or mechanical shrinking of the pressure vessel
- no mass input during discharging
- homogeneous conditions in the pressure vessel

Steady lines in the nomographs Nos 10.1 and 10.2:

- at the beginning of the discharge process, there exists only liquid water; there is no steam in the pressure vessel (steam contents $x_b = 0$)
- 1% water content of the steam output

The output steam mass m_{exW} is related to the initial water volume V_{Wb}.

Dotted lines in nomograph No 10.1:

If there is a mass of steam in the vessel at the start of the discharge (steam content $x_b > 0$), then the mass flow of steam output $\dot{m}_{exD}$ will always be related to the volume of steam in the vessel at the beginning of the process V_{Db}.

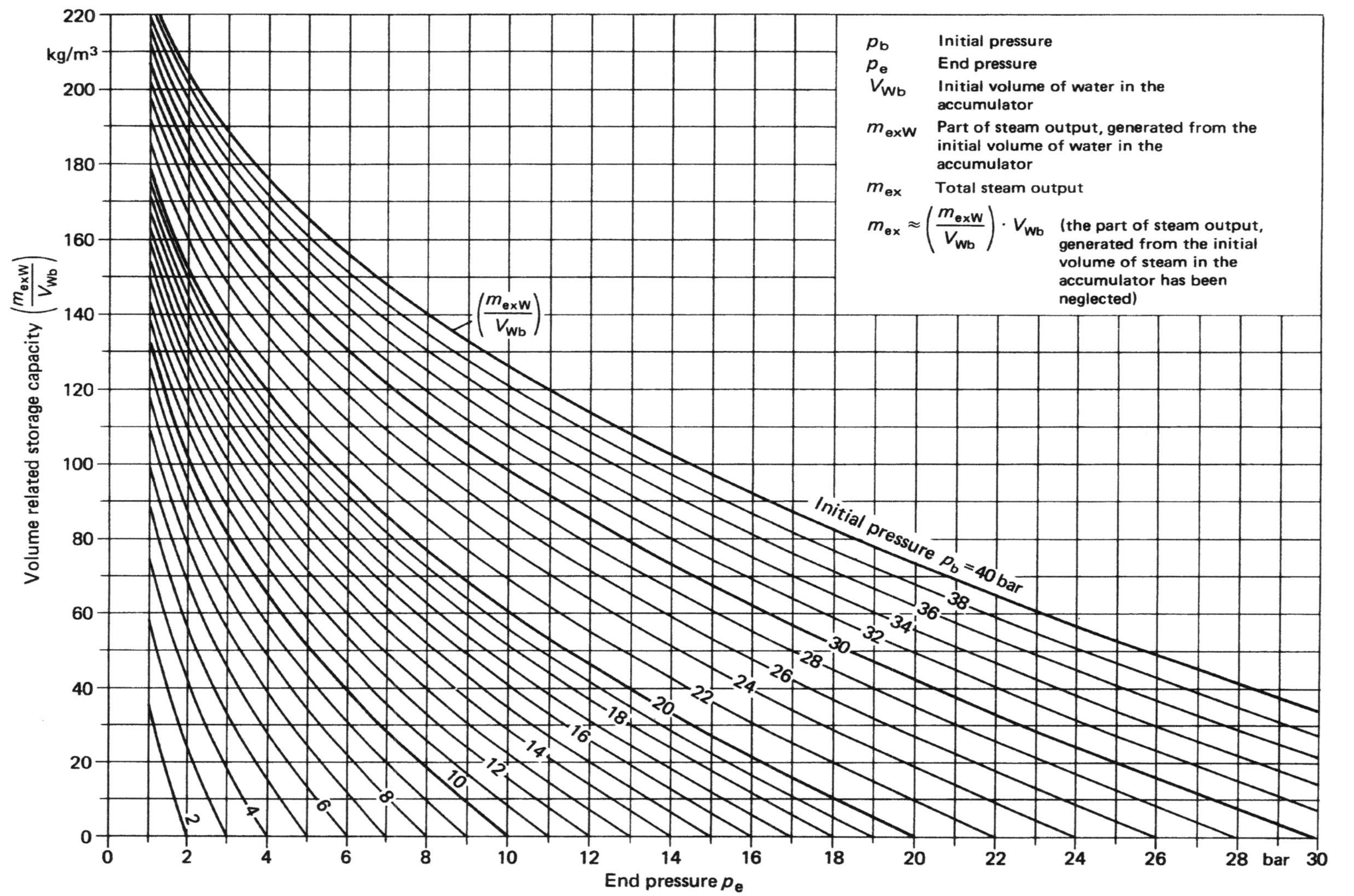

W. Bitterlich

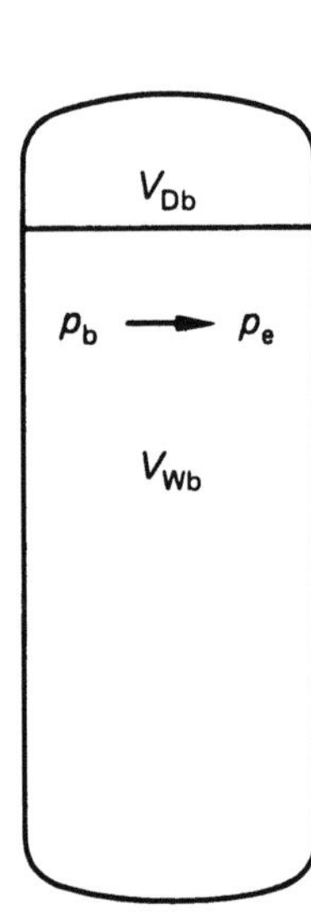
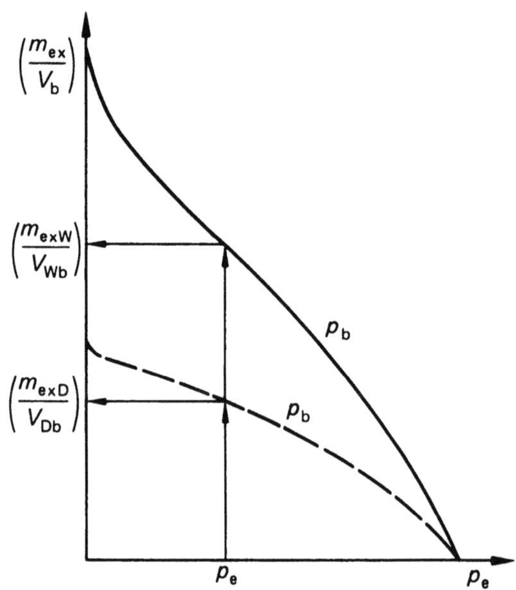

Example to 10.1:

Known values: $p_b = 160\ \text{bar}$ $V_{Wb} = 10\ \text{m}^3$

$p_e = 120\ \text{bar}$ $V_{Db} = 1\ \text{m}^3$

From worksheet No 10.1: $\left(\dfrac{m_{exW}}{V_{Wb}}\right) = 72\ \text{kg/m}^3$

$\left(\dfrac{m_{exD}}{V_{Db}}\right) = 36\ \text{kg/m}^3$

Calculation: $m_{ex} = 72\ \text{kg/m}^3 \cdot 10\ \text{m}^3 + 36\ \text{kg/m}^3 \cdot 1\ \text{m}^3 = 756\ \text{kg}$

Example to 10.2:

Known values: $p_b = 40\ \text{bar}$ $V_{Wb} = 10\ \text{m}^3$

$p_e = 30\ \text{bar}$ $V_{Db} = 1\ \text{m}^3$

From worksheet No 10.2: $\left(\dfrac{m_{exW}}{V_{Wb}}\right) = 35\ \text{kg/m}^3$

Calculation: $m_{ex} \approx 35\ \text{kg/m}^3 \cdot 10\ \text{m}^3 \approx 350\ \text{kg}$

A more exact determination with worksheet No 10.1 results in:

$$\left(\frac{m_{exD}}{V_{Db}}\right) = 5\ \text{kg/m}^3$$

$$m_{ex} = 350\ \text{kg} + 5\ \text{kg/m}^3 \cdot 1\ \text{m}^3 = 355\ \text{kg}$$

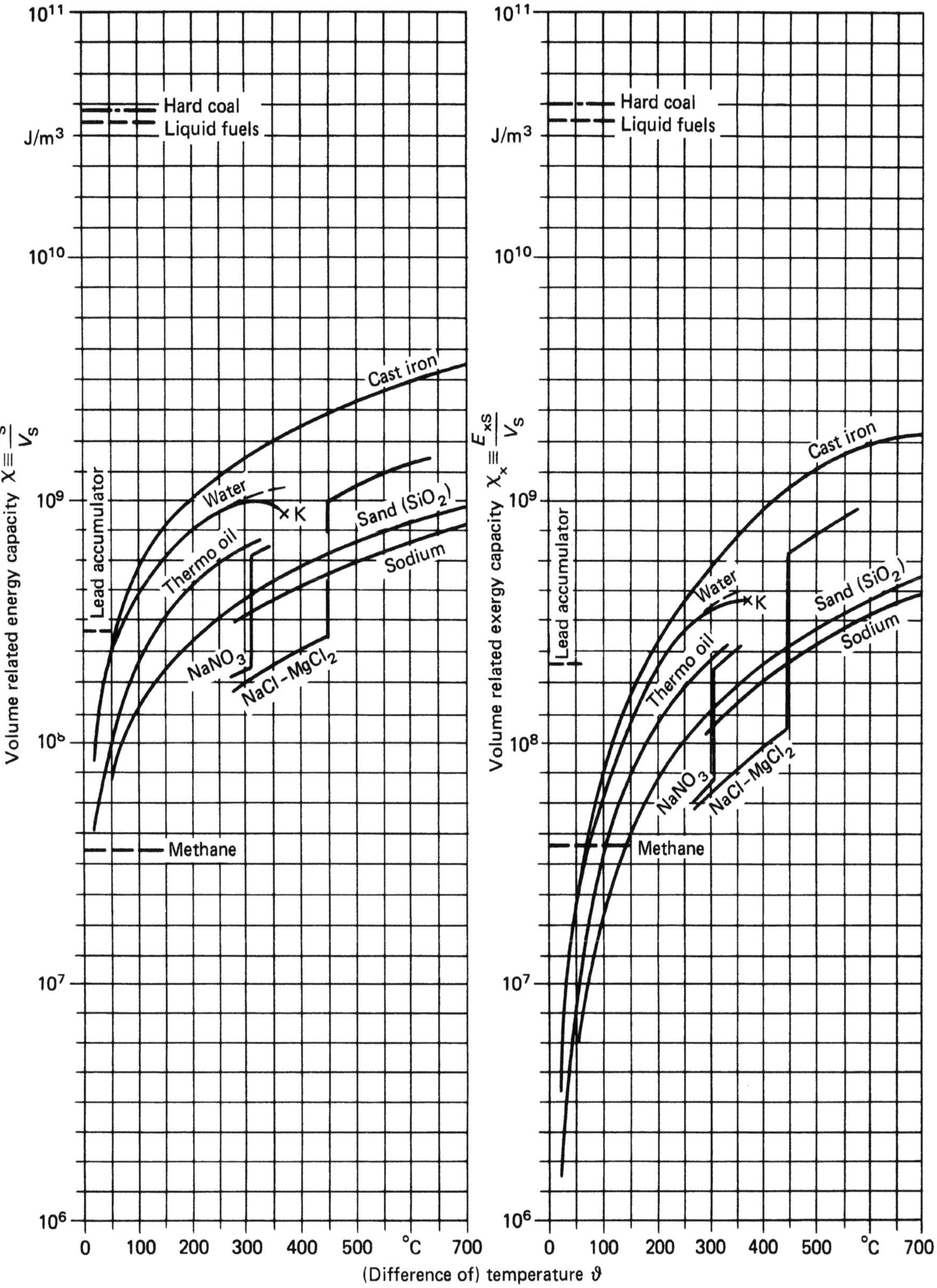

Bibliography

Bohn, T., W. Bitterlich: Grundlagen der Energie und Kraftwerkstechnik. Technischer Verlag Resch/Verlag TÜV Rheinland, Köln 1982

W. Bitterlich

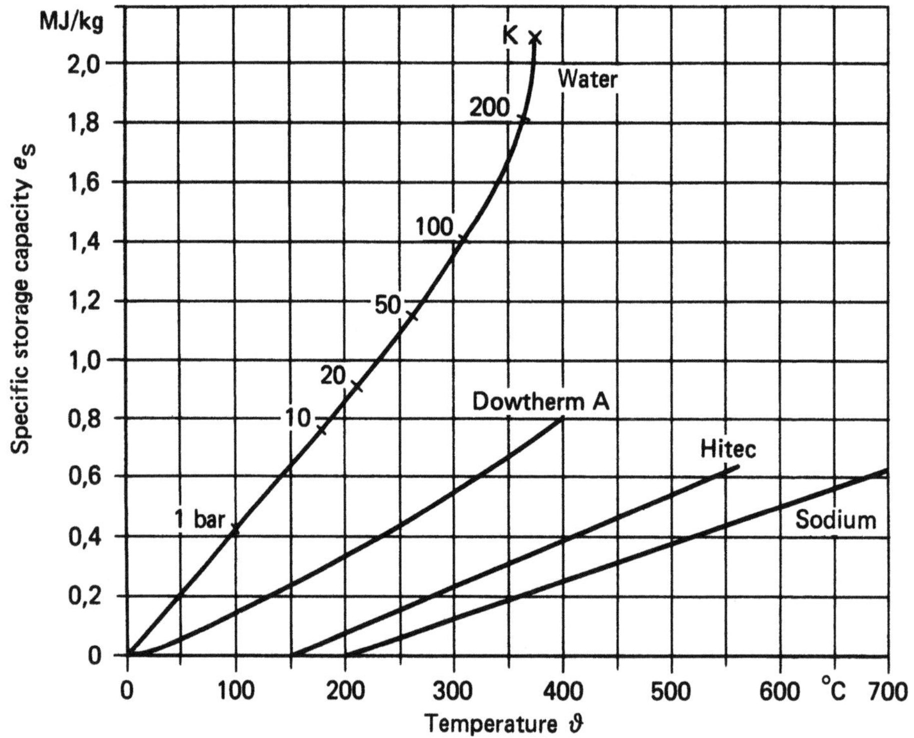

Specific storage capacity of selected storage media as a function of change in temperature

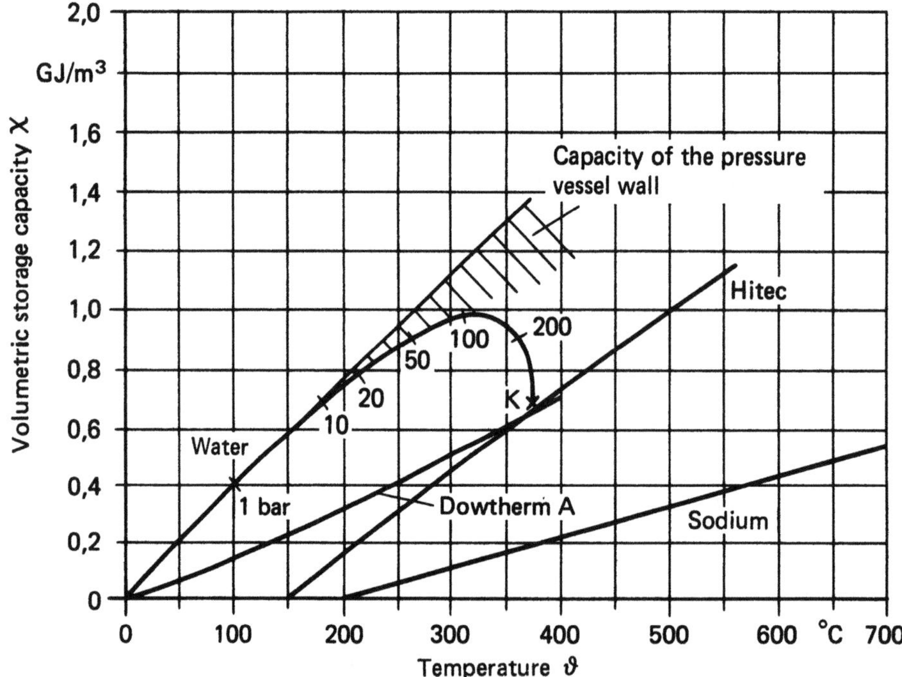

Volumetric storage capacity of selected storage media as a function of change in temperature

Bibliography

Thermische Energiespeicher für den Temperaturbereich 200 °C bis 500 °C. Institut für Kerntechnik und Energiewandlung e.V., Stuttgart. Forschungsvorhaben BMFT ET 4335 A, Report No IKE 5 TF-399-80, June 1980

$\vartheta = 0\ °C$
100
200
300
400
500
$\eta = 50\%$
60
70
80
90
(100)
$\Delta p = 1000\ Pa$
2000
3000
4000
5000
10000
70
50
m³/s
30
20
10
7
5
3
2
Volume flow $\dot{V}$ (for example air or flue gas)
20
30
40
50
70
100
200
kW
500
Power requirement P

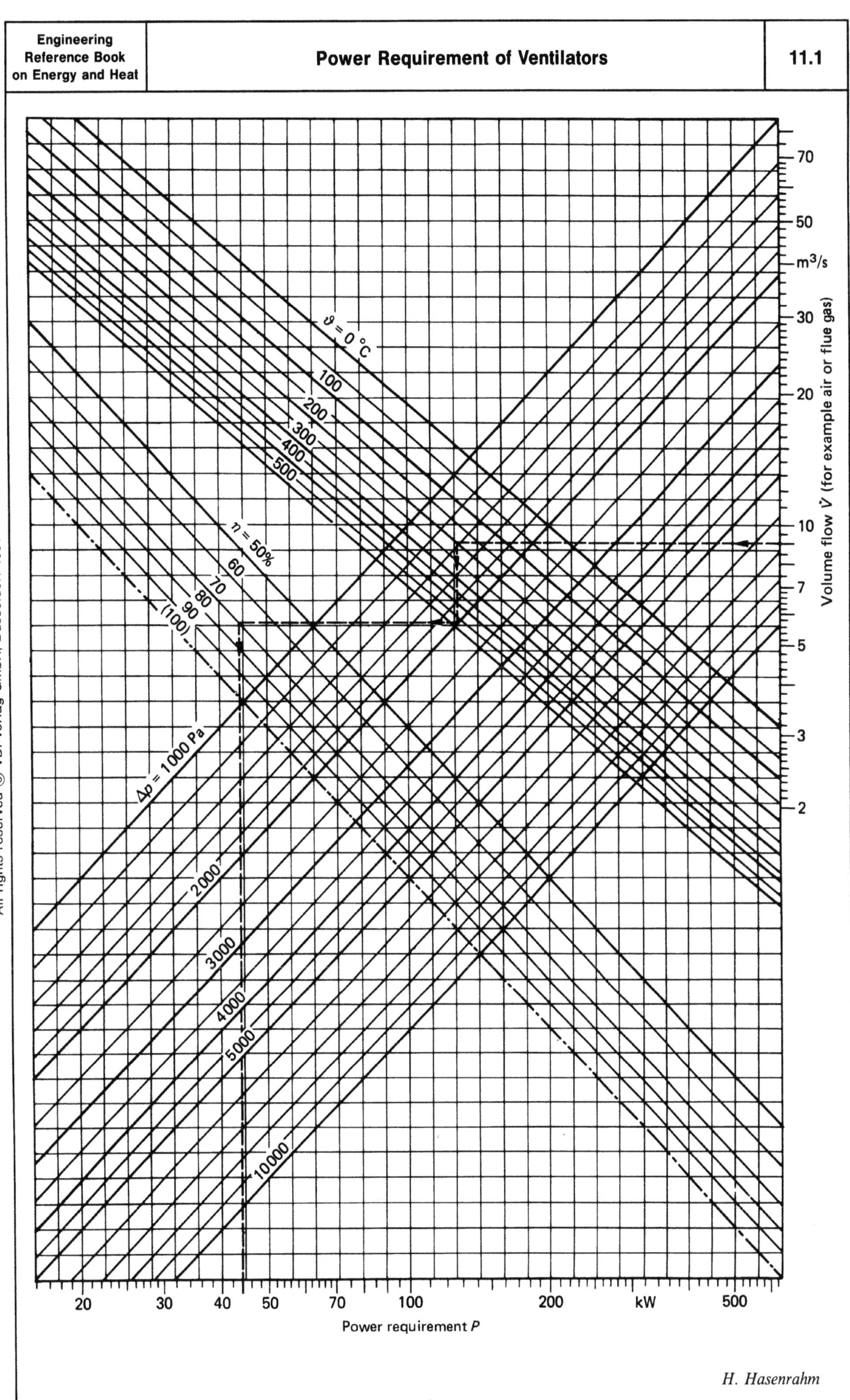

H. Hasenrahm

Example (air):

Known values:

Volume flow	$\dot{V}$	$= 9.0\ \mathrm{m^3/s}$
Temperature	ϑ	$= 200\,°\mathrm{C}$
Pressure rise	Δp	$= 2\,000\ \mathrm{Pa}\ (2\,000\ \mathrm{N/m^2})$
Efficiency of the ventilator	η	$= 0.70\ (70\%)$

Results in:

Power requirement $\qquad P \quad = 44\ \mathrm{kW}$

Basic equation (in above used units):

$$P = \frac{\dot{V}(1 + \vartheta/273)\,\Delta p}{10^3\,\eta}\,.$$

The worksheet is valid for an entrance pressure of $p_0 = 1.013$ bar. At a different entrance pressure p_1, the power requirement has to be multiplied by the factor p_0/p_1.

The terms "ventilators" and "compressors" have been defined in the explanation to Pos. 8414 of NIMEXE (Nomenclature for the foreign trade statistics of the EC-countries).

Specific compression work Y_S
1000
kJ/kg
600
400
300
200
100
80
60
40
30
20
10
8
6
4
3
2
1
$T_1 = 500$ K
400
300
250
200
150
100
$M^* = 100$
80
60
50
40
30
20
15
10
8
5
$\kappa = 1,5$
1,4
1,3
1,2
1,1
1,0
1,2
1,4
1,6
1,8
2,0
2,5
3,0
3,5
4,0
4,5
5,0
5,5
6,0
6,5
7,0
Pressure ratio p_2/p_1

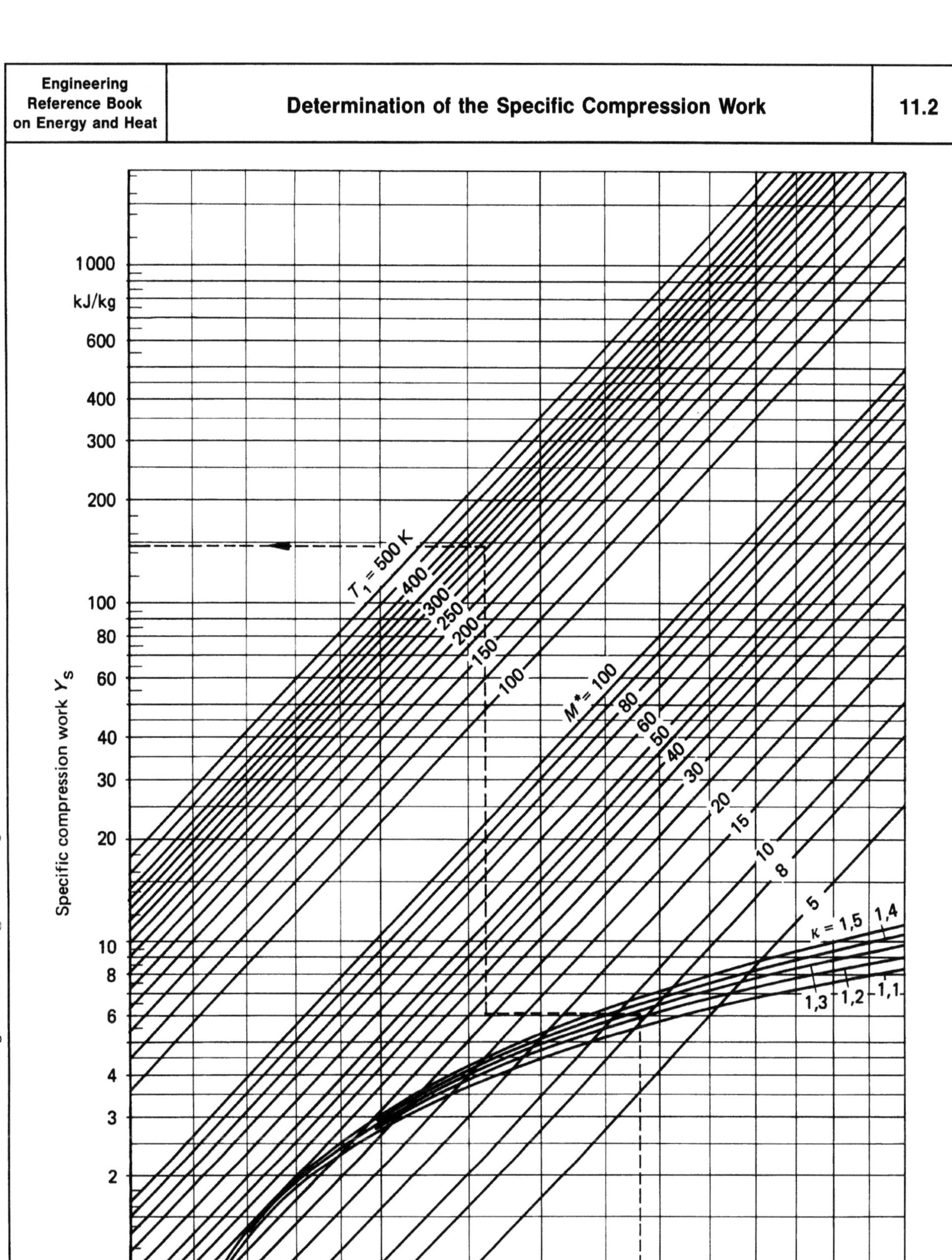

H. Hasenrahm

Explanation

The diagram is based on the equation:

$$Y_s = \frac{k}{k-1} \cdot Z \cdot R \cdot T_1 \left[p_2/p_1^{\frac{k-1}{k}} - 1 \right] \quad [\text{J/kg}]$$

The calculation of the isentropic compression work from the diagram can be divided into the following steps:

1) determination of the pressure ratio p_2/p_1
2) determination of the isentropic exponent k
3) determination of the molar mass M^*
4) determination of the inlet temperature T_1
5) determination of the isentropic compression work Y_s

    This value is valid for the uncooled compression process.

The molar mass M^* has been derived from the ideal gas constant R, multiplied by the compressibility factor Z: $M^* = 8\,314.41/Z \cdot R$.

Example:

Gas mixture of hydrocarbons and inert gases

Known values:

Molar mass	M^*	$= 26.5$
Pressure ratio	p_2/p_1	$= 3.8$
Isentropic exponent	k	$= 1.24$
Inlet temperature	T_1	$= 308$ K $(35\,^\circ\text{C})$

Result:

Specific compression work	Y_s	$= 147$ kJ/kg

Compression work

The compression of gaseous media can be described by the thermodynamic basic equation

$$Y_s = \frac{k}{k-1} \cdot Z \cdot R \cdot T_1 \left[p_2/p_1^{\frac{k-1}{k}} - 1 \right] \quad [\text{J/kg}]$$

The isentropic compression work is the theoretical compression without extraction or addition of heat (entropy $s = $ constant). It can be described by thermodynamic properties of the gas, as for example, isentropic exponent k, gas constant R and by the compressibility factor Z, the inlet temperature T_1 and the pressure ratio p_2/p_1, also called π. At uncooled compression, the equation describes a comparative process to determine the efficiency of the machine.

The compression work plus the power requirement due to the inner efficiency causes a temperature rise of the compressed gas. To reduce gas temperature and the compression work, and thus, the power consumption of the compressor at higher pressure ratios (dependent on the gas), it is useful to remove heat during compression. The power required is directly proportional to the inlet temperature T_1.

To compare the behaviour of a cooled compressor, the isothermal comparative process has been defined:

$$Y_T = Z \cdot R \cdot T \cdot \ln p_2/p_1 \quad [\text{J/kg}]$$

It is based on the idea, that the compression will take place at constant temperature.

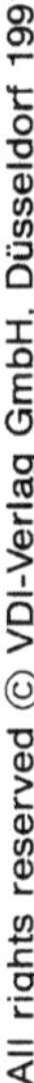

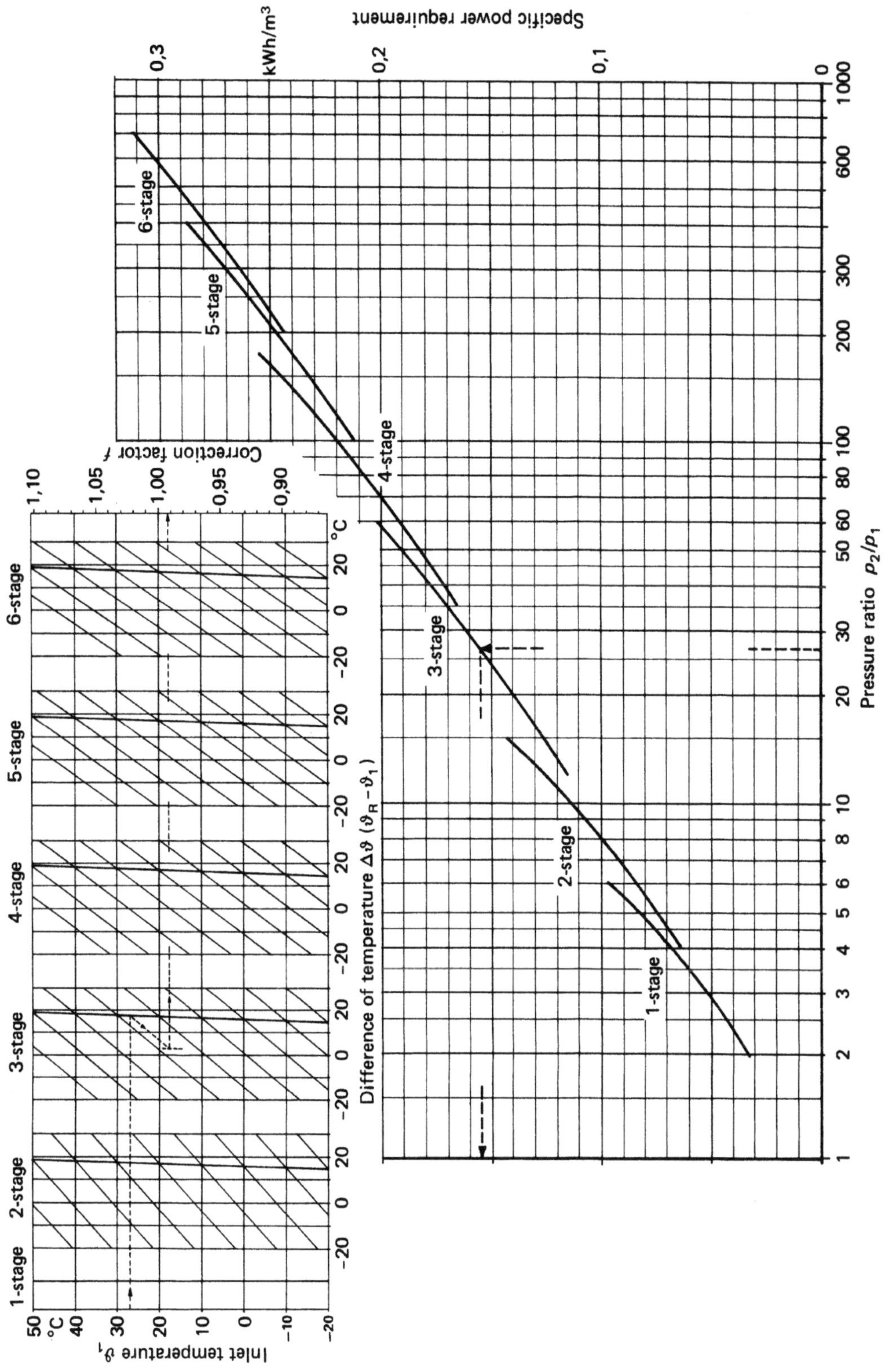
Specific power requirement
kWh/m³
0,3
0,2
0,1
0
Pressure ratio p_2/p_1
1000
600
300
200
100
80
60
50
30
20
10
8
6
5
4
3
2
1
6-stage
5-stage
4-stage
3-stage
2-stage
1-stage
Correction factor f
1,10
1,05
1,00
0,95
0,90
Difference of temperature $\Delta\vartheta$ $(\vartheta_R - \vartheta_1)$
°C
20
0
-20
Inlet temperature ϑ_1
50
°C
40
30
20
10
0
-10
-20

H. Hasenrahm

Explanation

The diagram is based on the following assumptions:

Inlet pressure $\quad\quad\quad\quad\quad p_1 = 1\,\text{bar}$
Inlet temperature $\quad\quad\quad\quad T_0 = 293\,\text{K}\ (20\,^\circ\text{C})$
Re-cooling temperature $\quad\quad T_R = 310\,\text{K}\ (37\,^\circ\text{C})$
Isentropic exponent $\quad\quad\quad k\ = 1.4$
Mechanical efficiency $\quad\quad\quad \eta_m = 0.92$

To determine the power requirement at the clutch, the specific power requirement has to be multiplied by the volume flow at standard conditions ($\dot{V}_N$: 1.013 bar, 273 K = 0 °C).

Example:

Medium: Nitrogen

Known values:

Volume flow at standard conditions $\quad \dot{V}_N \quad = 5\,000\,\text{m}^3/\text{h}$
Inlet pressure $\quad\quad\quad\quad\quad\quad\quad\quad\quad p_1 \quad = 4\,\text{bar}$
Outlet pressure $\quad\quad\quad\quad\quad\quad\quad\quad p_2 \quad = 108\,\text{bar}$
Pressure ratio $\quad\quad\quad\quad\quad\quad\quad\quad p_2/p_1 = 27$
Inlet temperature $\quad\quad\quad\quad\quad\quad\quad T_1 \quad = 300\,\text{K}\ (27\,^\circ\text{C})$
Re-cooling temperature $\quad\quad\quad\quad T_R \quad = 303\,\text{K}\ (30\,^\circ\text{C})$

Result:

Specific power requirement (from the diagram) = 0.155 kWh/m^3
Correction of temperature (auxiliary diagram) $\quad \vartheta_1 = 27\,^\circ\text{C},\ \Delta\vartheta = \vartheta_R - \vartheta_1 = 3\,^\circ\text{C}$
Correction factor (3-stage compressor) $\quad f = 0.992$
Power requirement at the clutch $\quad P_{Ku} = 0.155 \cdot 5\,000 \cdot 0.992\,\text{kW} = 768.8\,\text{kW}$

Compression with intermediate cooling

On isothermal compression, i.e. compression at constant temperature, the power requirement is at its lowest. It is an idealised process and technically not feasible. To approximate this idealisation, the pressure ratio is divided into several parts and the compressed gas is cooled down at the end of each stage. This requires additional compression work, because the cooling process takes place in a heat exchanger with an additional pressure drop. Therefore, the number of intermediate coolers has to be optimised by calculation. Some items of this calculation are: optimising the difference of temperature between the gas and the coolant in the intermediate cooler (this strongly influences the heat exchange area), additional power requirements caused by the additional pressure drop, comparative costs for different types of machinery and the costs of cooling water, which are often neglected and may be very high.

Small rotating displacement compressors up to a volume flow of about 1.2 m^3/s may operate with oil flood cooling. Here, most of the compression heat is removed by oil injected directly into the compression chamber, so that the discharge temperature of the gas is comparatively low. The hot oil is separated after the gas has left the compressor and is cooled down in a heat exchanger by a coolant. Then the cold oil again is injected into the compression chamber. This kind of cooling also causes additional power requirements, because there are displacement losses from the oil in the compression chamber and pressure drop caused by the oil separation.

Inlet volume flow $\dot V$
27,78 m³/s
25
22,22
19,44
16,67
13,89
11,11
8,33
5,56
4,17
2,78
1,39
10000 kW
5000
4000
3000
2000
1000
500
Pressure ratio p_2/p_1
30
25
20
15
10
5
4
Inlet volume flow $\dot V_1$
100·10³
80 m³/h
70
60
50
40
30
20
15
10
5

H. Hasenrahm

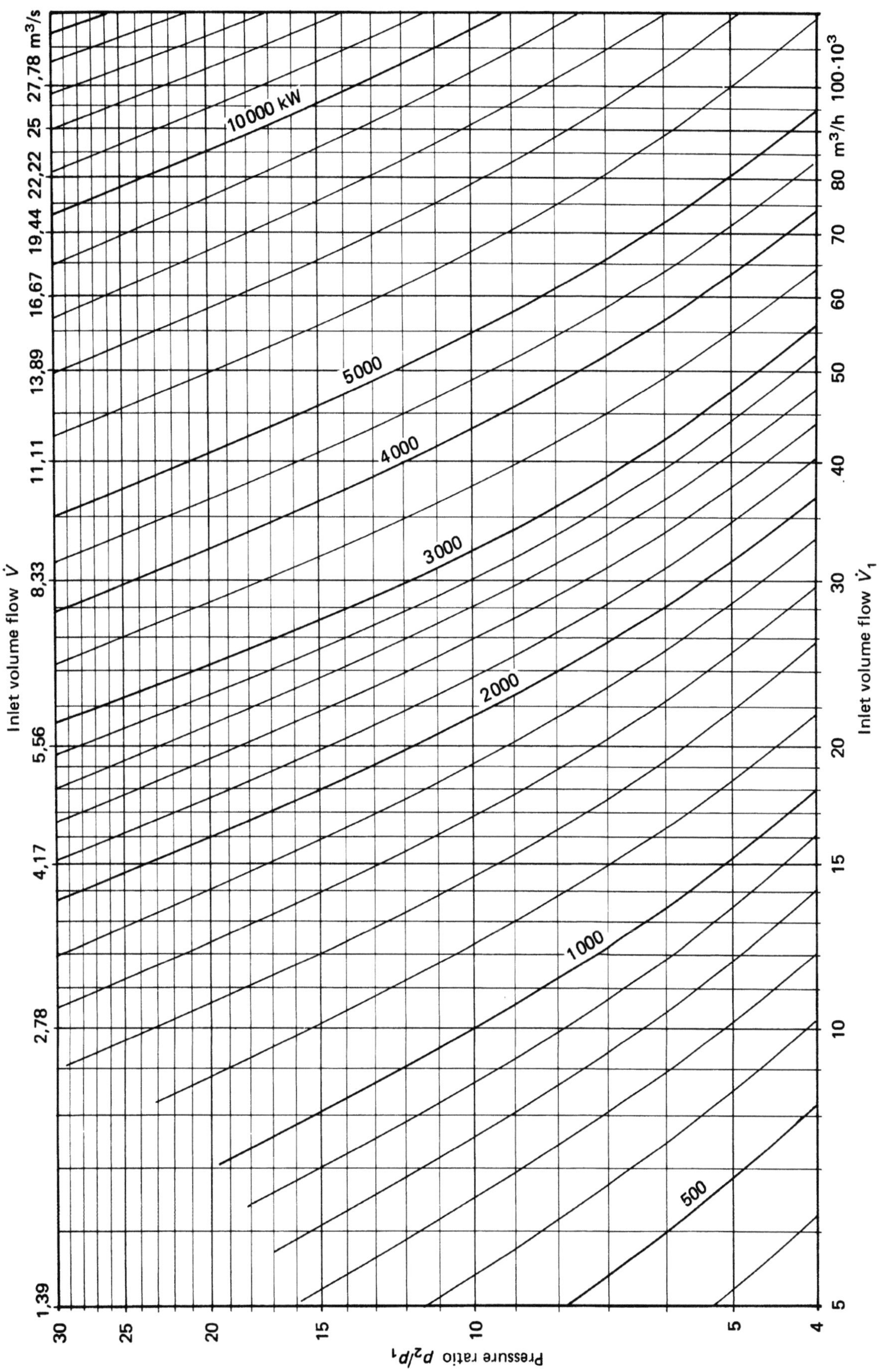

Explanation

The diagram is based on the following assumptions for the compression of air:

Inlet pressure	$p_1 = 1\,\text{bar}$
Inlet temperature	$T_1 = 288\,\text{K}\ (15\,°\text{C})$
Re-cooling temperature	$T_R = 298\,\text{K}\ (25\,°\text{C})$
Relative humidity	$\varphi = 70\%$
Gas constant	$R = 288.5\,\text{J/(kgK)}$

This diagram also can be used for gases, whose properties of media do slightly differ from those of the air. At differing inlet pressure p_1, the calculations are linear proportional to p_1.

The temperature difference between the inlet temperature T_1 and the re-cooling temperature does strongly influence the efficiency dependent on the number of intermediate cooling stages (rough standard value $4\,\text{K} \approx 1\%$, proportional).

Power requirement

The power requirement at the clutch of a compressor can be calculated from the specific compression work Y, the mass flow $\dot{m}$ and the efficiency η_{Ku}:

$$P_{Ku} = Y_{(s,\,T)} \cdot \dot{m}/\eta_{Ku} \quad [\text{W}]\,.$$

The efficiency is calculated from the inner compression losses, the inner and outer leakage losses and the mechanical losses (bearings, rod seals and, if it exists, gear box). To calculate the total power consumption, the motor requirements have to be added. At the calculation of the pressure ratio, if it exists, the pressure drop in suction filters (air and technical gases are seldom free of dust) and in silencers (compressors induce airborne and structure-borne sound to the suction and pressure line) has to be considered.

8
7
6
5
4
3
2
4,3
2,7
1 bar
1,65
°C
70
60
50
40
30
20
10
Re-cooling temperature
40
%
60
80
100
Relative humidity
1,2
1,0
0,8 bar
30
°C
20
10
0
Inlet temperature

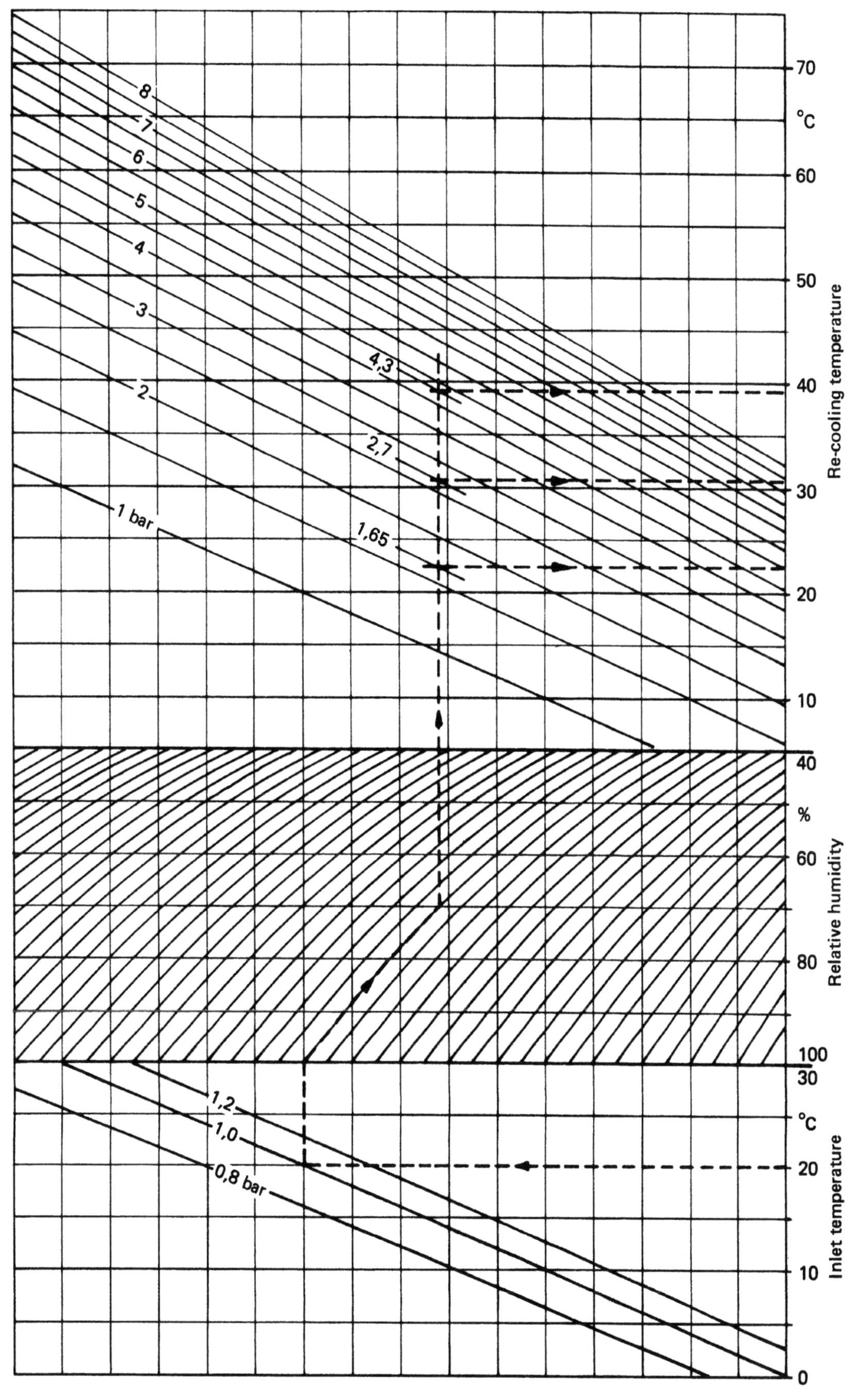

H. Hasenrahm

Explanation

With the diagram, one can calculate that limit re-cooling temperature in intermediate coolers of an air compressor, which avoids condensation of air.

Example:

Air of 1 bar, 20 °C and 70% relative humidity has to be compressed to 7.0 bar by a 4-stage compressor with 3 intermediate coolers. The pressures at the end of the stages in the intermediate coolers are 1.65, 2.7 and 4.3 bar. The dew point in the intermediate coolers occurs at the following re-cooling temperatures:

Intermediate cooler No. 1 (1.65 bar)   22.4 °C
Intermediate cooler No. 2 (2.7   bar)   30.7 °C
Intermediate cooler No. 3 (4.3   bar)   39.2 °C

Cooling

Because of temperature limits and for energy reasons, compressed air is cooled down. Usual coolants are the ambient air (heat exchange by forced convection) or, in the case of water cooling, raw water or circulating water. In the case of cooling with circulating water, a two circle system is possible: the primary system is cooled by treated water, which works as a transmission medium in a closed loop. It's heat is removed via heat exchangers in an open loop to the ambient air or to water with a worse quality.

The difference of temperature between the re-cooling temperature of the compressed air and the inlet temperature of the coolant does influence the *heat exchange area*. For environmental reasons, the rise in temperature of the coolant may be limited.

Dropping below the dew point

The ability of the inlet air to absorb steam decreases with increasing pressure. The specific mass of absorbed water can be calculated from:

$$x = R_{Gas}/R_{H_2O} \cdot p_D/(p - p_D) \quad [kg/kg];$$

p_D is the saturation pressure depending on the temperature.

At dropping below the dew point, condensate occurs in the cooler forming acid substances with impurities of the air. In this case, high alloy material has to be used for the heat exchanger and some parts of the compressor. At exact calculations, the appropriate dew point of the acid has to be considered. Compared with the condensation of pure water, it may change the beginning of condensation to higher temperatures.

Determination of the power requirement of pumps

Example:

Mass flow	$\dot{m}_w$	$= 33.3\ \text{kg/s}\ (=120\ \text{t/h})$
Temperature	ϑ	$= 300\,°\text{C}$
Pressure head of the pump	p	$= 160\ \text{bar}$
Efficiency	η	$= 0.78\ (78\%)$

Result:

Power requirement	P	$= 960\ \text{kW}$

Suggestion for the safety factors for the determination of the motor power: up to 7.5 kW about 20%, from 7.5 to 40 kW about 15%, from 40 kW about 10% additional power.

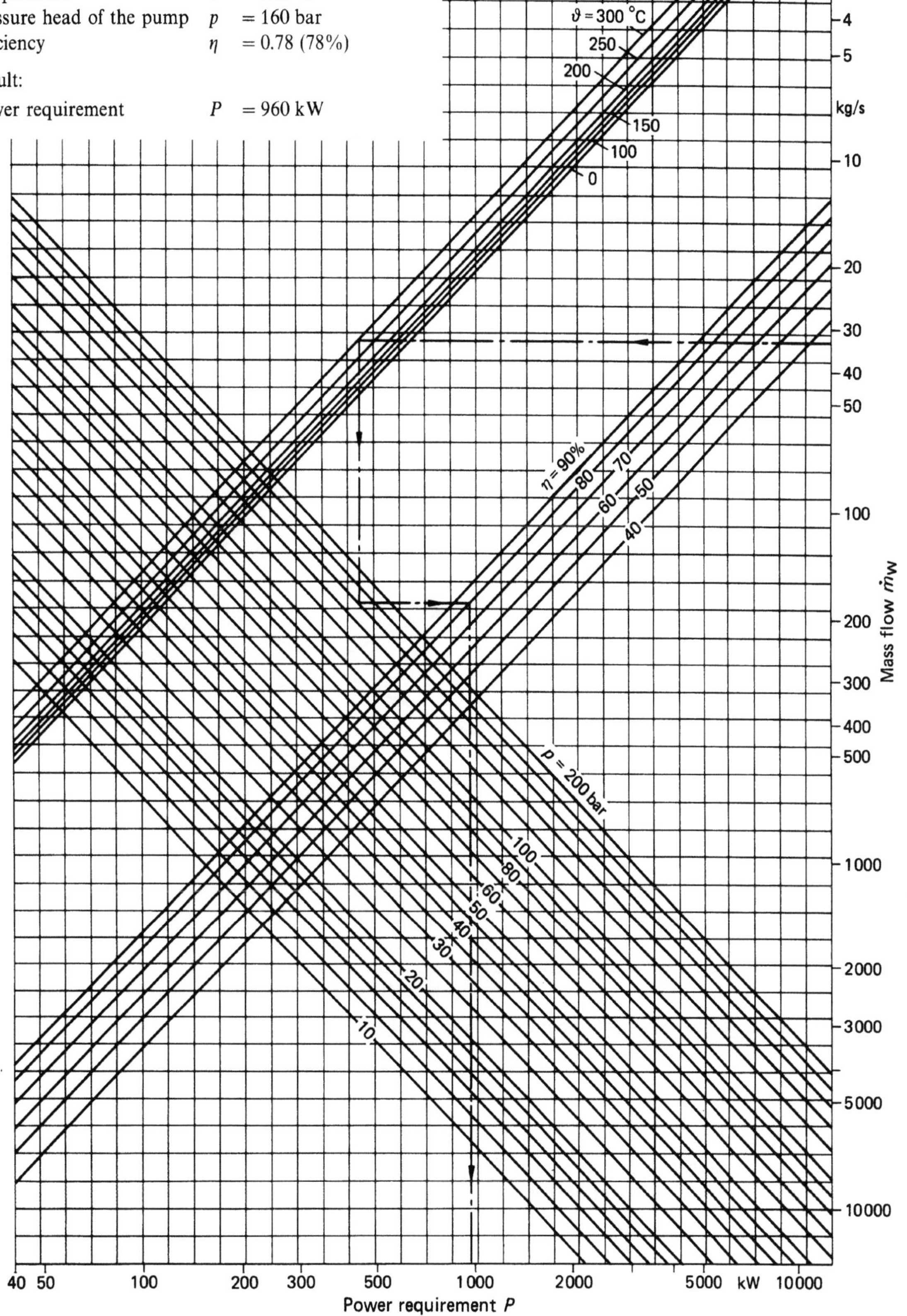

Basic equation (in above used units) $\qquad P = 100\,\dfrac{\dot{m}_w\,p}{\eta\,\varrho}$ $\qquad \varrho$ density of the feed water in kg/m^3 at the temperature ϑ

K. Gaffal

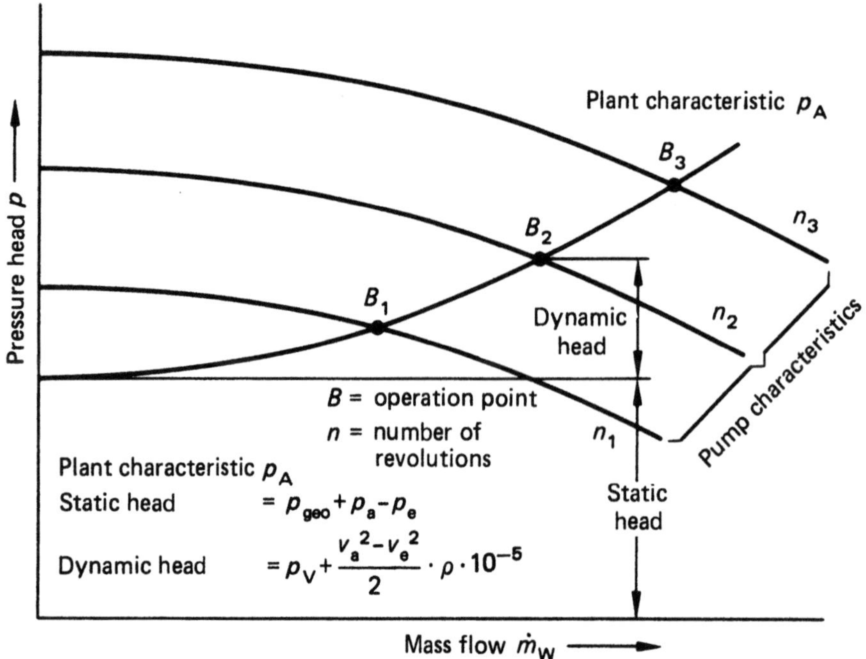

Change of the operation point on the plant characteristic p_A from B_1 to B_3 by increasing the numbers of revolution of the pump from n_1 to n_3.

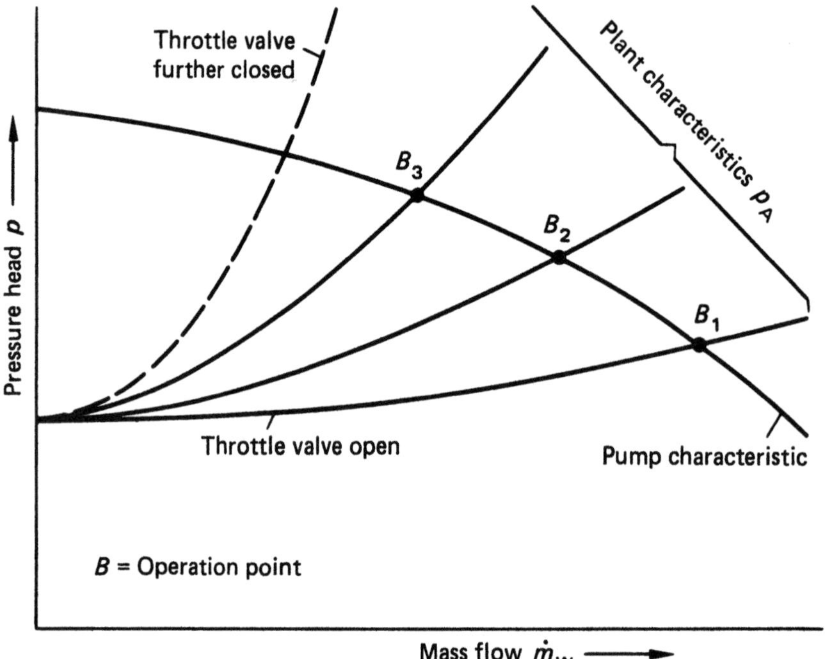

Change of the point of operation on the QH-line from B_1 to B_3 by throttling.

Definitions:

p_{geo} in bar	geodetic pressure difference
$p_a - p_e$ in bar	pressure difference between inlet and outlet cross section of the plant (if there is an open vessel this part has to be abolished)
v_a; v_e in m/s	flow velocity at the inlet resp. outlet cross section of the plant
p_V in bar	pressure drop in the pipe lines of the plant (see chapter No 9)
ϱ in kg/m³	density of the feed water

H
η
H_W,opt
H_Z
H_W
H_Z
η_W,opt
η_W
η_Z
η_Z
0,8 Q_W,opt   Q_W,opt   1,2 Q_W,opt
Q
f_Q,W
1,0 0,9 0,8 0,7 0,6 0,5 0,4
6,5
45
n_q,W 30 20 10
f_H,W
f_H,W
6,5
45
30 — 20 10
n_q,W
f_η,W
f_η,W
n_q,W   5 10 15 20 25 30 35 40 45
Curve No. 1 2 3 4 5 4 3 1
Curve No. 1 2 3 4 5
500 1/min
725 875
960
1450 1160
2000
3500 2900
6000
Number of revolutions n
Pressure head H_Z,Betr. , H_W,opt
400 m
200 300
100 150
60 80
30 40
15 20
8 10
4 6
2 3
1 1,5
Kinematic viscosity ν_Z , 1 m²/s
1,5
2 3
4 5
6 8
10 15
20 30·10⁻⁶
40 50
60 80
100 150
200 250
300 400
500 600
800 1000
1500 2000
2500 3000
4000
1 2 3 4 5 10 20 30 40 50 100 200 300 500 1000 2000 m³/h 5000 10000
0,3 0,5 1 2 3 4 5 10 20 30 40 50 100 200 300 500 1000 l/s 2000
Volume flow Q_Z,Betr. , Q_W,opt
K. Gaffal

Example:

Determination of the operation point

A mineral oil with a kinematic viscosity of $v_z = 500 \cdot 10^{-6}\,\mathrm{m^2/s}$ and a density $\varrho_z = 0.897\,\mathrm{kg/dm^3}$ has to be pumped.

The operation characteristics and the operation data of a pump pumping water are known:

$$Q_w = 31\,\mathrm{l/s}\ (= 111.6\,\mathrm{m^3/h})$$
$$H_w = 20\,\mathrm{m}$$
$$n = 1\,450\,\mathrm{l/min}$$

Index w = water
Index z = viscous liquid

To determine the new operating data on pumping mineral oil, the optimum pump data have to be calculated from the single operating characteristics. The following additional data have to be known:

Volume flow	$Q_{w,\,opt}$	31 [1]	l/s
Pressure head	$H_{w,\,opt}$	20 [1]	m
Efficiency	$\eta_{w,\,opt}$	0.78 [1]	–
Numbers of revolution	n	1 450	l/min
Kinematic viscosity	v_z	$500 \cdot 10^{-6}$	m²/s
Density	ϱ_z	0.897	kg/dm³
Acceler. due to gravity	g	9.81	m/s²

1) from the single operation characteristic

4 points on the new operation characteristic will be determined by the following calculation scheme:

$n_{q,\,w}$ from worksheet 11.9			27			1/min
$f_{Q,w}$	from work-		0.78			–
$f_{H,w}$	sheet No 11.7		0.83			–
$f_{\eta,w}$			0.49			–
Q/Q_{opt}		0	0.8	1.0	1.2	–
Q_w	from manufacturer's	0	24.8	31	37.2	l/s
H_w	data for 4 points on the operation	25	21.6	20	18.2	m
η_w	characteristics	0	0.74	0.78	0.73	–
$Q_z = Q_w \cdot f_{Q,w}$		0	19.3	24.2	29	l/s
$H_z =$		$= H_w$	$= H_w \cdot f_{H,w} \cdot 1.03$	$= H_w \cdot f_{H,w}$	$= H_w \cdot f_{H,w}$	
		25	2) 18.5	16.6	15.1	m
$\eta_z = \eta_w \cdot f\eta_w$		0	0.36	0.38	0.36	–
$P_z = \dfrac{\varrho_z \cdot g \cdot H_z \cdot Q_z}{\eta_z \cdot 1000}$			8.7	9.3	10.7	kW

By these data 4 points on the QH_z- and $Q\eta_z$ line and 3 points on the QP_z line are fixed

2) if H_z becomes greater than H_w is $H_z = H_w$

Determination of the pump size

At pumping mineral oil, the pump size has to be determined, which fulfils the following operation requirements:

Volume flow	$Q_{z,\,Betr}$	31	l/s
Pressure head	$H_{z,\,Betr}$	20	m
Kinematic viscosity	v_z	$500 \cdot 10^{-6}$	m²/s
Density	ϱ_z	0.897	kg/dm³

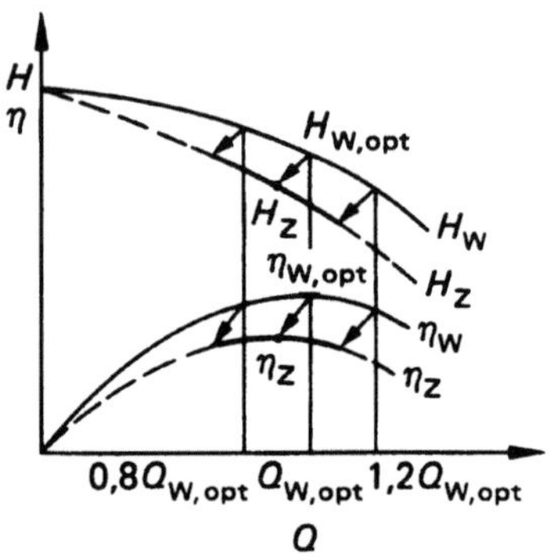

Graphic presentation of calculation

The above shown operating data can be recalculated according to water conditions by the following scheme of calculation and thus, the appropriate pump size can be determined.

n chosen		1 450	l/min
$n_{q,\,w}$ 3) from worksheet No 11.9		27	1/min
$f_{Q,z}$	from worksheet No 11.8	0.8	–
$f_{H,z}$		0.86	–
$Q_{w,\,Betr} = \dfrac{Q_{z,\,Betr}}{f_{Q,z}}$		38.8	l/s
$H_{w,\,Betr} = \dfrac{H_{z,\,Betr}}{f_{H,z}}$		23.3	m

3) with $\left.\begin{array}{l} Q_{z,\,Betr} = Q_{opt} \\ H_{z,\,Betr} = H_{opt} \end{array}\right\}$ approximately

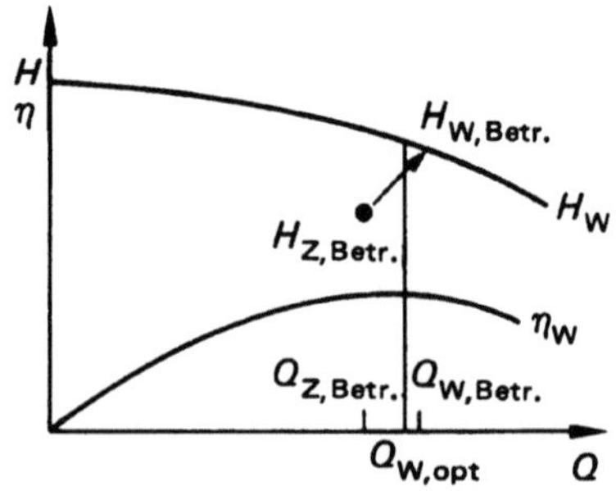

Graphic presentation of calculation

Known values: data for pumping viscous liquids; to be determined: data for pumping water

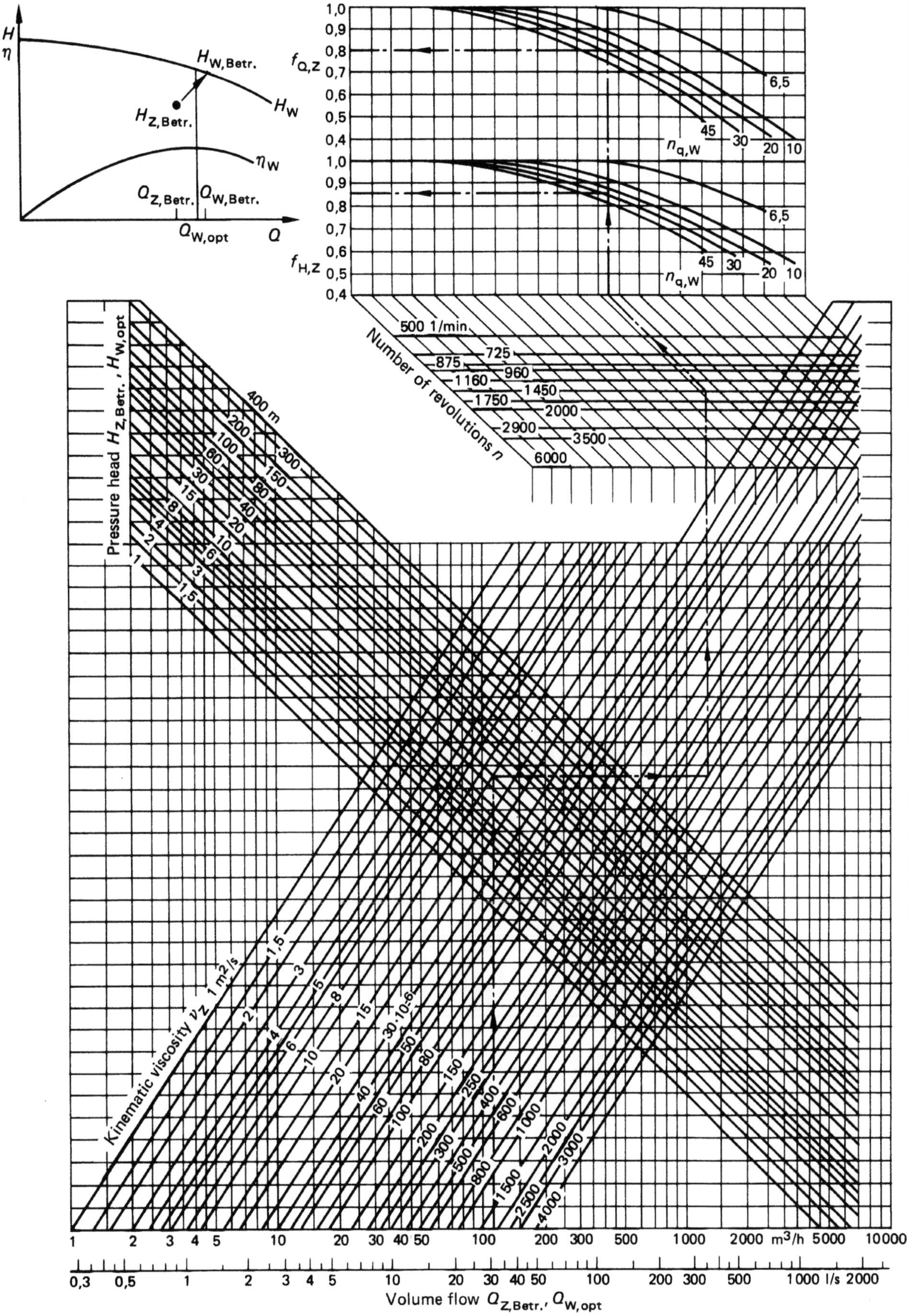

Volume flow $Q_{Z,Betr.}$, $Q_{W,opt}$

1 2 3 4 5 10 20 30 40 50 100 200 300 500 1000 2000 m³/h 5000 10000

0,3 0,5 1 2 3 4 5 10 20 30 40 50 100 200 300 500 1000 l/s 2000

Example for calculation see worksheet No 11.7

K. Gaffal

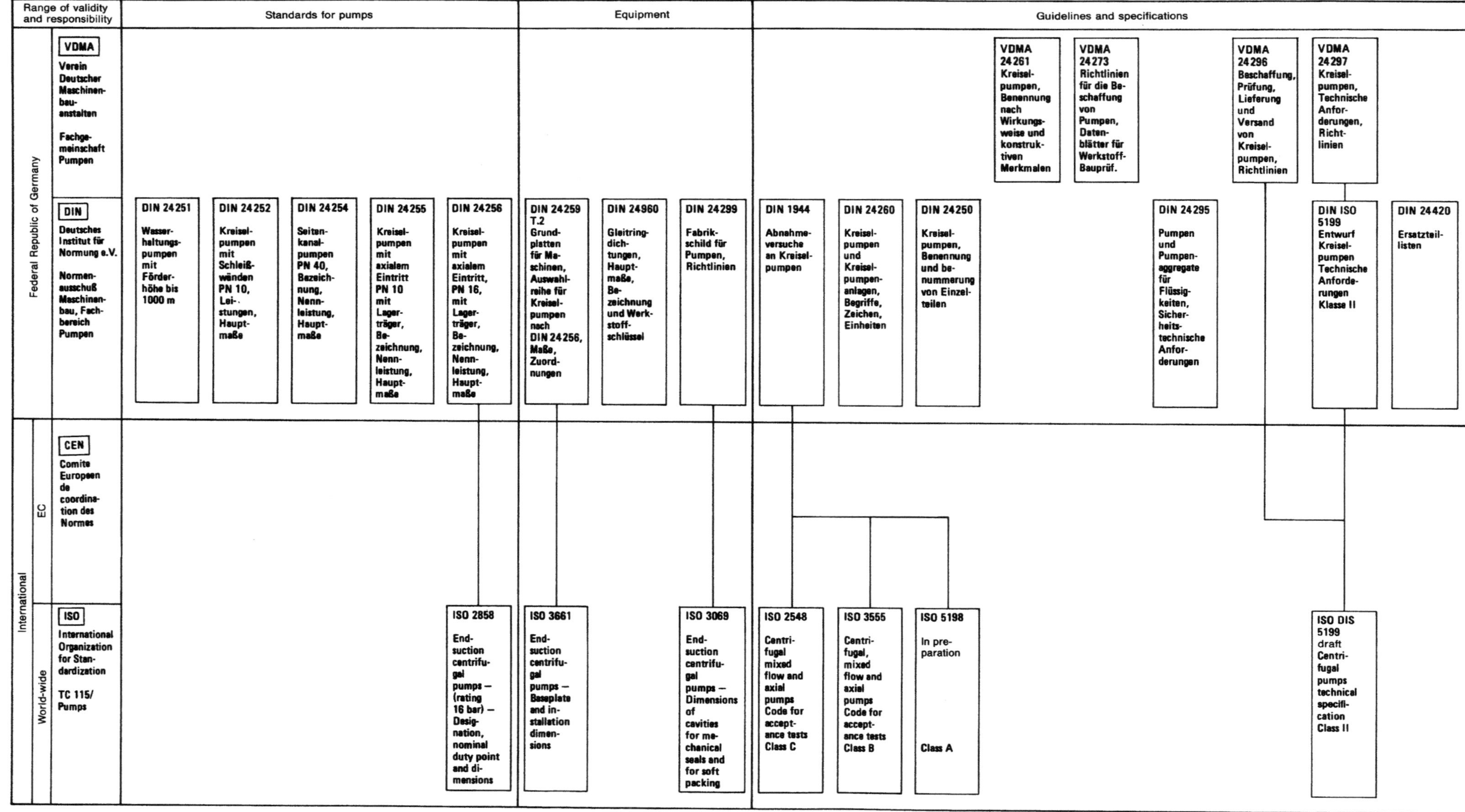

Range of validity and responsibility
Standards for pumps
Equipment
Guidelines and specifications

Federal Republic of Germany

VDMA
Verein Deutscher Maschinenbauanstalten
Fachgemeinschaft Pumpen

DIN
Deutsches Institut für Normung e.V.
Normenausschuß Maschinenbau, Fachbereich Pumpen

International
EC

CEN
Comite Europeen de coordination des Normes

World-wide

ISO
International Organization for Standardization
TC 115/ Pumps

DIN 24251 Wasserhaltungspumpen mit Förderhöhe bis 1000 m
DIN 24252 Kreiselpumpen mit Schleißwänden PN 10, Leistungen, Hauptmaße
DIN 24254 Seitenkanalpumpen PN 40, Bezeichnung, Nennleistung, Hauptmaße
DIN 24255 Kreiselpumpen mit axialem Eintritt PN 10 mit Lagerträger, Bezeichnung, Nennleistung, Hauptmaße
DIN 24256 Kreiselpumpen mit axialem Eintritt, PN 16, mit Lagerträger, Bezeichnung, Nennleistung, Hauptmaße
ISO 2858 End-suction centrifugal pumps — (rating 16 bar) — Designation, nominal duty point and dimensions

DIN 24259 T.2 Grundplatten für Maschinen, Auswahlreihe für Kreiselpumpen nach DIN 24256, Maße, Zuordnungen
DIN 24960 Gleitringdichtungen, Hauptmaße, Bezeichnung und Werkstoffschlüssel
DIN 24299 Fabrikschild für Pumpen, Richtlinien
ISO 3661 End-suction centrifugal pumps — Baseplate and installation dimensions
ISO 3069 End-suction centrifugal pumps — Dimensions of cavities for mechanical seals and for soft packing

VDMA 24261 Kreiselpumpen, Benennung nach Wirkungsweise und konstruktiven Merkmalen
VDMA 24273 Richtlinien für die Beschaffung von Pumpen, Datenblätter für Werkstoff-Bauprüf.
VDMA 24296 Beschaffung, Prüfung, Lieferung und Versand von Kreiselpumpen, Richtlinien
VDMA 24297 Kreiselpumpen, Technische Anforderungen, Richtlinien
DIN 1944 Abnahmeversuche an Kreiselpumpen
DIN 24260 Kreiselpumpen und Kreiselpumpenanlagen, Begriffe, Zeichen, Einheiten
DIN 24250 Kreiselpumpen, Benennung und benummerung von Einzelteilen
DIN 24295 Pumpen und Pumpenaggregate für Flüssigkeiten, Sicherheitstechnische Anforderungen
DIN ISO 5199 Entwurf Kreiselpumpen Technische Anforderungen Klasse II
DIN 24420 Ersatzteillisten
ISO 2548 Centrifugal mixed flow and axial pumps Code for acceptance tests Class C
ISO 3555 Centrifugal, mixed flow and axial pumps Code for acceptance tests Class B
ISO 5198 In preparation Class A
ISO DIS 5199 draft Centrifugal pumps technical specification Class II

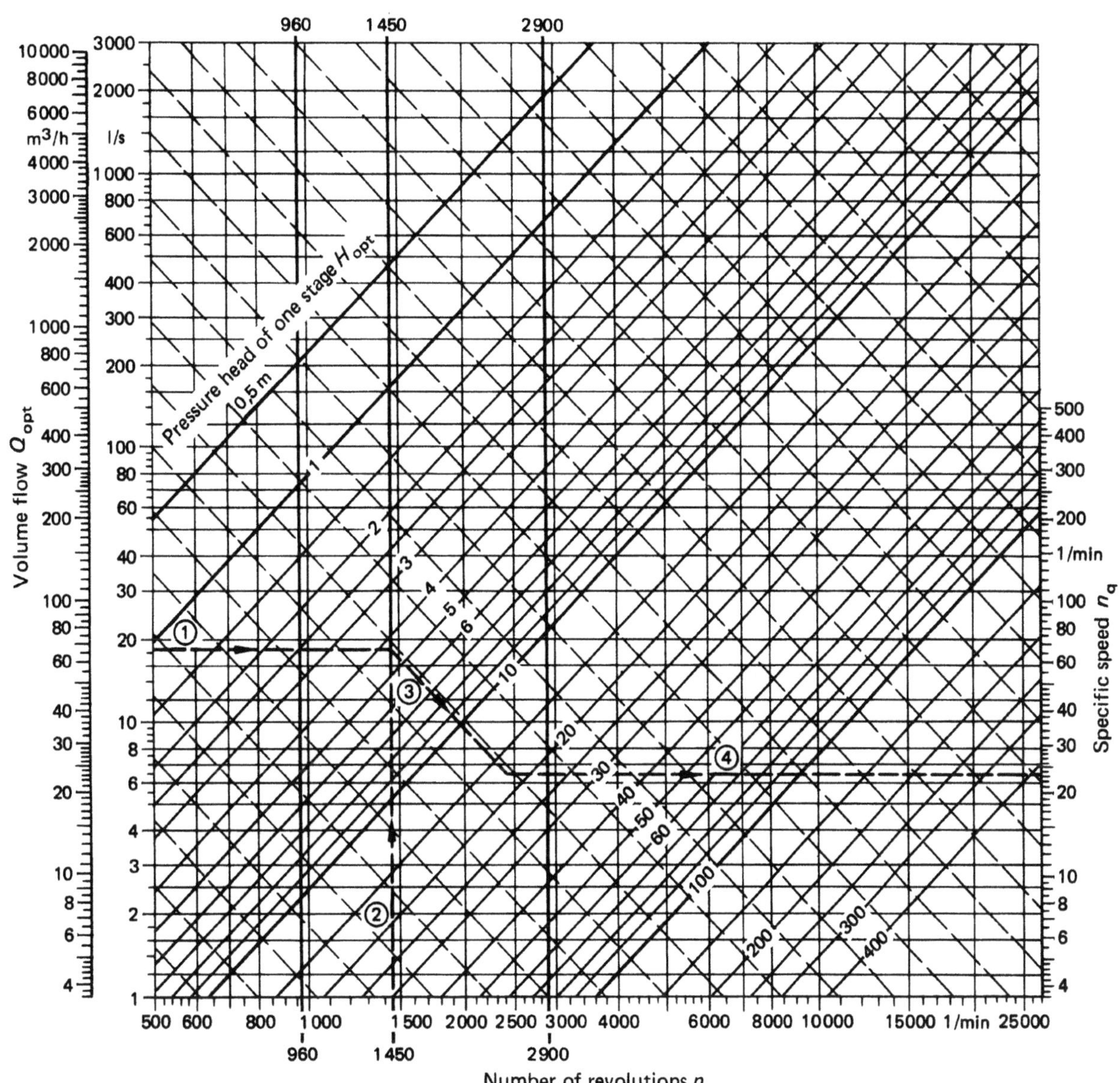

Equations	Units				
	Q_{opt}	H_{opt}	n	n_q	$g = 9.81$
$n_q = n \cdot \dfrac{\sqrt{Q_{opt}/1}}{(H_{opt}/1)^{3/4}}$	m^3/s	m	1/min	1/min	
$n_q = 333 \cdot n \cdot \dfrac{\sqrt{Q_{opt}}}{(g \cdot H_{opt})^{3/4}}$	m^3/s	m	1/s	1/min	m/s^2 DIN 24260
$n_q = 5.55 \cdot n \cdot \dfrac{\sqrt{Q_{opt}}}{(g \cdot H_{opt})^{3/4}}$	m^3/s	m	1/min	1/min	m/s^2

All equations have the same result.

In the case of multistage pumps, instead of the total head, the pressure head of one stage has to be used for calculation. If having pumps with two flooded rotors, only half of the volume flow has to be used for calculation.

Example: $Q_{opt} = 66$ m³/h = 18.3 l/s; $n = 1450$ l/min; $H_{opt} = 17.5$ m. Result: $n_q = 23$ l/min.

K. Gaffal

NPSH (net positive suction head) -values

Case a): Suction operation; pump is situated above liquid level

$$\text{NPSH}_{\text{vorh}} = \frac{p_e + p_b - p_D}{\varrho \cdot g} + \frac{v_e^2}{2g} - H_{v,s} - H_{s\,geo}.$$

At pumping cold liquids, for example, water with an open storage vessel with

$p_b = 1$ bar
$p_e = 0$ bar
$\varrho = 1\,000$ kg/m^3
$g = 10$ m/s^2 (a mistake of 2% occurs instead of using 9.81 m/s^2)

(The term $v_e^2/(2\,g)$ may be neglected because of low velocity heads in the suction resp. feed vessels)

the equation can be simplified for practical reasons to:

$$\text{NSPH}_{\text{vorh}} \approx 10 - H_{v,s} - H_{s\,geo}.$$

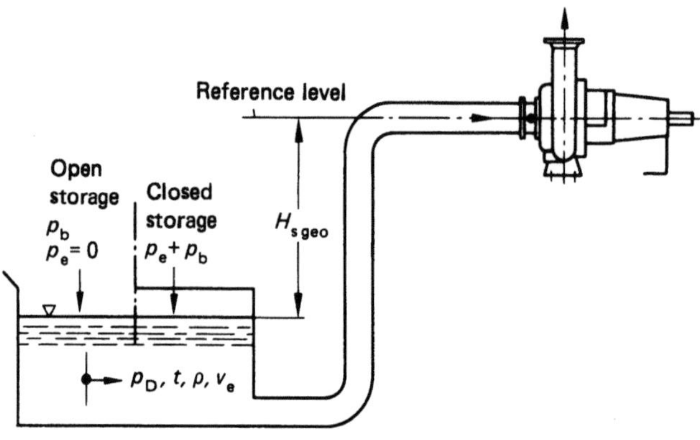

Case b): Pressure operation; pump is situated below liquid level

$$\text{NPSH}_{\text{vorh}} = \frac{p_e + p_b - p_D}{\varrho \cdot g} + \frac{v_e^2}{2g} - H_{v,s} - H_{z\,geo}.$$

Simplification for practical reasons under the same circumstances as in case a):

$$\text{NSPH}_{\text{vorh}} \approx 10 - H_{v,s} - H_{z\,geo}.$$

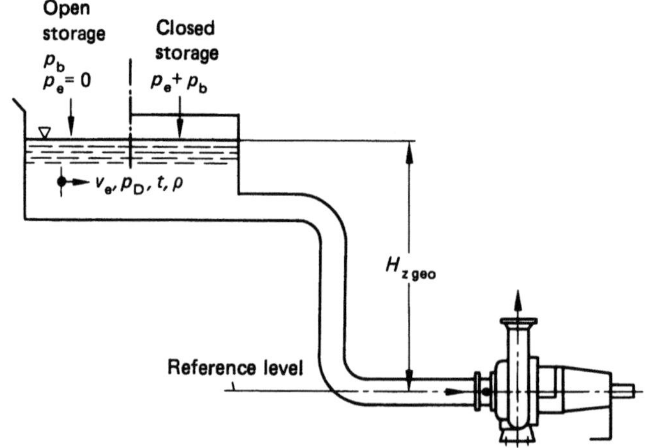

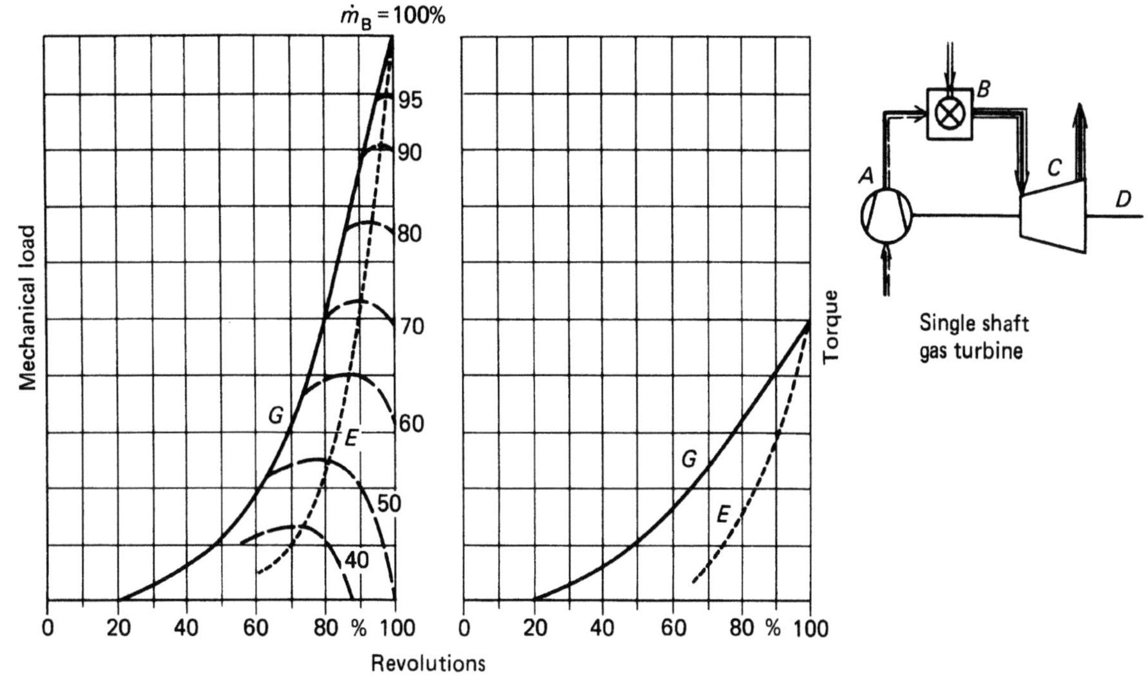

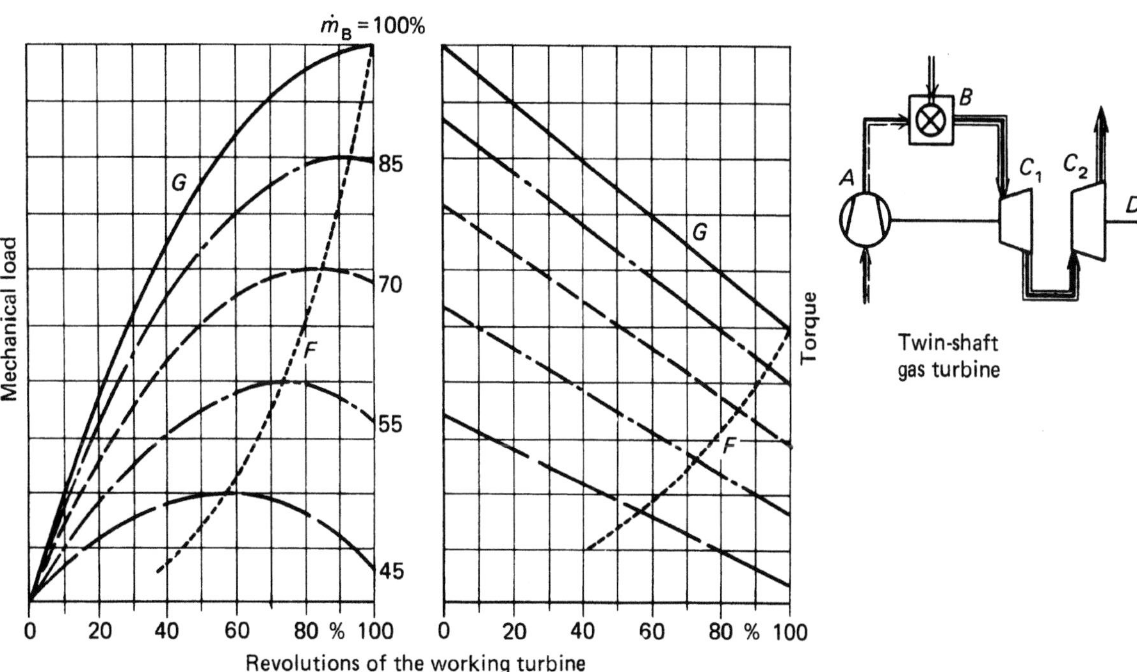

A	Compressor	D	Mechanical output
B	Heater (combustion chamber)	E	Optimum numbers of revolution
C	Turbine	F	Optimum speed of the working turbine
C_1	Compressor turbine	G	Limit of operating range
C_2	Working turbine	$\dot{m}_\text{B}$	Fuel mass flow

W. Hentschel, G. Oberländer

Explanation

In a cycle process of a gas turbine, the compressor load, as well as the mechanical load, is supplied by the energy from the expansion of the hot pressurized flue gas. Either turbine, compressor and driven assembly are joined by a single shaft, or the shaft is thus divided, so that the turbine part driving the compressor and the compressor are joined together, whilst the other turbine is used as the working engine. This is the reason for defining between single and multiple shaft gas turbines.

At a single shaft turbine compressor, turbine and driving assembly have the same number of revolutions. If there is a need for lowering the revolutions, the air flow rate and the pressure ratio of the compressor and thus, the torque and the output, will be reduced. If the driving revolutions and thus the air flow are kept constant, as for generator operations, greater changes in load can be controlled very quickly by varying the fuel supply of the combustion chamber without the need of changing the rotor speed.

At a twin shaft turbine, the driving revolutions is independent on the number of revolutions of the compressor with its driving turbine. If there is a need for a lower driving revolutions, the compressor and its driving turbine can keep their former revolutions and thus provide a constant energy of pressurized flue gas for the independent working turbine. Therefore, in the beginning, the decrease of output at lower revolutions is small and the torque increases strongly.

However, any change in load corresponds to a change in fuel supply, as well as a change of revolutions of the compressor and its driving turbine: in case of idling, it runs with lowest and in case of maximum load with highest number of revolutions by corresponding changes in air flow and pressure ratio. Therefore, not only the mass flow of fuel, but also the number of revolutions of the compressor and its driving turbine have to be changed with variations of load. Although short time peak load above the nominal load may be compensated elastically, greater changes in load causes strong deviations in revolutions and therefore are limited to certain ranges.

Because of this different operation performance, single shaft turbines preferably are used for driven devices with constant revolutions, as for example, generators.

Twin shaft turbines are used for driven devices with variable speed, as for example, vehicles or fluid engines. Besides the twin shaft turbine, there are other types of multiple shaft turbines, whose operation performance, depending on the special kind of assembly, is more or less equal to that of the twin shaft turbine.

In Germany, the following technical standards are valid for industrial gas turbines:

DIN 4340	Gas turbines	Definitions, names
DIN 4341	Gas turbines	Conditions of inspection of gas turbines
	Part 1	Basics
	Part 2	Examples of evaluation
DIN 4342	Gas turbines	Standard conditions, standard output, operating performance

The following technical standards are still in preparation:

DIN 4319	Thermodynamic and fluiddynamic vocabulary for steam and gas turbines
	Part 1 Basics
	Part 2 Definitions of types of operation and types of machinery

DIN 4343 Gas turbines Part 1   Measurement of flue gas emissions

DIN/ISO 4261 Gas turbines   Fuels for gas turbines

DIN 45 635 Part 65   Noise measurement on gas turbines

In Germany, the following technical rule is valid for industrial gas turbines:

Technische Anleitung zur Reinhaltung der Luft/TA-Luft dated 27. 2. 1986.

Every year, an up-to-date survey of gas turbines available, is published by the following technical journals:

Gas Turbine World, P.O. Box 447, Southport CT 06940, USA;
Turbomachinery International, P.O. Box 5550, Norwalk CT 06856-5550, USA.

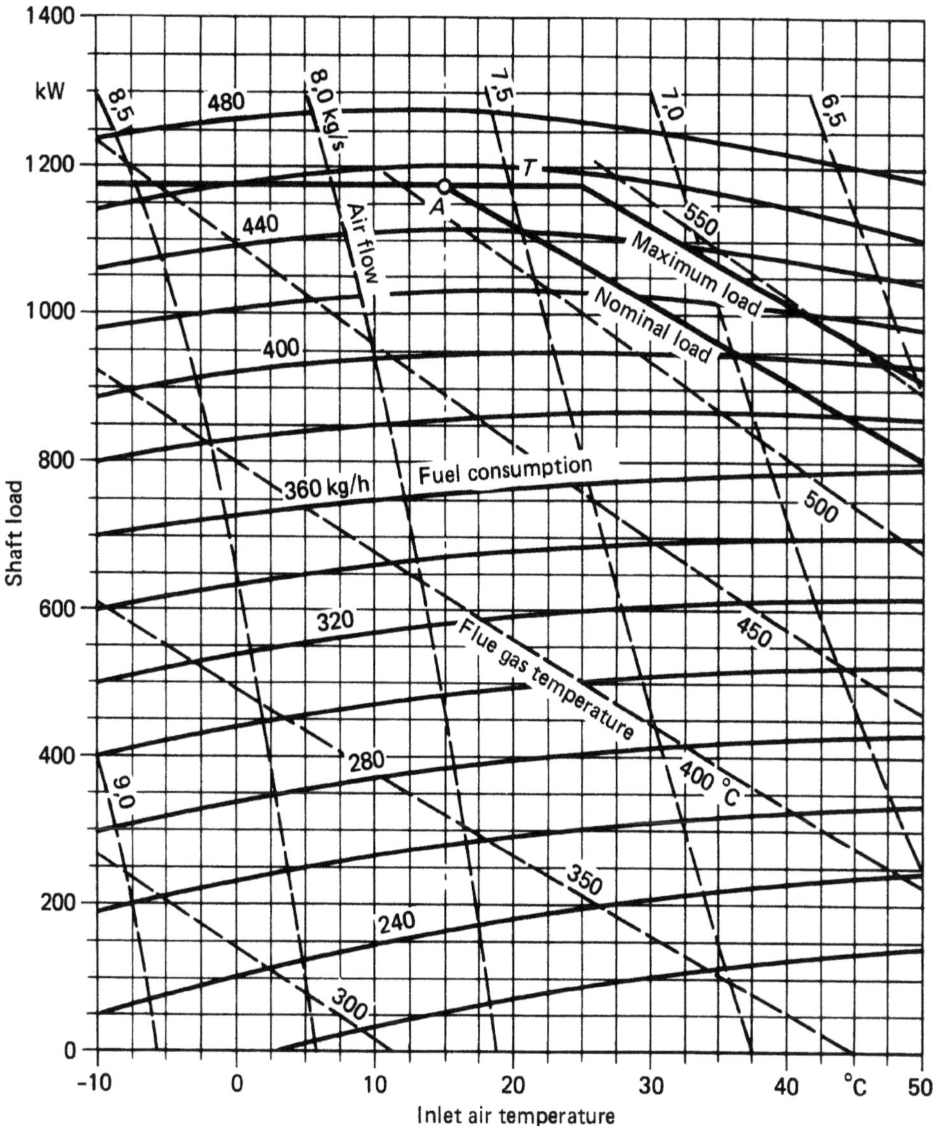

Characteristic diagram of a single shaft gas turbine
operating under standard conditions at nominal number of revolutions.

Condition of the surrounding air: absolute pressure   1.013 bar (sea level)
                                    temperature   15 °C
                                    relative humidity   60%
related to the mechanical output behind the gear box.

A   design operation at standard conditions
T   upper output limit ("topping")

Height correction:

Every 100 m of height an output reduction of 1.12% according to the density reduction of the air

Correction values for the pressure loss (dependent on the type of gas turbine):

Output loss per 10 mbar pressure loss

    in the air suction system   1.2 up to 2.5%
    in the flue gas system   0.5 up to 1.5%

Characteristic diagram and correction values have to be obtained from manufacturers specification.

W. Hentschel, G. Oberländer

Explanation

The output of a gas turbine is mainly influenced by the air flow, the pressure ratio and the drop in temperature. It is limited by the maximum permissible entrance temperature of the pressurized flue gas into the turbine and by its maximum number of revolutions.

At the same air temperature at compressor inlet point, the output limit is directly related to the pressure of the surrounding air resp. its density. It gains or decreases with the same ratio.

The air density, the pressure ratio of the compressor, as well as the available temperature drop, are influenced by changes of the temperature of the inlet air. These values influence mainly the output limits, which differ slightly depending on the type of turbine, but do have the same tendency. Here, a difference has to be made between an unlimited available continuous or basic load and a short time or peak load within limited conditions. To limit the size of driven machinery and gear box, it is usual to control the air inlet temperature up to a certain limit, because the output strongly increases with sinking air inlet temperature. Thus, it is possible to use the full turbine power up to these limits at low temperatures.

To examine each point of operation of a certain gas turbine, its characteristic diagram has to be used. In it its shaft output versus air inlet temperature is shown, and for each point of operation, its fuel consumption, air flow and flue gas temperature can be determined. These characteristic diagrams are valid for operations without losses at sea level. For differing heights, as well as for pressure losses at the air inlet and flue gas exit, they have to be corrected accordingly. Exact values of the influence of pressure losses have to be shown in the characteristic diagram of each gas turbine.

The power output is reduced by mechanical losses, mainly from the gear box and auxiliary engines. A part of it, including the bearing friction and lubrication, causes the rise in temperature of the lubricating oil and is in the range of 0.6% up to 1.8% of the consumed fuel energy. The lubricating oil consumption is within the range of 50 mg/kWh up to 80 mg/kWh.

The heat loss by radiation of the gas turbine surface is within the range of 0.4% up to 0.8% related to the fuel energy.

Mainly at operation with gaseous fuels, the continuous combustion in the combustion chamber of gas turbines favours a high combustion efficiency. Combustion losses increase the fuel consumption accordingly.

The greatest loss is the energy loss of the flue gas. At nominal load, according to the type of turbine, the flue gas temperature is within the range of 400 °C up to 600 °C. With sinking load the temperature will be lowered from 280 °C up to 400 °C. A part of this energy can be regained via cogeneration, as shown in worksheet No 12.4.

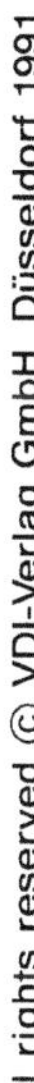

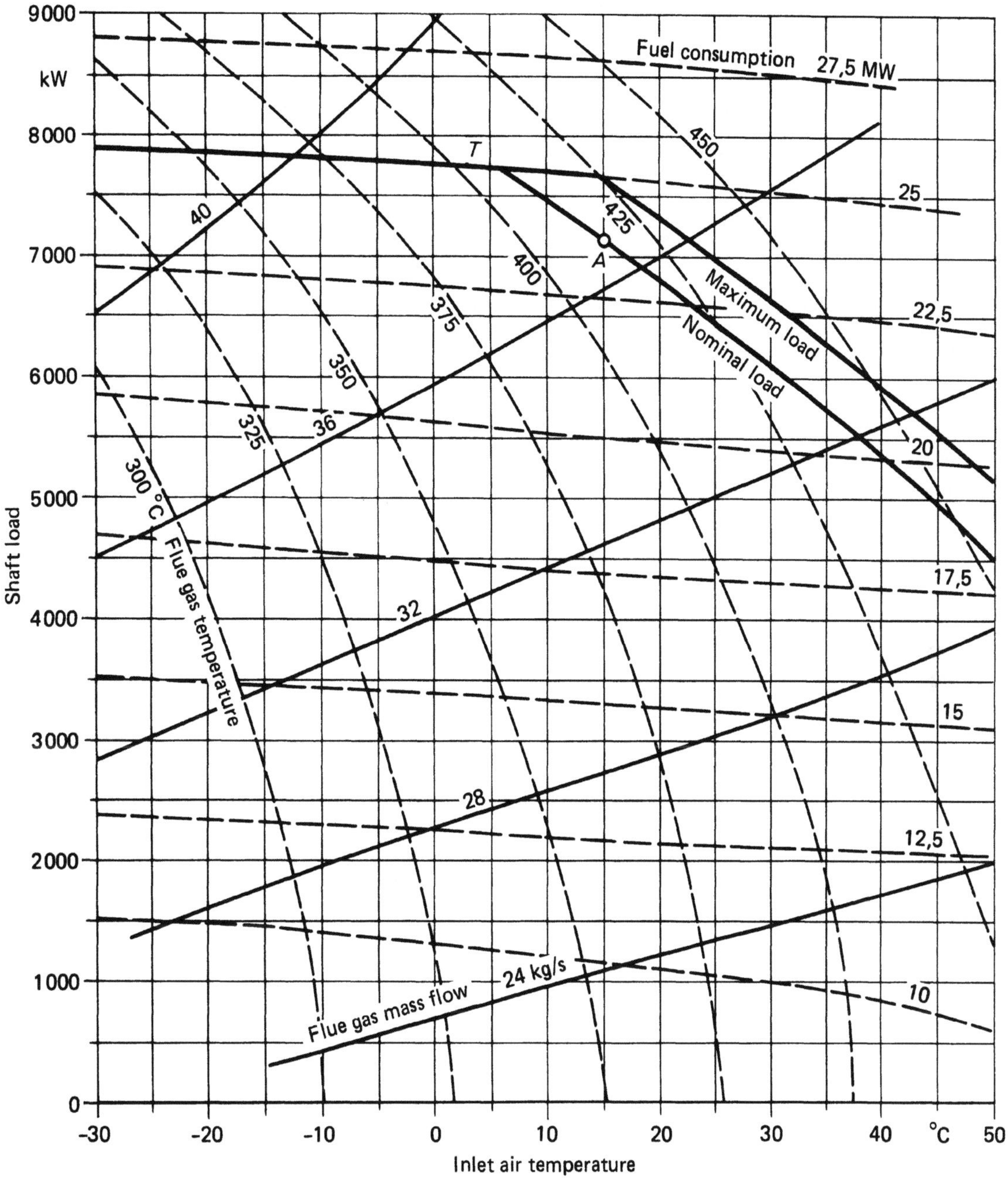

Characteristic diagram of a twin shaft gas turbine
operating under standard conditions at nominal number of revolutions.

Condition of the surrounding air: absolute pressure 1.013 bar (sea level)
temperature 15 °C
relative humidity 60 %
related to the mechanical output behind the gear box.

A   design operation at standard conditions
T   upper output limit ("topping")

Height correction:

Every 100 m of height an output reduction of 1.12 % according to the density reduction of the air

Correction values for the pressure loss (dependent on the type of gas turbine):

Output loss per 10 mbar pressure loss

in the air suction system   1.2 up to 2.5 %
in the flue gas system   0.5 up to 1.5 %

Characteristic diagram and correction values have to be obtained from manufacturer specifications.

W. Hentschel, G. Oberländer

Explanation

Besides the dynamic behaviour at changes of load, there are great differences in the
characteristic diagrams of single and twin shaft gas turbines. For both diagrams
it is common, that they depend on the appearance and the superposition of the
characteristic diagrams of the compressor and the turbine. For a closer examination
of a certain turbine, its characteristic diagrams and other manufacturer specifications have to be used. The following example shows the way of calculation for such
an examination.

E x a m p l e for the examination of the gas turbine with the characteristic diagram shown in the sheet before

Wanted load output	kW	....	according to contractors specification
Operation height	m	100	operation height A.S.L.
Standard temperature	°C	25	according to contractors specification
Pressure loss entrance	mbar	10	air inlet channel, silencer, filter
exit	mbar	20	exhaust gas channel, silencer, vessel
Efficiency of the generator at $\cos\varphi = 0.8$	%	96.3	according to manufacturer specification
Station consumption *)	kW	35	electrical pumps, ventilator, charger
Gear box loss	%	3	according to manufacturer specification
Type of turbine	–	....	according to manufacturer specification
Nominal output ISO (sea level, 15 °C)	kW	7150	characteristic diagram

Standard temperature	°C	25	design standard
Power output	kW	6450	characteristic diagram
Suction loss	kW	110	according to manufacturer specification
Exhaust loss	kW	100	according to manufacturer specification
Height correction factor	–	0.988	standard atmosphere for the required operation height
Corrected power output	kW	6165	
Gear box loss	kW	185	
Generator output	kW	5759	including gear box loss and generator efficiency
Terminal output of electric generator	kW	5724	including station consumption

Fuel consumption B	kW	22 200	characteristic diagram
Gain by pressure loss	kW	480	according to manufacturer specification
B corrected by pressure loss	kW	21 720	
Height correction factor	–	0.988	
B corrected by height	kW	21 459	
Manufacturer tolerance	%	±3	
Maximum fuel consumption	kW	22 103	guarantee value
Mass flow of fuel	kg/h	1 864	fuel oil EL, $H_u = 42{,}705$ MJ/kg
Overall electric efficiency	%	26.7	

*) The consumption of the plant does not include a possibly needed natural gas compressor

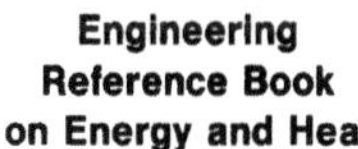

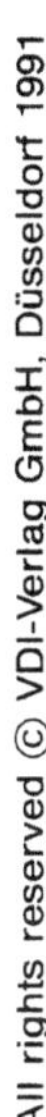

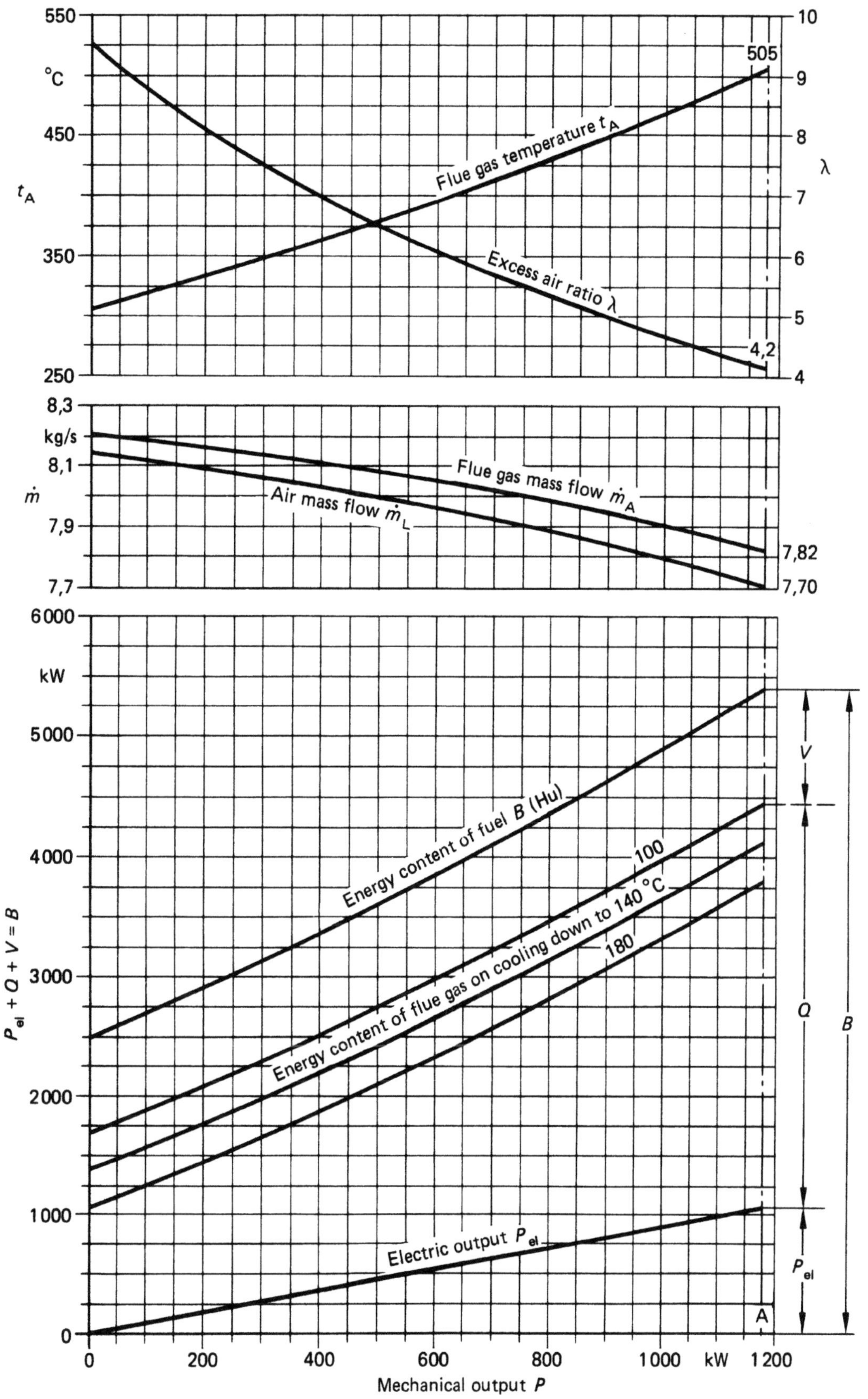

Energy proportions of a gas turbine according to the characteristic diagram No 12.2

W. Hentschel, G. Oberländer

Explanation

The drawing No 12.4 shows some typical energetic characteristic diagrams of a single shaft gas turbine with a thermal efficiency of about 20%. Turbines with a higher efficiency have losses in the same range; however, the proportions of the recoverable flue gas enthalpy and of the mechanical output change towards a bigger proportion of the mechanical output.

Examining a gas turbine being suitable for combined heat and power systems, its recoverable flue gas enthalpy has to be considered within the whole range of load. When fired with liquid fuel, the flue gas enthalpy can be recovered to a temperature of about 140 °C, using gaseous fuel to about 100 °C. (Diving below the dew point!)

Gas turbines, being used for combined heat and power systems, offer some considerable advantages:

- Its recoverable heat has only a single source: the flue gas. There is no cooling liquid. The heat of the lubricating oil and of the surface radiation can be neglected.
- It is possible to fire it with liquid as well as gaseous fuel, without any problems in changing the fuel at any point of operation.
- Its air pollution is very low, especially when fired with natural gas, and the Nox emissions can be further reduced by water injection into the combustion chamber.
- The lubricating oil consumption can be neglected.

Afterburner

The air ratio of gas turbines is about $\lambda = 4$. Thus, only 1/4 of the air oxygen will be consumed and the remaining 3/4 can be used to burn further fuel in an afterburner. This range is somewhat limited by the construction of boiler and burners, as well as the demand of low air pollution, so that the lowest excess air ratio should be $\lambda = 1.4$. Nevertheless, the additional gain by using an afterburner is very high. Comparing two plants with the same fuel consumption and with and without afterburner, the gain of the one with afterburner will surpass clearly the one without afterburner, because the additional fuel energy, used to fire the afterburner, can be nearly all recovered as heat, without changing the air flow and thus, the flue gas losses.

Combined gas turbine steam turbine plant

The steam, raised in a steam generator behind a gas turbine, can be used to drive a steam turbine, whose shaft is connected via a gear assembly to that of the gas turbine. There is additional usable heat from a section of the steam generator and the condenser.

Superposed gas turbine

Because of their quick start up and load control, gas turbines are suitable as superposed turbines for bigger steam turbines. Its hot flue gas can be heated up as described for the afterburner. There may be additional burners with separate air supply, which can also be used to fire the boiler. Besides the quick reaction of this plant to load changes, superposing of a gas turbine improves the thermal efficiency of the whole plant by using its flue gas enthalpy.

Enthalpy

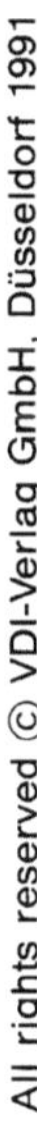

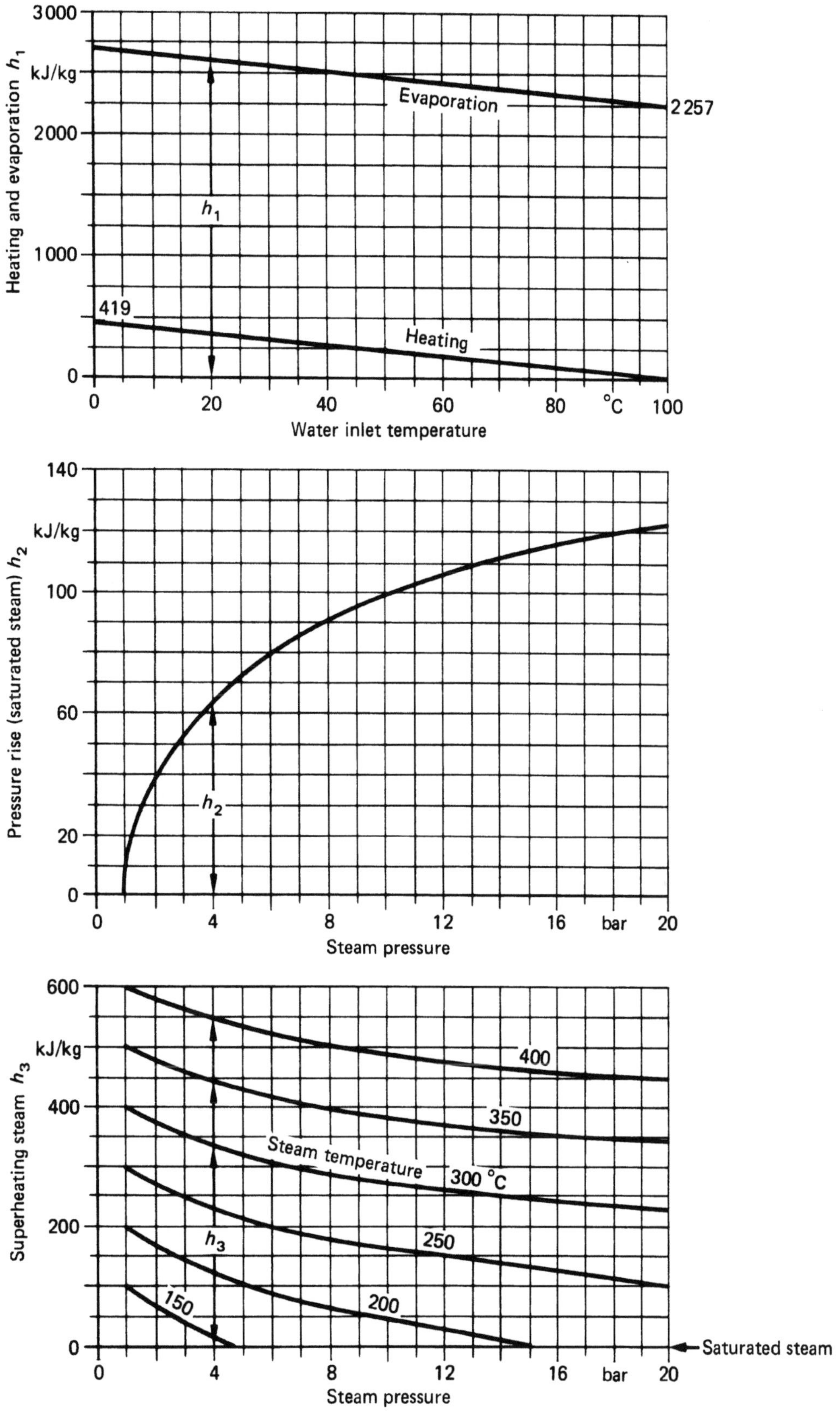

Diagrams for the calculation of the mass flow of steam for a known heat supply in kW

W. Hentschel, G. Oberländer

Explanation

The most common case of combined heat and power systems is to use the recoverable flue gas enthalpy to generate steam. With the diagrams in worksheet No 12.5, one can determine the mass flow of steam, which can be generated from a known enthalpy, depending on the values of steam temperature, steam pressure and of the water inlet temperature. It has to be considered, that the desired steam temperature should be at least 20 to 30 °C below the flue gas temperature.

The values of h_1, h_2 and h_3 can be calculated from the diagrams. Q means the recoverable enthalpy of the turbine flue gas. It can be either found in a diagram, as mentioned in worksheet No 12.4, or has to be calculated. The generated steam mass flow now can be calculated from:

$$\dot{m}_\mathrm{D} = \frac{3.6 \cdot Q}{h_1 + h_2 + h_3} \text{ in } [\text{t/h}] .$$

Example: Enthalpy calculation at the design point A in worksheet No 12.4

Water inlet	°C	40	$h_1 = 2\,500$ kJ/kg
Steam pressure	bar	6	$h_2 = 80$ kJ/kg
Steam temperature	°C	250	$h_3 = 200$ kJ/kg

Standard temperature	°C	15	basic assumption
Turbine output ISO	kW	1 177	according to manufacturer specification
Air flow	kg/s	7.7	characteristic diagram
Operating height	m	100	according to contractor specification
Height correction factor	–	0.988	standard atmosphere for the demanded operation height
Corrected flue gas mass flow	kg/s	7.73	air and fuel
Flue gas temperature	°C	505	characteristic diagram
Exit temperature of flue gas	°C	170	according to manufacturer specification
Usuable temperature difference	K	335	
$c_\mathrm{p} =$ at $\lambda = 4.2$	kJ/(kgK)	1.0781	dependent on λ and fuel
Flue gas enthalpy Q	kW	2 792	recoverable heat

Steam mass flow	t/h	3.6	from equation above
Steam temperature	°C	250	
Steam pressure	bar	6	= 5 bar gauge pressure

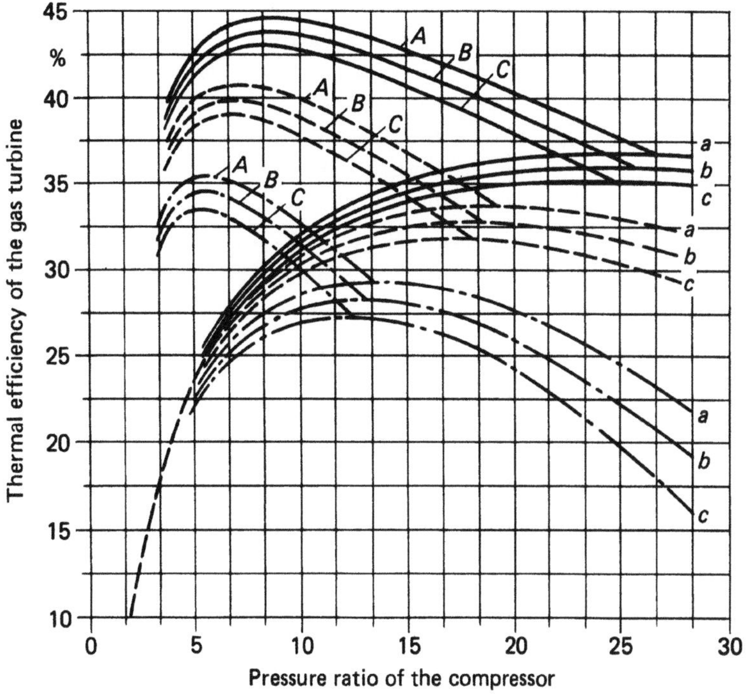

Turbine inlet temperature
——————— 1 300 °C
– – – – – 1 100 °C
— — — 900 °C

A, B, C with flue gas heat exchanger
a, b, c without flue gas heat exchanger

Efficiency of the compressor

A, a decreasing steadily from 87% at a pressure ratio of 3 to 84% at a pressure ratio of 27
B, b decreasing steadily from 85% at a pressure ratio of 3 to 82% at a pressure ratio of 27
C, c decreasing steadily from 83% at a pressure ratio of 3 to 80% at a pressure ratio of 27

Efficiency of the turbine

Decreasing steadily from 92% at a pressure ratio of 3 to 90% at a pressure ratio of 27

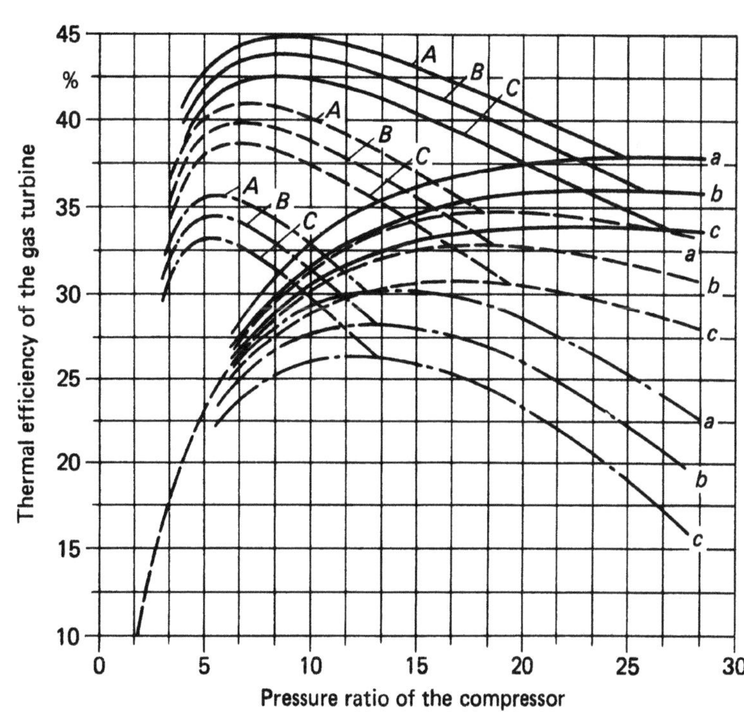

Turbine inlet temperature
——————— 1 300 °C
– – – – – 1 100 °C
— — — 900 °C

A, B, C with flue gas heat exchanger
a, b, c without flue gas heat exchanger

Efficiency of the compressor

Decreasing steadily from 85% at a pressure ratio of 3 to 82% at a pressure ratio of 27

Efficiency of the turbine

A, a decreasing steadily from 94% at a pressure ratio of 3 to 92% at a pressure ratio of 27
B, b decreasing steadily from 92% at a pressure ratio of 3 to 90% at a pressure ratio of 27
C, c decreasing steadily from 90% at a pressure ratio of 3 to 88% at a pressure ratio of 27

W. Hentschel,
G. Oberländer

Explanation

The power output of a gas turbine is the difference of the power generated by the turbine and the power consumed by the compressor. The higher the fluiddynamic efficiency of these components are, the higher is the thermal output of the gas turbine and thus, its thermal efficiency compared with the same conditions.

The pressure ratio of the compressor, the temperature of the pressurized flue gas at turbine inlet, the combustion efficiency and the pressure loss in the combustion chamber, do influence the thermal efficiency.

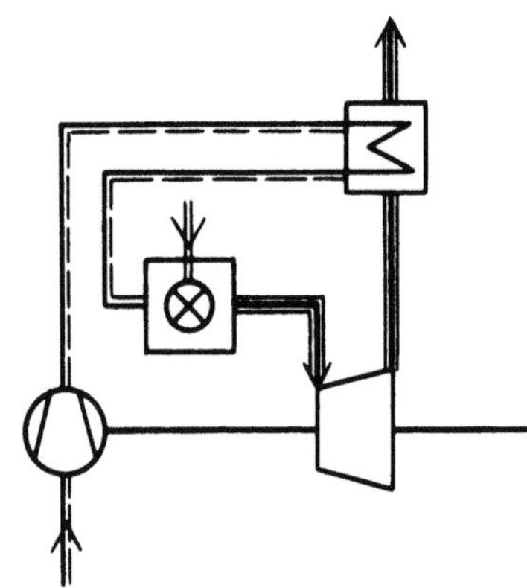

Gas turbine with flue gas heat exchanger for the internal use of the flue gas enthalpy

It is possible for some gas turbines to increase the temperature of the compressed air in front of the combustion chamber by means of a flue gas heat exchanger and thus save fuel to get the desired temperature at turbine entrance.

The exit temperature of the compressor increases proportionally with its pressure ratio. At the same temperature at turbine entrance, the turbine exit temperature is sinking with a rising pressure ratio. Only if the turbine exit temperature is much higher than the compressor exit temperature, the installation of a heat exchanger makes sense. At the same temperature level, heat exchange is not possible.

These interactions are shown in the diagrams of worksheet No 12.6. They are valid for gas turbines with and without heat exchanger and turbine entrance temperatures of 900, 1 100 and 1 300 °C, together with the assumed efficiencies and losses listed below. It shows, that for every gas turbine, there exists a region of optimum pressure ratio, as well as an upper limit for the use of a heat exchanger.

The following assumptions have been made for the diagrams:

Pressure loss

at compressor inlet	− 2%
between turbine exit and surroundings	− 7.3%
in the combustion chamber	− 4%
in the flue gas heat exchanger, cold side	− 2%
in the flue gas heat exchanger, hot side	− 4%

Thermal utilisation factor of the heat exchanger	−80%
Combustion efficiency of the combustion chamber	−99.5%
Mechanical efficiency	−96%
Air leakage	− 1.5%

Cooling air flow has not been considered.

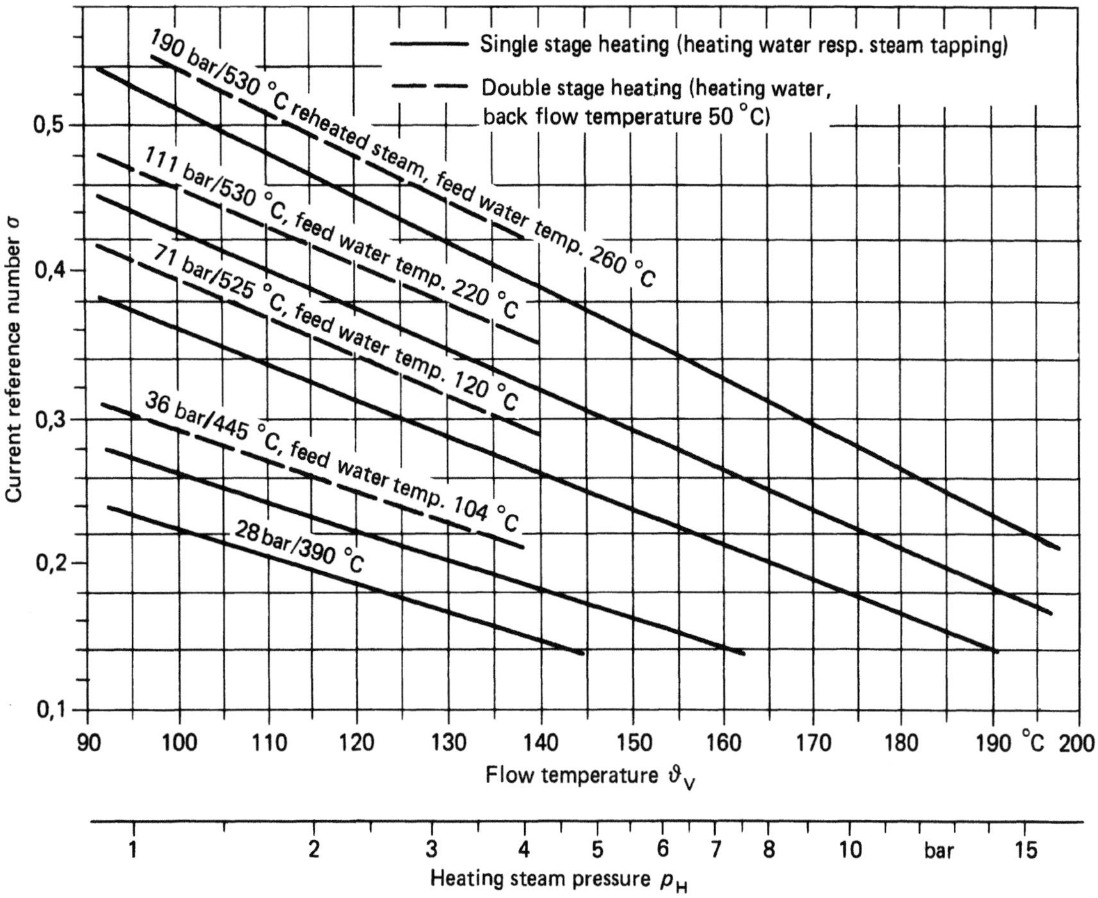
Single stage heating (heating water resp. steam tapping)
Double stage heating (heating water, back flow temperature 50 °C)
190 bar/530 °C reheated steam, feed water temp. 260 °C
111 bar/530 °C, feed water temp. 220 °C
71 bar/525 °C, feed water temp. 120 °C
36 bar/445 °C, feed water temp. 104 °C
28 bar/390 °C
Current reference number σ
Flow temperature ϑV
Heating steam pressure pH
bar

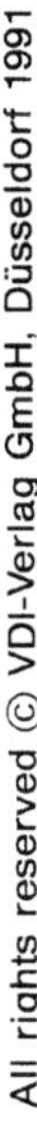

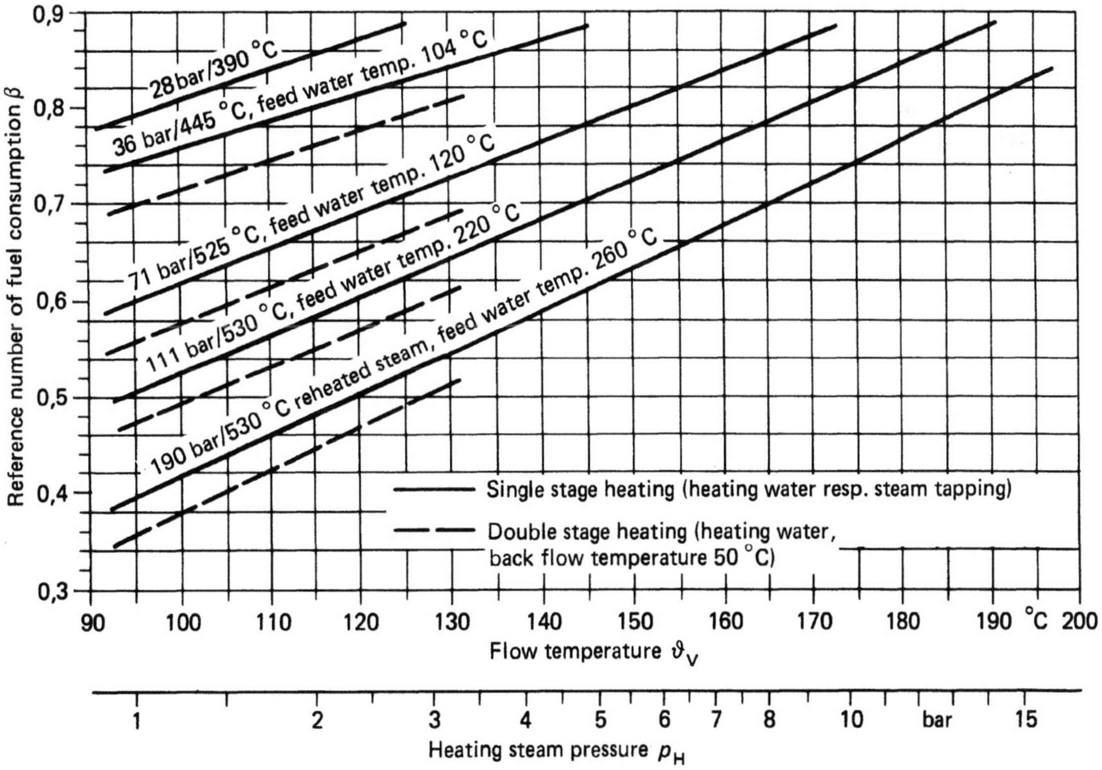
28 bar/390 °C
36 bar/445 °C, feed water temp. 104 °C
71 bar/525 °C, feed water temp. 120 °C
111 bar/530 °C, feed water temp. 220 °C
190 bar/530 °C reheated steam, feed water temp. 260 °C
Single stage heating (heating water resp. steam tapping)
Double stage heating (heating water, back flow temperature 50 °C)
Reference number of fuel consumption β
Flow temperature ϑV
Heating steam pressure pH
bar

J. Korek

Explanation

The diagrams in worksheet No 13.1 are related to back pressure steam cycle heat and power stations with reasonable feed water heating according to the live steam conditions. In the upper diagram of 13.1 the current reference number

$$\sigma = \frac{\text{generated electrical net output}}{\text{extracted enthalpy for heating}} = \frac{P_{\text{el}}}{\dot{Q}_{\text{H}}} \tag{1}$$

is shown. It depends on the heating steam pressure p_{H} of the hottest preheater stage of the heating water resp. the equivalent heating water temperature ϑ_{H}, regarding a temperature difference in the heat exchanger of 5 K. In the case of double stage heating, a back flow temperature of the heating water of 50 °C, with the same amount of heat flow from the blend steam and the back pressure steam has been assumed. By this, the water is nearly heated up to the same level.

The lines are valid for different live steam conditions. Except of the heat and power station with the 28 bar/390 °C live steam data, the blend steam is used for feed water heating. At the 190 bar heat and power station a reheating has been assumed.

The following further assumptions have been made:

Station requirement in current $\varepsilon = 0.05$ up to 0.07 (for p.f. furnaces and $p > 71$ bar, 0.05 up to 0.06 and for $p \leq 71$ bar 0.06 up to 0.07 is valid)

Generator efficiency $\eta_{\text{G}} = 0.96$ at 2 MVA and 0.98 at 120 MVA

Internal turbine efficiency $\eta_{\text{i}} = 0.82$ up to 0.86, growing with increasing steam flow

Mechanical turbine efficiency $\eta_{\text{m}} = 0.97$ at 2 MW; 0.98 at 25 MW; 0.992 at 100 MW.

Similar to the upper diagram, in the lower diagram of worksheet No 13.1 the fuel consumption reference number

$$\beta = \frac{\dot{Q}_{\text{Br}_{\text{HKW}}} - \dot{Q}_{\text{Br}_{\text{KW}}}}{\dot{Q}_{\text{H}}} = \sigma \left(\frac{1}{\eta_{\text{HKW}}} - \frac{1}{\eta_{\text{KW}}} \right) \tag{2}$$

is shown, depending on the heating steam pressure and the heating water temperature. With:

$\dot{Q}_{\text{Br}_{\text{HKW}}}$ fuel consumption of the heat and power station

$\dot{Q}_{\text{Br}_{\text{KW}}}$ fuel consumption of a compared condensation power station with the same electrical output

η_{HKW} electric efficiency of the heat and power station related to the fuel energy

η_{KW} electric efficiency of the compared condensation power station related to the fuel energy

From the energy balance:

$$\dot{Q}_{\text{Br}_{\text{HKW}}} \, \eta_{\text{DE}_{\text{HKW}}} + P_{\text{m}_{\text{P}}} - \dot{Q}_{\text{H}} - P_{\text{m}_{\text{T}}} = 0 \tag{3}$$

neglecting the mechanical load of the condensate pump $P_{\text{m}_{\text{P}}}$, together with equation (1) and (2), one gets the simple relation:

$$\beta \approx \frac{1}{\eta_{\text{DE}_{\text{HKW}}}} - \sigma \left[\frac{1}{\eta_{\text{KW}}} - \frac{1}{\eta_{\text{DE}_{\text{HKW}}} \, \eta_{\text{m}} \, \eta_{\text{G}} (1 - \varepsilon)} \right] \tag{4.}$$

For this worksheet $\eta_{\text{DE}_{\text{HKW}}} = 0.897$ and $\eta_{\text{KW}} = 0.38$ have been assumed. These values are valid for a coal fired power station.

Indexes:

Br	fuel
H	enthalpy
HKW	heat and power station
KW	power station (the comparison condensation power station)

0,20
0,15
0,10
0,05
0
Current loss SE
70
80
90
100
110
120
130
°C
140
Flow temperature ϑ_V
0,5
1
2
3
bar
4
Heating steam pressure p_H
Heating up of heating water
single stage
double stage
back flow temp. $\vartheta_R = 50\,°C$
Condenser pressure
0,12 bar
0,06 bar
0,03 bar

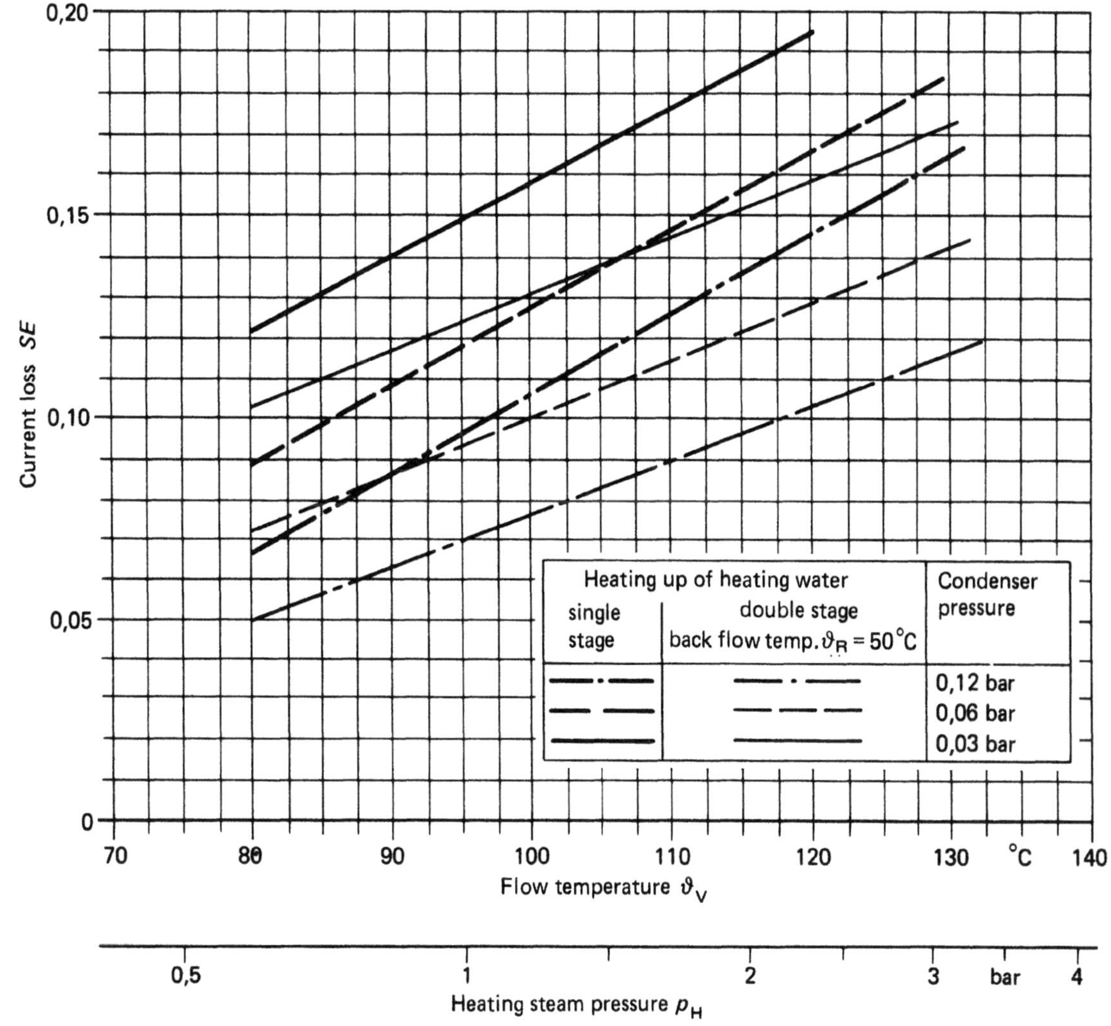

J. Korek

Explanation

The worksheet No 13.1 is related to condensation power stations with heat extraction. In the diagram the current loss from the extracted heating steam

$$SE = \frac{\text{reduced output of the power station because of the extracted heat}}{\text{extracted enthalpy flow for heating}}$$

$$= \frac{P_{el_{KW}} - P_{el_{HKW}}}{\dot{Q}_H}$$

$$= \frac{\Delta P_{m_T}\, \eta_m\, \eta_G (1 - \varepsilon)}{\dot{Q}_H}$$

is shown, depending on the heating steam pressure p_H of the hottest preheater stage of the heating water resp. the equivalent heating water temperature ϑ_H, regarding a temperature difference in the heat exchanger of 5 K.

With

$P_{el_{KW}}$ net electrical output of the condensation power station without heat extraction

$P_{el_{HKW}}$ net electrical output of the same condensation power station with heat extraction

ΔP_{m_T} reduced mechanical output of the turbine because of heat extraction

Further parameters are the condenser pressure p_0 and the number of heating water preheaters. In case of double stage heating, a back flow temperature of the heating water of 50 °C and the same amount of heat flow per stage has been assumed.

Each 10 K rise of the back flow temperature ϑ_R increases the current loss SE by 0.0054.

The following further assumptions have been made:

Steam condition after reheating: 40 bar at 530 °C

$\eta_i\ = 0.88$
$\eta_m = 0.995$
$\eta_G = 0.98$
$\varepsilon\ \ = 0.08$

Symbols and indexes see worksheet No 13.1

Temperature measurement technique

Range of validity

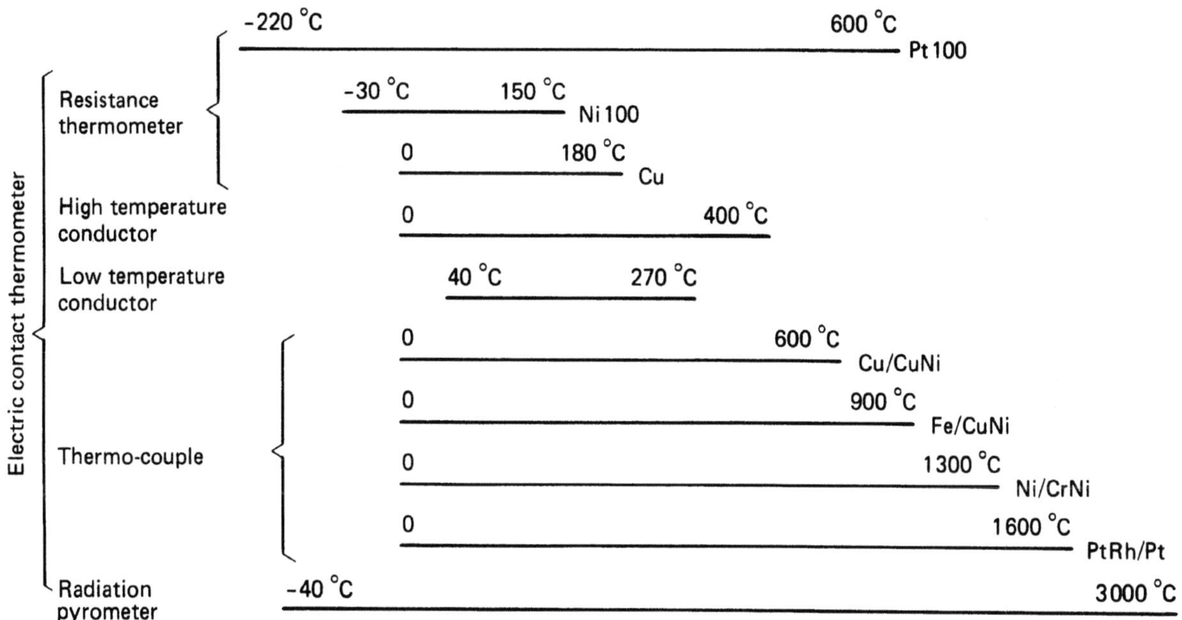

Electric contact thermometers can practically be used for all fluids and surfaces. If there are corrosion risks, this can be solved by various types and appropriate choice of the shielding material.

Ex-proof types are available.

DIN 43 710   Thermoelectric voltages, error limits
DIN 43 770   Survey
DIN 43 160   Thermometer, definitions

Field of application: energy and process engineering

For *radiation pyrometers*, the radiation emitted by the measured surface is a measure for the temperature (VDE/VDI guideline No 3511, Survey).

Field of application: steel mills, plastic industry

According to DIN 43 710

H. Hasenrahm

Pressure measurement technique

The most frequently pressure measurement device is the pressure gauge. It has a spring apparatus (tube, membrane or sylphone bellows). Normally there is only a single local indicator with a fixed range. Many types are able, to transmit electrical signals (for example 4 to 20 mA, limits) by means of auxiliary devices. The measuring accuracy is in the range of 0.2% and 6%, depending on the range of measurement and the measurement apparatus. The range of measurement is between 20 mbar and 1 000 bar.

Pressure transducer have adjustable ranges of measurement (for example 1 : 10), standardised output signals (for example 4 to 20 mA), its zero point can be suppressed (for example 60 to 100 bar) and with those, ranges of measurement of 1 mbar are possible. The measuring accuracy is in the range of 0.2% up to 0.4%.

Flow measurement technique

Industrial measurement technique uses flow meters of different kinds. The ranges of measurement vary from tubes of a few mm diameter (dosing) up to pipes of some meters in diameter (hydroelectric plant, waste water plant). Measuring accuracies, partly below 1%, will be achieved (weights and measures regulations), but mostly some percent are sufficient (automatic control, pipe fracture safety device). Many instruments do have a local indicator. All instruments (partly with auxiliary devices) are able to transmit electrical signals (for example 4 to 20 mA; impulses for remote control).

Kinds of flow measurement/mass flow measurement
(industrial measurement technique *))

Technology	Limits	Field of application
Volume meter		
Oval disk meter	Only pure	Chemical industry,
Cylindrical piston meter	liquids	food industry
Rotary piston gas meter	gases	gas industry
Dry gas meter	only	
Turbine meter	sensitive to fouling	gas and oil industry
Vortex meter	appropriate inlet has to be considered	gas and oil industry, heat suppliers
Mass flow meter		
Differential pressure meter	inlet problems, pure media	power stations, chemical industry
Floating particle meter	pure media, mostly small diameter	chemical industry, labs
Baffle plate meter	limited diameter	oil industry
Magnetic induction meter	liquids, medium has to be conductive	water suppliers, food and chemical industry
Ultrasonic meter	inlet problems	water suppliers, oil industry
Coriolis mass flow meter	limited diameter	food and chemical industry
Correlation mass flow meter	accuracy depends on correlation function	pneumatic conveying; bulk material measurement

*) Flow meters used only in house building technique or laboratory technique (e.g. thermal flow meters, laser anemometers) have not been considered

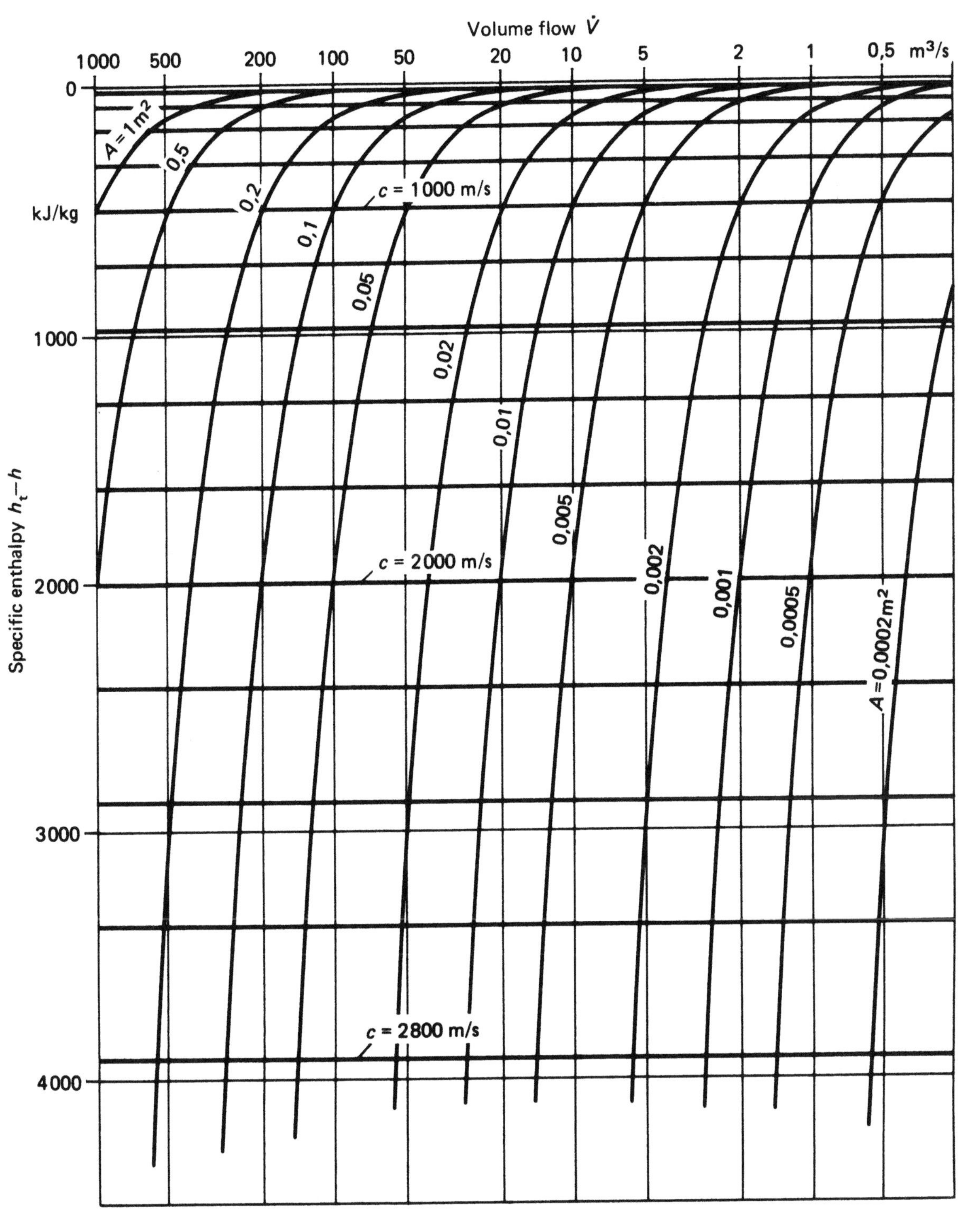

Volume flow $\dot{V}$
1000   500   200   100   50   20   10   5   2   1   0,5   m³/s
0
A = 1 m²
0,5
0,2
kJ/kg
0,1
c = 1000 m/s
0,05
0,02
1000
0,01
0,005
c = 2000 m/s
0,002
0,001
2000
0,0005
A = 0,0002 m²
3000
c = 2800 m/s
4000
Specific enthalpy $h_\tau - h$